COMMUNITY COLLEGE
OF DENVER
RED ROCKS CAMPUS

A BASIC/INTERMEDIATE COURSE FOR WATER SYSTEM OPERATORS

Volume 3
INTRODUCTION TO
Water
Distribution

PRINCIPLES and PRACTICES of WATER SUPPLY OPERATIONS

American Water Works Association
6666 W. Quincy Ave., Denver, Colorado 80235

ISBN 0-89867-188-4

Foreword

Introduction to Water Distribution is the third volume in a five-part handbook series designed for use in a comprehensive training program entitled "Principles and Practices of Water Supply Operations." Volume 3 examines the system facilities and procedures used to distribute treated water to consumers, with special attention to maintaining the quality of water in the system.

Other student volumes in the series include:

Volume 1	*Introduction to Water Sources and Transmission*
Volume 2	*Introduction to Water Treatment*
Volume 4	*Introduction to Water Quality Analyses*
Reference Handbook	*Basic Science Concepts and Applications*

Instructor guide and solutions manuals have been prepared with detailed lesson plan outlines, resource materials, and examination questions and answers for volumes 1 through 4.

Course content in the water supply operations series has been developed to meet training requirements of basic to intermediate grades of certification in water treatment and water distribution system operations. The modular format used throughout the series provides the flexibility needed to conduct both short- and long-term vocational training.

The reference handbook is a required companion textbook that is correlated with volumes 1 through 4 through footnote references. The purpose of a separate reference book is to provide the student with supplementary reading in the areas of mathematics, hydraulics, chemistry, and electricity.

The development of the training materials in the water supply operations series was made possible by funding through the US Environmental Protection Agency, Office of Drinking Water, under Grant Agreement No. T900632-01, awarded to the American Water Works Association.

Disclaimer

Several photographs and illustrative drawings that appear in this volume have been furnished through the courtesy of various product distributors and manufacturers. Any mention of trade name, commercial products, or services does not constitute endorsement or recommendation for use by the American Water Works Association or the US Environmental Protection Agency.

Acknowledgments

Publication of this volume was made possible through a grant from the US Environmental Protection Agency, Office of Drinking Water, under Grant Agreement No. T900632-01, as part of a national program strategy for providing a comprehensive curriculum in water treatment plant and water distribution system operations. John B. Mannion, Special Assistant for Communications and Training (currently Deputy Executive Director of the American Water Works Association Research Foundation), represented the Environmental Protection Agency, Office of Drinking Water, as project officer; and Bill D. Haskins, Director of Education, served as project manager for the American Water Works Association.

Original manuscript for this handbook was prepared under contract for AWWA by a consortium of training specialists and developers. The following are recognized for completing the difficult task of satisfying those critiquing their work:

John T. Quigley, Professor, Department of Engineering Professional Development, University of Wisconsin, Madison, Wis.

John E. Ekman, Circuit Instructor, Wisconsin VTAE Services District Consortium, Madison, Wis.

Guy W. Hansen, Environmental Specialist, Office of Operations and Maintenance, Wisconsin Department of Natural Resources, Rhinelander, Wis.

Ralph C. O'Connor Jr., Coordinator of Certification and Training, Office of Operations and Maintenance, Wisconsin Department of Natural Resources, Madison, Wis.

A special thanks to Ursula Wallace and Mary Kay Cousin, AWWA Technical Editors, who prevailed throughout the many manuscript revisions; Jane Olivier Benson, AWWA Associate Production Editor, for production of the handbook; Sheryl Tongue for typesetting and graphic art support; and Robert J. Love for graphic art support.

The completion of this publication has involved many key individuals who unselfishly contributed personal time in rewriting major portions of the manuscript and providing constructive suggestions. Many others volunteered their time in technical reviews or providing guidance in the development of working outlines for the modules that make up this handbook. To each, regardless of their level of involvement or support in making this training handbook possible, AWWA is indeed grateful. Completing this work exemplified that the strength and stature of AWWA lies in the professional commitment, integrity, and dedication of its members to be of service to others. To each of the following, a special thank you and acknowledgment for your help:

Gerald S. Allen, Vice President—Engineering, Pennsylvania Gas & Water Company, Wilkes-Barre, Pa.

David B. Anderson, President, Utility Technical Services, Inc., Englewood, Colo.

Kenneth J. Carl, Director of Municipal Surveys (Retired), Insurance Services Office, New York, N.Y.

James R. Daneker, Director of Training, BIF, A Unit of General Signal, West Warwick, R.I.

Jack O. Davis, Hydraulic Engineer, Denver Water Department, Denver, Colo.

David J. Hack, Water Specialist, Missouri Department of Natural Resources, Jefferson City, Mo.

John M. Hadden III, Superintendent of Distribution, Consolidated Mutual Water Company, Lakewood, Colo.

Jack W. Hoffbuhr, Deputy Director, Water Management Division, USEPA Region VIII, Denver, Colo.

Robert G. Jensen, Director of Administration, Denver Water Department, Denver, Colo.

Raymond L. Kocol, Engineer in Charge, Water Engineering Division, Department of Public Works, Milwaukee, Wis.

Jack E. Layne, Director of Engineering and Construction, Denver Water Department, Denver, Colo.

Ralph W. Leidholdt, Director of Technical Affairs and Chief Engineer, Directorate General of Water, Ministry of Electricity and Water, RUWI, Sultantate of Oman

Richard W. Looper, Water Service Superintendent, Colorado Springs Water Division, Department of Public Utilities, Colorado Springs, Colo.

Gerald L. Mahon, Director, St. Cloud Water Utility, St. Cloud, Minn.

Paul D. McQuade, Engineering Records/Drafting Supervisor, Denver Water Department, Denver, Colo.

James R. Myers, Corrosion Consultant, JRM Associates, Franklin, Ohio

Donald C. Renner, Superintendent of Utilities, Village of Arlington Heights, Arlington Heights, Ill.

Phillip S. Richardson, Assistant Supervisor of Process Control, Denver Water Department, Denver, Colo.

John A. Roller, Superintendent (Retired), Tacoma Water Division, Tacoma, Wash.

F.R. Seevers, Sales Representative, Mueller Co., Decatur, Ill.

Robert E. Vansant, Partner and Project Manager (Deceased), Black & Veatch, Kansas City, Mo.

Robert K. Weir, Director of Plant, Denver Water Department, Denver, Colo.

Vitte Yusas, Consultant, Madison, Wis.

One individual is especially deserving of recognition, not only for the major revisions made to the pipe installation, pumping, and instrumentation modules of this volume, but for his personal interest and dedication in overseeing the publication of the entire Water Supply Operations Training Handbook Series. For the numerous hours of voluntary overtime, appreciation is extended to John F. Rieman, AWWA Senior Technical Editor, for a job well done.

Introduction for the Student

Water distribution systems are made up of the pipe, valves, and pumps through which treated water is moved from the treatment plant to the homes, offices, and industries that consume the water. The distribution system also includes facilities to store treated water for use during periods when demand is greater than the treatment plant can supply, meters to determine how much water is being used, and fire hydrants to meet special service requirements.

Two important requirements of the distribution system are that it supply each user with a sufficient volume of water at adequate pressure and that it maintain the quality of the POTABLE* water delivered by the treatment plant.

The Operator's Role in Distribution

Some of the duties often assigned to water distribution operators include

- Maintaining pipes, valves, pumps, and other facilities to ensure a continued flow of potable water
- Monitoring and operating valves and pumps to vary the amount of water supplied as the demand varies
- Installing new connections to existing lines to supply water service to new customers
- Sampling water to ensure that its quality is maintained
- Investigating possible cross connections to nonpotable water, to ensure that the nonpotable liquid cannot flow into the potable system
- Maintaining system maps and records.

To perform the most basic of these duties, an operator entering the water distribution field should have good judgment, strong mechanical skills, and an education in the fundamentals of mathematics, hydraulics, and electricity. The more advanced and responsible operator will need training and experience in the operation, maintenance, repair, and replacement of distribution system equipment.

Topics Covered in the Text

This text includes modules that describe in detail the construction, operation, and maintenance of the major components of the distribution system, including pipes, valves, services and meters, fire hydrants, pumps and prime movers, storage facilities, and instrumentation and control equipment. A module on tapping explains the procedures used to add large and small connections to an

---

*Words set in SMALL CAPITAL LETTERS are glossary terms. Definitions for these terms can be found in the glossary at the end of this volume.

existing water line; and a module on cross connections details the opportunities for connection of nonpotable systems to the treated-water supply, with descriptions of the protective measures and equipment appropriate to such situations.

A module is also included describing the maps and records that are critical to efficient distribution system operation. The final module of the book suggests guidelines for effective public relations, since the distribution system operator will frequently be in the public eye.

Other Resources

Additional readings on the topics covered in the text are listed in the supplementary readings at the end of each module. The operator will also benefit from reading the other volumes in this series, especially the reference handbook, *Basic Science Concepts and Applications,* which is cited in footnotes throughout the book; and Volume 4, *Introduction to Water Quality Analyses,* which includes detailed discussions of the sampling and testing procedures used to monitor the quality of water within the distribution system.

Table of Contents

Water Distribution

Module 1

Pipe Installation and Maintenance

There are four types of piping systems in water utilities: transmission lines, in-plant piping systems, distribution mains, and service lines. TRANSMISSION LINES* carry water from a source of supply to a treatment plant or from a treatment plant or pumping station to a distribution system. Raw-water transmission and transmission lines are discussed in detail in Volume 1, *Water Sources and Transmission*, Module 4, Water Transmission. IN-PLANT PIPING SYSTEMS are the pipes located in pump stations and treatment plants. They are discussed in Volume 2, *Water Treatment,* and in this volume in relationship to the facilities in which they are found. DISTRIBUTION MAINS are the pipelines that carry water from transmission lines and distribute it throughout a community. This module deals with distribution mains. Service lines, or SERVICES, are small-diameter pipes that run from the distribution mains to the customer premises. Their selection, installation, and maintenance are discussed in Module 5, Services and Meters.

Water distribution system operators are involved extensively in the repair and maintenance of water mains. They may also become involved in installation and inspection of new water mains. This module begins with an overview of the concerns related to planning water systems and water-system extensions, additions, and replacements. This is followed by a discussion of system installation, repair, and maintenance.

Because practices and procedures vary from one part of the country to another, the operator is encouraged to become familiar with the site-specific conditions that influence local distribution system operations. Local conditions

---

*Words set in SMALL CAPITAL LETTERS are glossary terms. Definitions for these terms can be found in the glossary at the end of the volume.

may affect selection of materials, installation procedures, and maintenance requirements. Product literature and the supplementary readings listed at the conclusion of this module will be helpful in illustrating approved distribution system practices in different regions.

After completing this module you should be able to

- Identify and describe three distribution system configurations.

- Be familiar with the criteria used to size water mains to meet customer demand and fire-flow requirements.

- Discuss the general characteristics of commonly used pipe materials, giving selection criteria and standards of performance.

- Use *C* values to determine friction head loss given parameters for flow rate and pipe material, size, age, and length.

- Explain the types of corrosion problems that are found in a water distribution piping system and outline what preventive measures should be taken for each.

- Describe the analytical and evaluative procedures used to determine the bacteriological quality of water in the distribution system, as well as the rate of corrosion and scale formation; recommend ways to deal with any water-quality problems discovered.

- Explain the procedures for performing leakage and pressure tests and prepare troubleshooting guidelines that can be used if a section of pipe fails a leak test.

- Identify three methods for disinfecting water mains and note when each method might be used; explain what procedures are required before a water main can be returned to service after disinfection.

- Explain several procedures for maintaining or improving the carrying capacity of the distribution system.

- Define the term *unaccounted-for water*, including a list of contributing factors; explain the difference between a leak-detection program and a water audit.

- Recommend safety precautions that should be observed during trenching operations and installation of pipe, including measures for work-area protection and traffic control.

1-1. System Planning

In most communities, system planning and design is done by the city engineer or a consulting engineer. The operator, however, is often the person who knows the most about the existing system. For that reason, the operator's opinions and suggestions may have an important effect on final planning decisions. Good system planning should consider all of the following topics.

Policy Considerations

Certain general planning considerations related to new system development or system expansion require policy decisions about factors such as future growth, cost, financing methods, and ordinances. Although these decisions are usually made by the utility or city governing board, the operator should be aware of the general concerns.

Future growth.   System planning must include a long-range program for distribution system growth. Demand forecasts should be made covering a minimum of 10 years in the future. The projected data should include population growth and domestic water usage, expected growth of commercial and industrial water usage, and fire-protection requirements. With this information, future pumping, storage, flow, and pressure needs can be projected.

Cost.   When planning for a specific addition or extension, estimates of cost should be included. The estimates are often grouped or divided into these categories—materials, labor, overhead, contingencies, profit, engineering and inspection, legal expenses, and land acquisition.

The estimates of water-main costs may also be stated on a "per linear foot (metre)" basis for excavation of rock and earth; installation of mains, valves, and hydrants; and backfilling and repair. The cost of construction, inspection, final flushing, disinfection, and testing of any special sections should also be included.

Financing methods.   The system planning or engineering report may also contain an evaluation of alternative methods of financing, with recommendations as to the preferable method. Common methods to be considered are government grants and loans, private capital, revenue bonds, water/sewer charges, or any combination of these methods. Changes in water rates or rate structure should also be addressed at this time.

Ordinances.   If local ordinances regarding private-well abandonment, elimination of cross connections, and responsibilities regarding thawing of frozen services are inadequate or do not exist, then it is often desirable to have the planning report cover these matters and include basic recommendations for development of adequate ordinances.

System Layout

Configuration.   Distribution system layout is usually designed to fit one of three configurations: arterial-loop system, grid system, or tree system.

The ARTERIAL-LOOP SYSTEM is shown in Figure 1-1. All major demand areas should be served by an arterial-loop system. High-demand areas are served by distribution mains tied to the arterial loop to form the second configuration, a complete GRID SYSTEM without dead ends (Figure 1-2). Areas where an adequate water supply must be maintained at all times for health or fire-control purposes should be tied to two arterial mains where possible. Minor distribution lines or mains make up the secondary system, which is the major portion of the grid; they supply fire hydrants, and domestic and commercial consumers.

The third type of configuration, the TREE SYSTEM (Figure 1-3), is not recommended because few loops are formed, making it difficult to supply a continuous flow of good-quality water to all parts of the system.

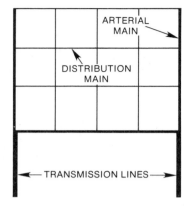

Figure 1-1. Arterial-Loop System

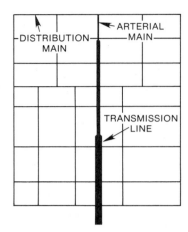

Figure 1-2. Grid System

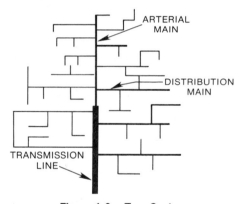

Figure 1-3. Tree System

Mapping.   As an initial step in system planning, a basic system map should be developed to show the following information:

- The existing system (including main size, water pressure, valves, and hydrants)
- Existing and planned streets
- Areas outside the system designated for future expansion
- Ground-level elevations, contour lines, and topographic features
- Existing underground utility services, including sanitary sewers; storm sewers; water, gas, and steam lines; and underground electric, telephone, and television cables
- Population densities (present and projected)
- Normal water consumption (present and projected)
- Proposed additions or changes to the system.

Sectional maps (50 or 100 ft/in. [0.5 or 1.0 m/mm]) should be drawn of the system to show more detail, including valve and hydrant locations. These maps are usually based on the water system's engineering records. Module 10, Maps, Drawings, and Records, discusses mapping and the use of sectional maps in more detail.

Valving. Shut-off valves should be provided so that areas within the system can be isolated for repairs or maintenance. Valves should be located at regular intervals and at all branches from the arterial mains to minimize interruptions of service. Where line intersections exist in a grid, no more than one branch (preferably none) should be without a valve. The system should be planned so that most of the flow will be maintained if any one section of the system is taken out of service.

Air-relief valves may be required at high points, and blowoff valves may be required at low points. (Refer to Module 3, Valves, for further discussion of the selection, location, and operation of valves.) Where cross connections are made between the domestic supply and any other supply, a backflow preventer that has approval of the state health department is recommended and is usually required by state regulations.

1-2. Sizing Mains

The size of a main determines its carrying capacity. Main sizes must be selected to provide the flow (capacity) to meet domestic, commercial, and industrial demands in the area to be served, as well as provide for fire-flow requirements at the necessary pressure.

Quantity requirements. Domestic usage requirements for a service area can be determined either from past records or from general usage figures like those shown in Table 1-1, with estimates corrected as necessary to reflect local conditions. This data should then be adjusted for commercial, industrial, and projected growth factors, to ensure that the system's design capacity will meet future demand.

The determining factor in sizing mains, storage facilities, and pumping facilities for communities with a population less than 50,000 is usually the need for fire protection. Fire-flow requirements for each community are set by an organization representing the fire insurance underwriters, called the Insurance Services Office (ISO). This group determines the minimum flow that the system must be able to maintain for a specified period of time in order for the community to achieve a certain fire-protection rating. Fire insurance rates are then based, in part, on this classification.

There are some small rural water systems designed to serve only domestic water needs. Fire protection is not provided, and fire-flow requirements are not considered in the design of these systems. Most systems, however, do provide water supply capacity for fire protection. The water delivery, storage, or pumping requirements can be increased tremendously when needs for fire flow are considered.

Fire insurance underwriters recommend that no main in the distribution system be less than 6 in. (150 mm) in diameter. For sound design, the next size

Table 1-1.   Water Used for Public Supplies, by States, 1975*

State	Gallons per Capita per Day	State	Gallons per Capita per Day
Alabama	210	New Mexico	236
Alaska	442	New York	154
Arizona	211	North Carolina	169
Arkansas	139	North Dakota	130
California	185	Ohio	167
Colorado	200	Oklahoma	130
Connecticut	134	Oregon	190
Delaware	171	Pennsylvania	178
Florida	168	Rhode Island	128
Georgia	158	South Carolina	242
Hawaii	228	South Dakota	126
Idaho	236	Tennessee	130
Illinois	199	Texas	176
Indiana	146	Utah	331
Iowa	146	Vermont	150
Kansas	170	Virginia	119
Kentucky	101	Washington	256
Louisiana	152	West Virginia	154
Maine	143	Wisconsin	156
Maryland	147	Wyoming	191
Massachusetts	145	District of Columbia	215
Michigan	168	Puerto Rico—Virgin Islands	125
Minnesota	135		
Mississippi	120		
Missouri	158		
Montana	267		
Nebraska	248		
Nevada	321		
New Hampshire	115		
New Jersey	145	United States	168

*Estimated Use of Water in the U.S. in 1975. Geological Survey Circular 765 (1977).

larger main than that indicated by calculations should be used. Fire insurance experts also suggest that minimum pipe sizes be governed, in part, by the type of area to be served. High-value districts should have minimum pipe sizes of 8 and 12 in. (200 and 300 mm); residential areas should have minimum sizes of 6 and 8 in. (150 and 200 mm). Smaller sizes should be used only when they complete a desirable grid. These recommendations may, however, be difficult to follow in practice. The nearest ISO should be consulted if more specific information on fire-flow requirements or a fire insurance rating is needed.

Pressure requirements.   In areas requiring high fire-flow capacity, minimum static pressure required at all fire hydrants is generally 35 psi (240 kPa) or higher. Pressure under required fire-flow conditions should not drop below 20 psi (140 kPa). An insurance company may require that both a minimum water pressure and a minimum flow requirement be maintained to the building they insure.

Velocity requirements.   Velocity of flow is also a factor in determining the capacity of pipes and, therefore, the required pipe size. Velocities should normally be 5 fps (1.5 m/s) or less, because of high friction losses that occur at greater velocities.[1] This may be difficult to obtain under normal operating conditions, and velocities can significantly exceed this guideline under fire-flow conditions.

Network analysis.   Pipe carrying capacity depends on pipe size, pressure, flow velocity, and head loss resulting from friction. Friction factors include roughness of the pipe, flow velocity, and pipe diameter. The required pipe size can be calculated when the other requirements and characteristics are known. The Hazen–Williams formula is often used for this purpose. Charts, nomographs, and special slide rules using this formula, or based on it, have been developed for various sizes and types of pipe to help in selecting proper pipe size.[2]

When the distribution system or system expansion is extensive, it will probably be necessary to analyze the system and balance the flow among all areas in relation to demand. This analysis requires a plot of pressures and flows at points throughout the system and may be accomplished using either the Hardy–Cross or Newton–Raphson Methods. Today most system evaluations are performed using computerized modeling techniques. Table 1-2 illustrates some of the data obtained from a computerized analysis of one system's flow characteristics.

1-3. Material Selection

The operator should be consulted regarding the type of pipe selected for extensions or system repair. In some systems, the operator is solely responsible for making this selection. Many factors should be considered. For example, what qualities should a pipe have? What types of pipe are available? Is one type better than another? Are all types acceptable? What type of water main does the community have now? Why was it selected? Are the materials for the existing pipe and the proposed new pipe compatible? The following paragraphs discuss general requirements for pipe, then conclude with brief descriptions of several piping materials commonly used in water distribution systems.

Pipe Qualities

Strength.   Pipe must have the strength to withstand the EXTERNAL LOAD (SUPERIMPOSED LOAD) exerted on it. The external load is made up of the BACKFILL, the material that is filled back into the trench after the pipe has been laid, and the TRAFFIC LOAD, the weight or impact of the traffic passing over the line. The pipe must also be able to resist damage from impact during installation. The ability to withstand external forces such as these is often referred to as a measure of the

---

[1] *Basic Science Concepts and Applications*, Hydraulics Section, Head Loss (Friction Head Loss).
[2] *Basic Science Concepts and Applications*, Mathematics Section, Graphs and Tables (Graphs—Nomographs).

Table 1-2.   Results of Computer Analysis of a Distribution System

Fowler Program - Menomonie Example Maximum Day Demand

Number of Iterations - 9

Diameter	Upstream Node	Downstream Node	Pipe	Flow	Head Loss	Velocity	Upstream HGL	Downstream HGL
4.00	1	2	1	1.29475	.00585	.03306	1034.55	1034.54
12.00	1	3	2	-14.59475	.00208	-.04141	1034.55	1034.55
8.00	2	3	3	-21.87044	.00752	-.13961	1034.54	1034.55
8.00	3	4	4	-73.56519	.21394	-.46961	1034.55	1034.77
4.00	2	4	5	-14.43481	.22145	-.36858	1034.54	1034.77
6.00	4	5	6	-92.80000	4.00571	-1.05315	1034.77	1038.77
6.00	5	15	7	-41.84239	.27153	-.47485	1038.77	1039.04
6.00	15	16	8	-30.50758	.07563	-.34622	1039.04	1039.12
6.00	15	14	9	-15.53481	.01896	-.17630	1039.04	1039.06
6.00	14	6	10	.70674	.00012	.00802	1039.06	1039.06
6.00	6	7	11	-11.99326	.01678	-.13611	1039.06	1039.08
6.00	7	8	12	-9.23175	.01860	-.10477	1039.08	1039.10
6.00	12	8	13	2.65467	.00062	.03013	1039.10	1039.10
6.00	9	8	14	11.37708	.04565	.12911	1039.14	1039.10
6.00	9	11	15	-4.83075	.00311	-.05482	1039.14	1039.15
6.00	10	11	16	32.93697	.11985	.37379	1039.27	1039.15
6.00	11	12	17	23.90622	.04815	.27130	1039.15	1039.10
6.00	12	7	18	10.76151	.01922	.12213	1039.10	1039.08
6.00	12	13	19	6.29004	.00305	.07138	1039.10	1039.09
6.00	13	14	20	19.44155	.03284	.22064	1039.09	1039.06
6.00	16	13	21	16.35151	.02383	.18557	1039.12	1039.09
6.00	16	10	22	-51.05909	.14721	-.57945	1039.12	1039.27
6.00	10	9	23	28.24633	.12296	.32056	1039.27	1039.14
6.00	17	10	24	147.14239	3.65777	1.66987	1039.27	1042.92
6.00	17	33	25	98.20100	2.80066	1.11445	1042.92	1040.12
6.00	5	35	26	-71.05761	.36198	-.80641	1038.77	1039.13
6.00	35	34	27	-82.75761	.42003	-.93919	1039.13	1039.55
6.00	34	33	28	-90.75761	.56948	-1.02998	1039.55	1040.12
4.00	33	32	29	-2.65661	.00593	-.06784	1040.12	1040.13
6.00	32	29	30	-32.55259	.06396	-.36943	1040.13	1040.19
8.00	32	31	31	6.29598	.00712	.07145	1040.13	1040.12
6.00	31	30	32	-4.90402	.00063	-.03131	1040.12	1040.12
6.00	29	30	33	21.70402	.07046	.24631	1040.19	1040.12
6.00	29	28	34	-60.65661	.27003	-.68837	1040.19	1040.46
12.00	27	28	35	12.80574	.00058	.03633	1040.46	1040.46
12.00	26	27	36	-12.41563	.00018	-.03523	1040.46	1040.46
12.00	26	23	37	24.61206	.00109	.06983	1040.46	1040.46
12.00	24	23	38	-20.30000	.00061	-.05759	1040.46	1040.46

CRUSHING STRENGTH of a pipe. External load is figured in pounds per linear foot (lb/lin ft [kg/m (linear)] or pounds per square inch (psi [kPa]).

Pipe must also be able to withstand the INTERNAL LOAD, or HYDROSTATIC PRESSURE, within the pipe that results from a combination of REGULAR OPERATING DELIVERY PRESSURE and SURGE PRESSURE. Internal load is also figured in psi (kPa). Normal water pressure depends on local conditions and requirements, but is usually in the 40–100 psi (280–690 kPa) range. Surge is a momentary increase in pressure, usually caused by a sudden change in velocity or direction of water flow. Surge is also known as WATER HAMMER, and may be caused by the rapid opening or closing of valves or the sudden starting or stopping of pumps. Water hammer results from shock or pressure waves that travel rapidly through the pipe. These waves can cause extensive damage, such as ruptured pipe and damaged fittings. Figure 1-4 illustrates the shock waves during a water-hammer condition.

Two terms used to describe the strength of the pipe are TENSILE STRENGTH and FLEXURAL STRENGTH. Tensile strength is the resistance a material has to "longitudinal," or lengthwise pull, before it fails; flexural strength is the ability of the material to bend or flex without breaking. Pipe shear breakage or beam breakage may occur when a force exerted on a pipe is stronger than its tensile or flexural strength.

A shear break occurs when the earth shifts. Beam breakage, which resembles a shear break, may occur when a pipe is unevenly supported along its length. A pipe that is resting on a narrow unyielding support point, like a rock or a wall, may break like a beam does when overloaded. Figure 1-5 shows these effects.

Figure 1-4.   Shock Waves During Water Hammer

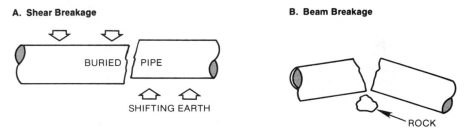

A. Shear Breakage

B. Beam Breakage

BURIED   PIPE

SHIFTING EARTH

ROCK

Figure 1-5.   Shear and Beam Breakage

Pipe should be carefully chosen to ensure that its pressure rating is adequate for handling the pressures in a specific system. Load factors can be calculated using various formulas and tables found in current AWWA standards. A safety factor of 2.5 to 4 times the design load is normally added to the calculated value. When replacing a section of pipe, the new piece must have a pressure rating equal to or greater than the piece being replaced.

Specific requirements or standards for all kinds of pipe have been established and published by AWWA to ensure adequate and consistent quality of water mains. Other agencies that have established standards for pipe include federal and state governments; Underwriters Laboratories, Inc.; National Sanitation Foundation (NSF); American Society for Testing and Materials (ASTM); and the manufacturers themselves. These specifications, which cover the method of design, manufacture, and installation in detail, should be used when selecting pipe.

Durability.   Durability is the degree to which a pipe will provide satisfactory and economical service under the conditions of use. It implies long life, toughness, and the ability to maintain tight joints with little or no maintenance.

Corrosion resistance.   Corrosion resistance, which is closely related to durability, refers to a pipe's resistance to aggressive soils and waters. Troublesome chemical reactions often occur between the pipe wall and its surroundings, adversely affecting the quality of the water the pipe is carrying, as well as the capacity and service life of the pipe. Water pipe should not react with the water it carries and it should resist corrosion both inside and out. Special measures or materials may be necessary to ensure corrosion resistance. This topic will be dealt with more extensively later in this module.

Smooth inner surface.   A smooth inner surface will ensure maximum flow capacity for the water pipe. The ease with which water passes through the pipe is measured by a factor called the FLOW COEFFICIENT, the SMOOTHNESS COEFFICIENT, or the ROUGHNESS COEFFICIENT. Pipe is rated by this coefficient, which is usually referred to as the C VALUE. The higher the C value of the pipe, the smoother it is.

A. Smooth Pipe

B. Tuberculated Pipe

Courtesy J-M Manufacturing Co., Inc.   *Courtesy of Girard Industries*

Figure 1-6.   Effect of Tuberculation

Table 1-3. _C_ Values of Various Pipe Materials

Pipe Material	C Value
Asbestos–cement	140+
Cast iron (new)	140+
Bitumastic enamel, centrifugally applied	140+
Cement lined, centrifugal method, actual diameter	140+
Pit cast, tar dipped	140
Cast iron (20 years old)	100
Tar dipped (inactive water)	130
Bitumastic enamel (inactive water)	135
Cement lined	140
Tuberculated	100*
Concrete (quality pipe)	140+
Ductile iron (cement lined)†	140+
Plastic	140+
Steel	140+
Wood stave (smooth)	120

*Less with high degree of tuberculation.
†Greater capacity because of larger internal diameter for normal outside diameters.

Figure 1-6 illustrates the difference between smooth pipe and pipe with an interior surface roughened by TUBERCULES. The hydraulics section in _Basic Science Concepts and Applications_ deals more extensively with the _C_ value.[3]

Table 1-3 shows the _C_ values for various pipe materials, ages, and conditions. New pipe generally has a _C_ value of 140 or greater. Old pipe may have a _C_ value of 100 or less. The table also shows that a smooth inner surface may be preserved by using pipe lined with cement, tar, epoxy, or plastic.

Economy.   Several items are involved in determining whether or not the pipe selected is the most economical choice. For example, is the size and type of pipe being considered readily available? Are the necessary tees, elbows, and couplings readily available? Can the needed supplies be purchased locally or from a company with a local representative? Special orders, custom-made pipe or fittings, long delivery times, difficulty in obtaining extra material, or returning damaged or defective pieces will most likely increase the cost and the construction time for the project.

Another factor to consider is installation costs. Will the time and money spent on installation be increased or decreased because of the type of pipe? Some installation features that must be considered include pipe section length, weight, coating, lining, strength, joint type, cost, flex, and ease of installation. Abnormal installation procedures must be used in case of unusual topsoil or conditions, uneven terrain, high ground water or bedrock, river or highway crossings, proximity to sewer lines, proximity to other utility services, available sizes, and ease of tapping.

Installation is a major part of the cost of a project, whether the utility does the work itself or contracts for it to be done. The projected cost as well as the

---

[3] _Basic Science Concepts and Applications_, Hydraulics Section, Head Loss (Friction Head Loss).

additional unplanned-for costs or savings will vary significantly for the same job based on factors such as amount and quality of advance planning and engineering, reliability of delivery times, and weather conditions.

Tapping and repair.   The pipe selected should be easy to tap for installation of SERVICE CONNECTIONS, valves, and additional pipe. It should hold the service connection firmly without cracking, breaking, or leaking. It should be easily replaceable or, at least, repairable.

Water quality maintenance.   Maintenance of water quality must be considered relative to the characteristics of the water being provided. No taste, odor, chemicals, or other undesirable qualities should be imparted to the water.

Corrosion Resistance

Once installed, water pipe is in continuous close contact with the soil surrounding it and the water moving through it. Either or both may cause problems that will affect the performance and life of distribution system pipe, as well as the quality of the water being delivered to the customers. It is, therefore, important to know what conditions exist and what future problems could occur when selecting water-main pipe.

CORROSION is generally defined as an electrochemical reaction that deteriorates a metal or an alloy. For the purposes of this discussion, corrosion also includes the dissolving of other distribution system materials (for example, asbestos–cement pipe) through contact with water.

Corrosion is a natural phenomenon. Metals are normally found in their stable, oxidized form in nature. Iron ores, for example, are found as iron oxides. These oxides are chemically reduced in the refining process to useful metal. In the presence of oxygen and water or under certain soil and electrical conditions refined iron tends to return to its more stable form of iron oxide, or rust. The qualities of some waters and the conditions found in some soils are especially favorable to corrosion.

There are several types of corrosion to be considered in a water distribution piping system:

- Internal electrochemical corrosion—caused by aggressive water flowing through the pipes

- External corrosion—caused by the chemical and electrical conditions of the soil in which the pipe is buried

- Galvanic corrosion—caused by connecting components made of dissimilar metals, such as connecting galvanized steel pipe to a copper water line without an insulating union

- Stray-current corrosion—caused by uncontrolled DC electrical currents flowing in the soil.

Internal electrochemical corrosion.   It is often assumed that the quality of the treated water entering the distribution system will not change before it reaches the customer. However, many natural and treated waters are corrosive—they dissolve certain pipe materials or furnish substances that react with the material.

These corrosive waters cause deterioration of the distribution system and domestic plumbing systems. This deterioration or corrosion can, in turn, cause the quality of the water delivered to the customer to be significantly degraded. Corrosion can also reduce flow capacity and shorten the life span of the distribution system.

Figure 1-7 illustrates a corrosion cell, showing how corrosion can cause pitting and the formation of rough tubercules. Even slight corrosion and TUBERCULATION can have a major effect on the smoothness of a pipe's surface, seriously reducing its carrying capacity. Major corrosion can weaken pipes, and corrosion products may break off of mains and clog services or cause water quality problems and customer complaints.

The topic of internal corrosion is complex. The principal concern with internal corrosion has been its effect on unlined cast-iron mains, steel tanks, and other metal surfaces in the distribution system. As a result, unlined cast-iron pipe is no longer installed in new systems. Internal corrosion, although now mainly a problem in existing pipe, must still be a consideration in the selection of pipe for new and replacement purposes.

For more information on internal pipe corrosion, refer to Volume 2, *Water Treatment*, Module 9, Stabilization, and the chemistry section in *Basic Science Concepts and Applications.*[4]

External corrosion. The soil that will surround a pipe should also be considered during pipe selection. Soil characteristics, like water characteristics, vary greatly from one area of the country to another. Soil characteristics can also change greatly from one part of a community to another. The frequency and severity of external-corrosion problems in distribution systems vary with the type of soil, pipe location, and type of pipe. Generally, nonmetallic cement pipe and plastic pipe will not experience external-corrosion problems. Asbestos-cement pipe may suffer external corrosion under certain soil conditions. Steel and cast-iron pipes will require some kind of protection from most soils. The remainder of this discussion deals with metallic pipe.

The electrochemical corrosion process that takes place on the outside of metal pipe is very similar to the corrosion process that takes place on the inside of the pipe. External pipe corrosion involves a chemical reaction and a simultaneous flow of electrical current. Differences in the soil and the pipe surface create potential differences on the buried metal. These differences form a corrosion cell, which has an anodic (negative) area and a cathodic (positive) area. A direct electrical current flows from the anodic area to the cathodic area through the soil and returns to the anodic area through the metal.

The surface impurities, nicks, and grains that create corrosion cells are usually present in metal pipe. Oxygen differentials in the soil can also create corrosion cells. Although some corrosion will occur when almost any metal pipe is buried, severe corrosion can be serious enough to cause leakage, breaks, and otherwise shorten the service life of the pipe.

---

[4] *Basic Science Concepts and Applications*, Chemistry Section, Treatment Processes (Scaling and Corrosion Control).

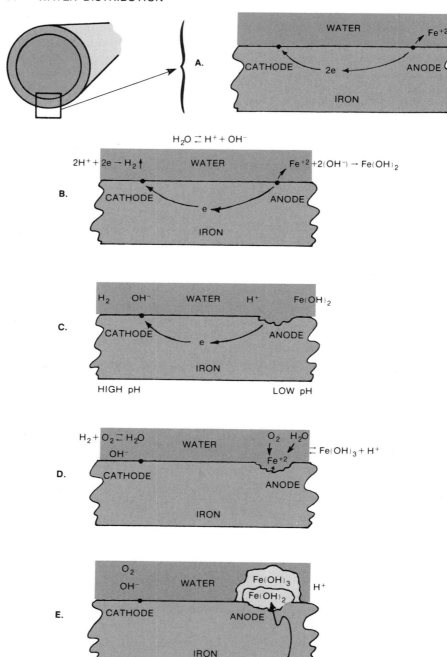

Figure 1-7.   Chemical and Electrical Reactions in a Corrosion Cell

Corrosion most frequently takes the form of pits in an otherwise relatively undisturbed pipe surface. These pits occur at the anodic areas and may eventually penetrate the pipe wall; cathodic areas are generally protected in this process. This type of corrosion will occur on steel, cast-iron, and ductile-iron pipe, as well as on copper and stainless-steel pipe. Although most underground corrosion destroys a relatively small area of the total pipeline, it requires replacement of the whole pipe.

Factors affecting external corrosion.   The degree of corrosion and the size of the area affected by corrosion depend on the quantity, size, and intensity of the corrosion cells. This, in turn, depends on the corrosivity of the soil. Some soil characteristics and conditions that are likely to increase the rate and amount of corrosion include:

- High moisture content
- Poor aeration
- Fine texture (for example, clay or silty materials)
- Low electrical resistivity
- High organic-material content
- High chloride and/or sulfate content, which increases electrical conductivity
- Highly acid or alkaline pH, depending on the metal or alloy
- Presence of sulfide
- Presence of anaerobic bacteria.

Most of these characteristics affect the electrical resistivity of a soil, which is a fair measure of a soil's corrosivity relative to other soils. The amount of metal removed from the pipe wall is proportional to the magnitude of the current flowing in the corrosion cell. The amount of current is, in turn, inversely proportional to the electrical resistivity of the soil or directly proportional to the soil conductivity. Therefore, a soil of low resistivity or high conductivity fosters corrosion currents and increases the amount of metal oxidized.

Soil resistivity and conductivity are measurable. Increased corrosion of steel can generally be expected when soil resistivity is below 10,000 $\Omega \cdot$cm. Significant corrosion of steel and iron is highly probable when soil resistivity falls below 2000 $\Omega \cdot$cm.

The results of several studies have shown that certain microorganisms can accelerate corrosion reactions and enhance tuberculation. Iron bacteria or iron oxidizers can have these effects under ANAEROBIC conditions. Corrosion can also be caused or aided by sulfate-reducing bacteria, again under anaerobic conditions. Sulfate-reducing bacteria affect corrosion by altering the environment so that it becomes more favorable for corrosion reactions.

Preventing external corrosion.   Several solutions to the problem of corrosion have been tried. The most effective solution will vary depending on the pipe material and the chemical and electrical conditions of the surrounding soil.

Methods that have been found effective under certain conditions include:

- Specifying extra thickness for pipe walls
- Bitumastic or coal-tar coatings
- Wrapping in polyethylene plastic sleeves
- Cathodic protection.

Adding extra thickness to a pipe as a corrosion allowance may seem to be an easy solution, but it is not an effective long-term remedy. Pipe corrosion is not uniform, and specifying a corrosion allowance for buried pipe only postpones the problem. In most cases, the extra cost is prohibitive compared to the cost of other corrosion-prevention methods.

Applying a hot bitumastic or coal-tar coating to the pipe exterior will also probably not prevent or reduce soil corrosion. When pinholes or breaks occur in the coating, bare metal at these locations will be anodic to the surrounding coated pipe and corrosion will be concentrated in these areas. This is also true for pipe such as steel that can be obtained with a cement-mortar coating. This coating will, however, have some "self-healing" properties for small breaks. Coatings coupled with cathodic protection (discussed below) are the best corrosion control for steel in aggressive soils, and this approach is also used successfully for ductile iron.

Coated pipe must be inspected and handled carefully to ensure that the coating is not damaged. Perforations of the pipe wall often occur sooner in coated steel pipe than in bare pipe because the corrosion is concentrated within small pits instead of spread over the entire surface of the pipe.

One means of corrosion control for gray-iron or ductile-iron pipe in corrosive soils is encasing the pipe in a polyethylene sleeve. Figure 1-8 shows this approach.

CATHODIC PROTECTION may also be used to control external pipe corrosion. Cathodic protection stops corrosion by canceling out, or reversing, the corrosion currents that flow from the anodic area of a corrosion cell. There are two principal methods for applying cathodic protection—sacrificial anodes and impressed-current systems.

Sacrificial anodes, also called galvanic anodes, consist of magnesium alloy or zinc castings that are connected to the pipe through insulated lead wires, as shown in Figure 1-9. Magnesium and zinc are anodic to steel and iron, so a galvanic cell is formed that works opposite to the corrosion being prevented. As a result, the magnesium or zinc castings corrode (sacrifice their metal), and the pipe, acting as a cathode, is protected.

Impressed-current cathodic-protection systems use an external source of DC power that makes the structure to be protected cathodic with respect to some other metal in the ground. In most cases, rectifiers are used to convert AC power to DC, and a bank of graphite or specially formulated cast-iron rods is connected into the circuit as the anode (corroding) member of the cell, with the distribution system component being the protected cathode (Figure 1-10).

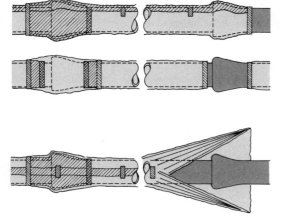

Method A: One length of polyethylene tube for each length of pipe, overlapped at joint.

Method B: Separate pieces of polyethylene tube for barrel of pipe and for joints. Tube over joints overlaps tube encasing barrel.

Method C: Pipeline completely wrapped with flat polyethylene sheet.

Figure 1-8.   Polyethylene Encasement of Ductile-Iron Pipe

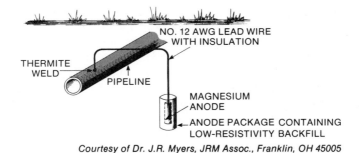

Courtesy of Dr. J.R. Myers, JRM Assoc., Franklin, OH 45005

Figure 1-9.   Cathodic Protection Using Sacrificial Anodes

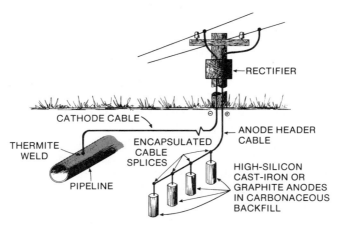

Courtesy of Dr. J.R. Myers, JRM Assoc., Franklin, OH 45005

Figure 1-10.   Cathodic Protection Using Impressed Current

Table 1-4.   Galvanic Series for Metals Used in Water Systems

Corroded End (Anode)	MOST ACTIVE
Magnesium	+
Magnesium alloys	
Zinc	
Aluminum	
Cadmium	
Mild steel	
Wrought (black) iron	
Cast iron	
Lead–tin solders	
Lead	
Tin	
Brass	
Copper	
Stainless steel	−
Protected End (Cathode)	**LEAST ACTIVE**

(Corrosion Potential increases from LEAST ACTIVE upward to MOST ACTIVE)

Cathodic protection is a very complicated and technical area. A competent consultant should be involved early in any problem analysis or system revision.

Galvanic corrosion.   GALVANIC CORROSION, another form of corrosion an operator may encounter, occurs when two electrochemically dissimilar metals are connected. This is often the case where a service is connected to a main or where other APPURTENANCES are connected into the distribution system. The two metals form a corrosion cell or, more exactly, a GALVANIC CELL. This can result in loss of the anodic metal and protection of the cathodic metal. A practical galvanic series for common pipe and appurtenance materials is shown in Table 1-4. Each metal is anodic to, and may be corroded by, any metal below it. The greater the separation between any two metals in the series, the greater the potential for corrosion and (barring other factors) the more rapid the corrosion process.

Corrosion resulting from the action of a galvanic cell generally appears as a pitting of the anodic metal. The extent of pitting decreases with an increase in the distance from the junction of the two metals. Factors that determine the severity of this corrosion are the relative amounts of surface area of each metal involved and the resistivity of soil around the piping. A large area of cathodic metal connected to a small area of anodic metal may result in rapid corrosion of the anodic metal, as illustrated schematically in Figure 1-11. The reverse situation, a small cathode coupled to a large anode, is shown in Figure 1-12. Although the large total area of the anode helps reduce the severity of corrosion in this situation, rapid corrosion can still occur at the point where the anode (service line and main) connects to the cathode (valve). Other factors being equal, the rate of corrosion caused by a BIMETALLIC (two-metal) connection will generally be greater in low-resistivity soils than in high-resistivity soils.

Corrosion-control methods for bimetallic connections are designed to break the electrical circuit of the galvanic cell. Dielectric barriers, or insulators, are the method most commonly used to stop the flow of corrosion current through the

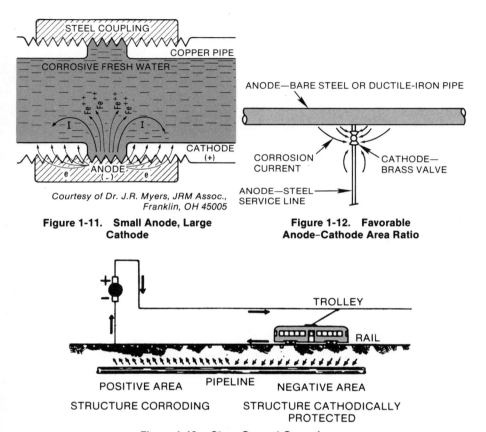

Courtesy of Dr. J.R. Myers, JRM Assoc.,
Franklin, OH 45005

**Figure 1-11.   Small Anode, Large
Cathode**

**Figure 1-12.   Favorable
Anode–Cathode Area Ratio**

Figure 1-13.   Stray-Current Corrosion

metal. Coating a bimetallic connection reduces the amount of corrosion current passing through the soil. If only one metal can be coated, it is best to coat the cathode, since coating the anode may create an unfavorable area relationship, with a relatively small anode (exposed through pinholes in the coating) and a large cathode. The best solution is to insulate the service from the main. Cathodic protection can also be used to control galvanic corrosion.

Stray-current corrosion.   Stray-current corrosion is caused by DC current that leaves its intended circuit, collects on a pipeline, and discharges into the soil, as shown in Figure 1-13. This is called interference, and can be caused by rectifier-operated cathodic-protection systems, transit systems (such as subways using DC current) and other sources of DC current.

Corrosion caused by stray-current discharge often appears as deep pits concentrated in a relatively small area on the pipe. If a main or service has this appearance, particularly at the crossing of a cathodically protected pipeline or transit facility, electrical tests should be conducted to determine if interference is the cause of the problem. Since many liquid-fuel and natural-gas pipelines are required to have cathodic protection, these lines become potential sources of stray current.

The cathodic-protection systems for pipelines should be designed and installed to minimize any adverse effects on nearby underground metallic structures. Interference can be by the installation of a bond wire between the intended circuit and the main. The wire provides a metallic path for the current to leave the main and eliminates the stray-current corrosion. Other methods that may be effective for reducing stray-current corrosion in some situations include coatings, polyethylene encasement, cathodic protection, and the strategic placement of insulators.

In dealing with stray-current corrosion, it is frequently necessary to ensure the electrical continuity of the affected pipe. Stray-current effects can only be measured and counteracted if the pipeline is electrically continuous. Pipelines to be installed in areas of high stray currents, as well as pipelines to be cathodically protected, should have bonds at all nonwelded joints.

Like cathodic protection, stray-current corrosion is a very complicated area. Consultants and trained personnel, available through engineering firms specializing in cathodic protection, may be required to help design a system that causes minimum interference with other metallic structures. Also, there are coordinating committees throughout the United States to aid in resolving stray-current problems.

General guidelines for corrosion prevention. Decisions on the need for and type of corrosion prevention for buried piping are based on conditions unique to each situation. Some areas may require specialized knowledge, but a few guidelines can be established. For piping projects in new service areas, soil tests should be used to estimate the corrosivity of the soil. Factors such as bimetallic connections, pipe material, pipe coating or covering, and cathodic protection should be reviewed at the planning stage.

For expansions or replacements in present service areas, experience and leak records are good indicators of corrosion problems to be anticipated. A careful examination of both interior and exterior conditions of piping removed from service is helpful in determining the extent of corrosion that is taking place. A number of procedures can be used to determine the corrosion or internal scaling rate from pipe specimens, including calculations of the depth of pits in the pipe or loss of weight of the pipe section. Experienced corrosion engineers should be consulted for design of corrosion-control systems.

Pipe Materials

This section discusses the most commonly used types of pipe materials and couplings or joints. Pipe description, characteristics, available sizes, normal use, and advantages or disadvantages are briefly noted in Table 1-5. It should also be noted that state and local regulations may govern the preferred types of pipe to be used for water distribution systems. Information on pipe design, manufacture,

Table 1-5. Transmission and Distribution Pipeline Materials

Material	Common Sizes Diam. in.	Normal Max. Working Pressure psi	Advantages	Disadvantages
Asbestos–cement	4–35	200	Good flow characteristics light weight, easy handling, low maintenance	Low flexural strength in small sizes, more subject to impact damage, difficult to locate underground, deteriorates when used with certain aggressive soils or waters
Cast iron (ductile, cement lined)	4–30	350	Durable, strong, high flexural strength, lighter weight than cast iron, greater carrying capacity for same external diameter, fracture resistant, easily tapped	Subject to electrolysis and attack from acid and alkali soils, heavy to handle
Concrete (reinforced)	12–168	50	Durable with low maintenance, good corrosion resistance, good flow characteristics, resists backfill and external loads	May deteriorate in alkali soil if cement type is improper or in acid soil if not protected
Concrete (prestressed)	16–120	250	Durable, low maintenance, good corrosion resistance, good flow characteristics, resists backfill and external loads	Same as above
Steel	4–120	High	Light weight, easy to install, high tensile strength, low cost, good hydraulically when lined, adapted to locations where some movement may occur	Subject to electrolysis; external corrosion in acid or alkali soil; poor corrosion resistance unless properly lined, coated, and wrapped; low resistance to external pressure in larger sizes; air-vacuum valves imperative in large sizes; subject to tuberculation when unlined
Polyvinyl chloride	4–36	200	Light weight, easy to install, excellent resistance to corrosion, good flow characteristics, high tensile and impact strength	Difficult to locate underground, requires special care when tapping, susceptible to damage during handling

difficult & costly to tap.

testing, and shipping is included in the following, where pertinent. The distribution system pipe materials discussed here include:

- Gray cast-iron pipe (CIP)
- Ductile-iron pipe (DIP)
- Steel pipe (SP) and reinforced concrete pipe (RCP)
- Asbestos–cement pipe (ACP)
- Plastic pipe, such as polyvinyl chloride pipe (PVC).

Standard specifications for the design, manufacture, and installation of pressure pipe have been developed by the American Water Works Association (AWWA), American National Standards Institute (ANSI), and American Society for Testing and Materials (ASTM). These guidelines have been established by experimentation, testing, and experience in practice. Most regulatory agencies use these guidelines to set their own specifications, which operators may be required to follow. Pipe manufacturers also publish product literature that is useful in pipe selection and installation. Table 1-5 summarizes the advantages and disadvantages of pipeline materials, and Table 1-6 summarizes related data and comments for pipe joints and their applications.

Gray and ductile cast-iron pipe.   Cast-iron pipe has been the standard for water distribution systems in the United States for many years. There are more miles of this pipe in use today than any other pipe. Two types of cast iron are commonly found in distribution systems: gray cast-iron pipe (CIP) and ductile-iron pipe (DIP). (In this text, CIP is used as the abbreviation for gray cast-iron pipe; it does not stand for cast-iron pipe in general.)

Gray cast-iron pipe is strong but brittle, usually offers a long service life, and is reasonably maintenance free. Ductile-iron pipe resembles CIP in appearance and has many of the same characteristics. It differs from CIP in that the graphite is distributed in the metal in spheroidal or nodular form—that is, in ball-shape form—rather than in flake form, as shown in Figure 1-14. This is achieved by adding a material called an innoculant, usually magnesium, to the molten iron. Ductile-iron pipe is much stronger, tougher, and more ductile than CIP. Gray cast-iron pipe is no longer manufactured in the United States, due to the increased reliability of DIP. Gray cast-iron is still used in the manufacture of valves and fittings.

Although unlined cast iron has a certain resistance to corrosion, aggressive waters can cause the pipe to lose carrying capacity through corrosion and tuberculation. The development of a process for lining pipe with a thin coating of cement mortar made it possible to minimize tuberculation and maintain the carrying capacity of the pipe. The cement-mortar lining is about ⅛-in. (3-mm) thick and adheres closely to the pipe. The lined pipe may be cut or tapped without damage to the lining. Ductile-iron pipe internal lining is normally cement mortar. Bituminous external coating and polyethylene wraps are methods commonly used to reduce external corrosion.

Table 1-6.   Pipe Joints and Their Applications

Pipe Material	Type of Joint	Application
Asbestos–cement	Coupling with rubber ring gasket	All locations
Cast iron	Push-on	General use
	Mechanical	Soft soils where settlement is anticipated or where flexibility is required
	Bell and spigot	Only in stable soil where settlement is not excessive
	Flanged	Where valves or fittings are to be attached in vaults or above grade
	Flexible ball	River crossings
Concrete	Galvanized-steel ring, bell-and-spigot types, or their variations with rubber gaskets and cement fills	All locations
Plastic	Solvent weld	In small lines
	Bell and spigot with rubber O-rings	In large sizes
Steel	Mechanical-type couplings	Pipes less than 24 in. ID, especially with coal–tar linings
	Welded joints	Pipes greater than 24 in. ID with inside coatings
	Flanged joints	Where valves or fittings are to be attached
	Expansion joints (stuffing-box type)	At points to relieve strain on welded joints

A. Cast Iron (Graphite Flakes)

B. Ductile Iron (Graphite Nodules)

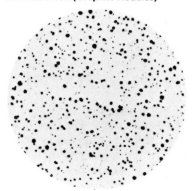

Courtesy of Ductile Iron Pipe Research Assoc.

Figure 1-14.   Microphotographs of Cast Iron and Ductile Iron

Ductile-iron pipe is available with standard wall thicknesses (referred to as THICKNESS CLASSES) in diameters of 4 in. (100 mm) and larger. The standard lengths are 18 and 20 ft (5.5 and 6.1 m). Ductile-iron pipe is strong and can withstand the working pressures found in distribution systems. It is also durable and can be cut and tapped in the field.

Advantages of DIP include:

- Durability, strength, flexural strength
- Good carrying capacity for a given external diameter
- Fracture resistance
- Ease of tapping.

Disadvantages include:

- Potential for electrical and chemical corrosion if not protected from acidic or alkaline soils or aggressive waters
- Relatively great weight, which increases the difficulty of installation.

CIP and DIP joints.   Gray cast-iron and ductile-iron joints of the following types have been used to join pipe lengths together (listed in order of development):

- Flanged joints
- Bell-and-spigot joints (no longer installed)
- Mechanical joints
- Ball-and-socket or submarine joints
- Push-on joints.

Flanged joints consist of two machined surfaces that are tightly bolted together with a gasket between them to prevent leakage. These joints are easy to make and require no special tools. They are used aboveground in water plants, pump houses, and other places where rigidity, self-restraint, and tightness are required. Figure 1-15 shows a flanged-joint cross section and an installation. Flanged joints will not flex and are not normally used underground.

Leaded bell-and-spigot joints (Figure 1-16) were commonly used until the 1920s. They have been gradually replaced by joints requiring less skill and time to make. Generally, these joints, if leaking, should be replaced with a mechanical joint.

Mechanical joints, which gradually replaced the bell-and-spigot joint, are shown in Figure 1-17. This joint is made by bolting a movable follower ring on the spigot to a flange on the bell and compressing a rubber gasket to form a tight seal. The mechanical joint is a little more expensive than previously mentioned joints, but is easily made and requires no special skill. Since the bell-and-spigot ends need not fit tightly, each joint can be made to deflect slightly.

Ball-and-socket joints, shown in Figure 1-18, are special-purpose joints most commonly used for submerged installations. Their great advantage is that they provide for large deflections (up to 15°). This makes them very useful for pipe lines laid across mountainous terrain or under rivers. Boltless flexible pipe joints, designed on the ball-and-socket principle, are also available.

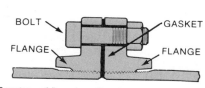

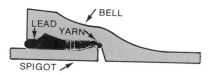

Courtesy of American Cast Iron Pipe Company

Courtesy of U.S. Pipe

Figure 1-15. Flange Joints

A. Partially Completed

B. Completed

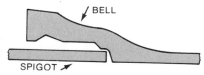

Figure 1-16. Leaded Bell-and-Spigot Joint

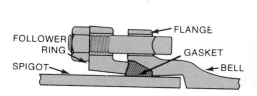

*Courtesy of American
Cast Iron Pipe Company*

*Courtesy of Pacific States
Cast Iron Pipe Company*

Figure 1-17. Mechanical Joint

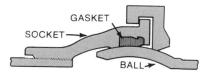

Courtesy of American Cast Iron Pipe Company

Figure 1-18. Ball-and-Socket Joint

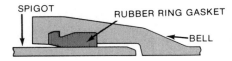

Courtesy of American Cast Iron Pipe Company   Courtesy of J-M Manufacturing Co., Inc.

Figure 1-19.   Push-On Joint

Push-on joints, the most recently developed pipe joints, are the most popular joints in water distribution system installations today. The joint consists of a bell, with a specially designed recess to accept a rubber ring gasket, and a beveled-end spigot. The joint offers ease of installation, and when made up, the rubber ring gasket is compressed to produce water tightness and locked in place against further displacement. Push-on joints are available in several designs. In addition to ease of installation and water tightness, the joint permits deflections of 3° to 5°, depending on the design, a fact that makes it possible to install pipe on a curve. Small diameters may be assembled by hand; larger sizes usually require mechanical aids. Figure 1-19 shows a push-on joint.

Steel pipe and reinforced concrete pipe.   Steel pipe and reinforced concrete pipe are sometimes used as large feeder mains in water distribution systems. These pipe materials are described in Volume 1, *Water Sources and Transmission*, Module 4, Water Transmission.

Asbestos–cement pipe (ACP).   Asbestos–cement pipe was introduced into the United States around 1930. It is now widely used in water systems, especially in areas where metallic pipe is subject to corrosion.

Asbestos–cement pipe is made of asbestos fibers, silica sand, and portland cement in a ratio of about 20:30:50. The asbestos fibers provide much of the strength of the cement in the pipe and reduce the content of uncombined calcium hydroxide to less than 0.5 percent, thus producing a relatively stable and chemically resistant compound. Asbestos–cement pipe is available for service in working pressure classes of 100, 150, and 200 psi (690, 1030, and 1380 kPa). It is commonly supplied in standard sizes from 4 to 42 in. (100 to 1070 mm) in each class and lengths of 10 to 13 ft (3 to 4 m). In the presence of aggressive soils and certain adverse water conditions, ACP should not be used. In any case, appropriate manufacturer's recommendations and AWWA standards should be consulted before installation. For use with soft waters, pipe-production standards should be specified with a lower limit of uncombined calcium hydroxide to prevent leaching.

Advantages of ACP include:

- Light weight

- Low initial cost

- Nonmetallic construction (It is nonconductive and, therefore, immune to electrolysis and tuberculation.)

- Smooth bore (The initial *C* factor is 140 plus, and it often stays high. Water pressure and volume remain constant.).

Disadvantages of ACP include:

- Breakage (This pipe breaks easily if it is flexed in handling or during uneven settling in the pipe trench. It must be handled and installed carefully. Proper bedding in the trench is very important in preventing flex and breakage after installation. Operators must also be particularly careful when excavating around ACP, because it can be easily punctured by excavating tools.)

- Nonmetallic construction (Asbestos–cement pipe cannot be located with conventional pipe locators or thawed electrically. [Metal tape or wire can be installed running along the pipe to allow location with pipe locators that sense metal.])

- Special care required during tapping (Corporation stops often must be installed with a tapping saddle due to possible splitting of the pipe.)

- Limited sizes (Short lengths require more couplings.)

- Special safety precautions required during installation (When pipe is cut or machined during installation, masks must be worn to prevent inhalation of asbestos dust, which has been identified as a cause of lung cancer.).

Asbestos–cement pipe is joined by sleeved couplings also made of asbestos–cement, as shown in Figure 1-20. Couplings should be marked according to the pressure of the pipe with which they are to be used. The couplings consist of a sleeve with two interior rubber rings that provide for a watertight seal. The joint provides for some deflection without leakage. The joint is normally jacked or pulled onto the pipe. Except for the couplings, fittings for ACP are generally cast iron.

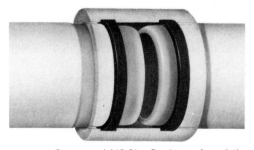

Courtesy of A/C Pipe Producers Association

Figure 1-20. Asbestos–Cement Sleeve Coupling

Plastic pipe. Plastic pipe is being used increasingly in the water utility industry. Plastic pipe was first introduced in the United States around 1940. All plastic pipe now used in water distribution should carry the seal of approval of the National Sanitation Foundation for potable water use and meet applicable AWWA or other industry standards.

Plastic pipe materials include polyvinyl chloride (PVC), polyethylene (PE), polybutylene (PB), and acrylonitrile-butadiene-styrene (ABS). Only PVC is used widely for distribution pipe. AWWA standards cover PVC pipe in sizes from 4 to 12 in. (100 to 300 mm) in diameter.

Polyvinyl chloride pipe is a semirigid pipe manufactured by an extrusion process. The fittings are made by a mold process. It is available in diameters up to 36 in. (0.9 m); lengths of 20–40 ft (6–12 m); and various types, grades, and pressure ratings. Within a given nominal pipe diameter, there are several equivalent systems for specifying internal and external diameters. Plastic pipe is normally specified by a sizing system other than the common iron pipe size (IPS) system. The AWWA standard is based on outside diameter the same as DIP. A new sizing system, termed the standard dimension ratio–pressure rated (SDR–PR) system, is a ratio of the outside pipe diameter to wall thickness. This system recognizes the strength properties of plastic and allows pipe of one pressure rating to be available in various sizes. Check with local state regulations to determine what types and sizes of plastic pipe are approved.

Advantages of PVC pipe include:

- Exceptionally smooth interior (It has superior flow efficiency and a C value of at least 150.)
- Chemical inertness (It usually will not react with water on the inside or soil on the outside. No extra linings, coating, or other special protection are needed to prevent corrosion.)
- Very light weight (It is cheaper and easier to ship, handle, and install. It is also flexible to some degree, permitting deflection as the pipe is laid.)
- Ease of handling and cutting.

Disadvantages of PVC pipe include:

- Susceptibility to damage during handling (Plastic pipe's strength can be affected by surface scratches or falling rocks during excavation and/or backfilling. It is recommended that the pipe be repaired or replaced if a scratch is deeper than 10 percent of the pipe wall thickness.)
- Nonmetallic construction (The pipe cannot be located with conventional pipe locators unless metal wire or tape is installed with the pipe, and frozen pipe cannot be thawed electrically.)
- Limited ability to hold corporation stops (A saddle should be used on any tap larger than 1 in. [25 mm].)
- Susceptible to damage from the ultraviolet radiation in sunlight when left uncovered for long periods.

Pipe made of PVC may be joined by a bell-and-spigot push-on joint or by a solvent-weld joint. In the push-on method, a rubber O-ring set in a groove in the bell end provides the necessary seal. In the solvent method, an NSF-approved solvent is used, as recommended, which causes a sleeve-type coupling and the plain-end pipe to be cemented together, creating a watertight seal.

1-4. Pipe Installation

The proper installation of pipe will enable the utility to reduce or minimize maintenance problems and aid in public health protection. Problems with an improperly installed main can result in temporary loss of service to a substantial number of consumers. Pipe cannot be buried and forgotten; it must be both carefully made and carefully installed.

Several operations in the installation of distribution lines are the same regardless of the pipe material used. These operations will be discussed here without reference to specific materials, except as may be necessary for clarification. Refer to AWWA standards and manuals for specific installation procedures for individual materials.

Pipe Handling

Inspection. Pipe normally receives a final inspection before leaving the factory. It can, however, be damaged in transit and should be inspected before it is unloaded or as it is unloaded. All materials—pipe, rings, gaskets, and fittings—should also be checked against a tally sheet for proper size, class, number, etc., before accepting delivery. Pipe should be carefully checked to see that it is not chipped or cracked. Any missing, damaged, or improper materials should be acknowledged by the driver, in writing, and reported to the shipping company. Damaged material should be saved and a claim made according to the shipping company's instructions.

Unloading. It is generally best to use mechanical equipment to unload pipe to protect it from damage. Care should be taken to prevent abuse and damage to the pipe, no matter what method of unloading is used. Pipe may be unloaded using a derrick and cable with hooks or a sling, or by skids and snubbing ropes, as shown in Figure 1-21. Pipe should not be dropped or allowed to strike other pipe. Lined and coated pipe must be handled carefully to avoid damaging the protective material. A forklift can damage the pipe coating or lining. Hooks for handling lined pipe should be rubber covered. Care should be exercised to avoid damage to plastic pipe when working in cold weather. Pipe should never be unloaded by pushing it off the truck.

When the pipe is unloaded at the jobsite, it should be placed as near as possible to where it is to be used to avoid excessive handling. Pipe is normally placed alongside the trench on the side opposite the excavated material. It should never be moved by the direct use of a metal blade or bucket, such as those found on a backhoe. All pipe, fittings, and gasket material should be protected from contamination and kept as clean as possible.

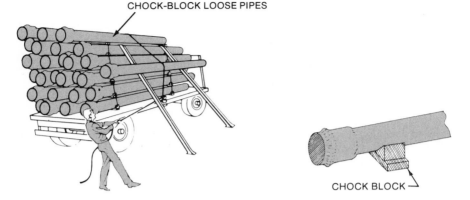

CHOCK-BLOCK LOOSE PIPES

CHOCK BLOCK

Figure 1-21. Unloading PVC Pipe With Snubbing Ropes

Stacking. If the pipe is to be stockpiled, stacking should be done in accordance with the manufacturer's directions. The following general recommendations should be observed:

- Stockpiles should be built on a flat base, off the ground to minimize contamination (Figure 1-22).
- The bottom layer should be supported uniformly along the barrel of the pipe so that the bells do not touch the ground.
- When stacking pipe, bell ends of bell-and-spigot pipe should project over the end of the barrels in alternate layers.
- Pipe of the same sizes and classes should be grouped together. Short lengths, fittings, and adapters should be placed in separate piles.
- Stack height should be controlled within the limits of safety and practicality.
- PVC pipe should not be stacked loose over 3-ft (0.9-m) high. It should be protected from the ultraviolet radiation in sunlight by covering with a tarpaulin or other opaque material. Air should be able to circulate beneath the cover, and plastic sheets should not be used.
- Pipe should be stacked so that it will not roll downgrade.

Stringing. If pipes are to be distributed (strung) along the trench, the following procedures should be observed:

- Lay or place pipe as near to the trench as possible to avoid excess handling. Secure the pipe so it cannot roll.
- If the trench is open, it is best to string pipe on the side opposite the spoil bank wherever possible, so that pipe can be moved easily to the edge of the trench for lowering into position.
- If the trench is not open, determine which side excavated earth will be thrown to, and then string the pipe out on the opposite side, leaving room for the excavating equipment.

Courtesy of Ductile Iron Pipe Research Assoc.

Figure 1-22. Pipe Stockpiles Kept Off Ground Surface

- Place pipe so as to protect it from traffic and heavy equipment. Safeguard pipe from the effects of any blasting that may be done.
- Face the bells in the direction the work is to proceed.
- Take care that pipe does not roll into the trench.
- If there is a danger of vandalism or other damage, string out only enough pipe for one day's laying.
- Cover the pipe ends to prevent contamination.

Regardless of whether pipe is strung or stacked, all rubber gaskets and lubricant should be stored at a central point and distributed as needed. Keep them clean and away from oil, grease, and excessive heat. The best way is to store them in a cool, dry place in their original cartons.

Excavation

Preliminaries.   Detailed plans of the project should be prepared by the project engineer. These plans should include alignment, grade, and depth specifications. The plans should be submitted to the agencies that have jurisdiction over the project area; approval from appropriate state, county, city, or other organizations should be obtained. Any question of access should be settled. The same plans should be submitted to and approved by the affected gas, sewer, and other utilities.

Before digging, the operator must notify other utilities of the impending excavation, and underground lines should be physically located and marked by the responsible utility. Utility companies are usually willing to do this, provided they are given adequate notice.

The excavation site should be properly marked with barricades, flashing lights, warning or detour signs, and whatever else is needed to protect the work crew and the public. Work-area protection is discussed in the Safety section of this module. It is also a good idea to mark the work area in advance with signs to let the public know what will be done, why it is being done, and that the utility is sorry for any inconvenience.

Selection and use of appropriate excavation equipment is an important consideration. Machines that are too large will cause unnecessary damage. Machines that are too small will move the job too slowly, which is not only uneconomical but may also cause problems such as settlement due to frozen backfill or side-slope instability. Bucket widths should be checked to minimize excavation and maintain side support. Most buckets can be changed very quickly, and an appropriate size can be rented, if necessary.

Most trench work today is done by hydraulic backhoes (Figure 1-23), because they are easy to operate and provide excellent control. Other trenching machinery may be useful where maneuver room is available and the soil contains little rock. Tight spaces and rock or other underground obstructions may require the use of a backhoe.

There are a variety of machines available for cutting hardened surfaces. Asphalt is cut with a piece of equipment called a clay spade or a stomper. Concrete is cut with large power saws. The edges of all cuts should be smooth. A jack hammer is also often used, but it does not leave a very smooth cut. All concrete and asphalt debris must be hauled away before the excavation starts—it cannot be used as fill.

Trench depth.   Several factors control the depth and width of trench excavation. Climatic conditions, the load the pipe will be carrying, ground-water conditions, and soil characteristics are some of the technical factors that must be considered. Practical considerations include pipe size, available equipment, economics, and surface-restoration requirements.

Because of the expense of excavation, the trench should be as shallow and as narrow as conditions permit. The depth of the trench or, more specifically, the depth of cover over the pipe, usually depends on the depth of frost penetration in colder climates and on the surface-load conditions in warmer climates. The cover

Courtesy of American Cast Iron Pipe Company
Figure 1-23.   Hydraulic Backhoe

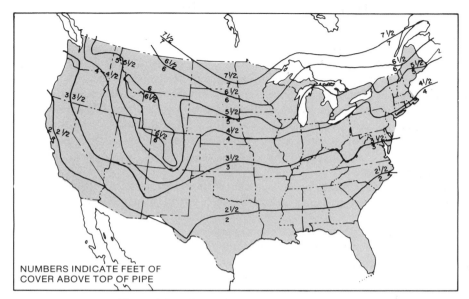

Figure 1-24.   Recommended Depth of Cover

must compensate for these factors. Excessively deep trenches, however, add expense and trouble; they must be shored and braced, and any water that collects in them must be pumped out.

During an extremely severe winter, frost may penetrate to twice the average winter frost depth. Regardless of the depth of frost penetration, structural requirements dictate 2 to 2½ ft (0.6 to 0.8 m) of minimum cover (Figure 1-24).

Points to consider in areas where freezing may be a problem include future changes in ground level, frost penetration under roads and driveways, elevation of service lines, looped mains, and insulation.

Future changes in ground-surface elevations due to road construction or erosion may significantly decrease the desirable amount of pipe cover and cause freezing problems. Frost penetration is significantly greater under roads and driveways than under areas not plowed or traveled. Service lines are often installed at a higher elevation (more shallow depth) than the main, resulting in more freezing problems. The smaller the diameter of the line, the more likely it will freeze (depending on flow). Looped mains, where water can flow, will have fewer freezing problems.

If adequate depth is not possible, the line should be insulated. Closed-cell styrofoam insulation board, 2 in. (50 mm) thick and 2 to 4 ft (0.6 to 1.2 m) wide, depending on pipe and trench size, works well. The insulation should be placed on stable, flat fill, a few inches above the pipe, for as long a distance as necessary. Bead-board type sheets should not be used; they are less expensive but have low insulating value and will absorb water, thereby further reducing their effectiveness.

Trench width.   The width of the trench is governed by safety considerations, economics, and the need to minimize the external loading on the pipe. Generally, the trench width should be no more than 1 to 2 ft (0.3 to 0.6 m) greater than the outside diameter of the pipe. This permits proper installation of the pipe and gives the worker enough room to make up the joints and tamp the backfill under and around the pipe. Furthermore, the superimposed load can increase markedly with trench width; wide trenches should be avoided for small-diameter pipe, particularly in hard clay soils. Refer to Tables 1-7, 1-8, and 1-9 for appropriate trench widths for ductile-iron, asbestos–cement, and PVC pipe. For special conditions, such as heavy-load areas, contact the manufacturer for recommendations.

When pipe must be laid on a curve, the maximum permissible deflection of the joint limits the degree of curvature. Trench width for curved pipelines will be wider than usual.

Trenching.   As the trenching process proceeds, excavated soil should be piled on one side of the trench, between the trench and the traffic, but far enough away from the trench so that it will not fall back into it or significantly increase the weight on the trench wall, which will increase the likelihood of cave-in. There must be enough room for workers to walk alongside the trench.

Tools and materials should be stored so that they do not directly damage trees, lawns, etc. Stored material should also be kept back from the wall of the trench to prevent cave-ins. It is also important to consider what might happen during a rainstorm or extended exposure to the elements.

Table 1-7.   Trench Widths for Ductile-Iron Mains

Nominal Pipe Size in.	Trench Width in.	Nominal Pipe Size in.	Trench Width in.
4	28	20	44
6	30	24	48
8	32	30	54
10	34	36	60
12	36	42	66
14	38	48	72
16	40	54	78
18	42		

Table 1-8.   Trench Widths for AC Pipe

Pipe Diameter in.	Trench Width Minimum in.	Trench Width Maximum in.
4	18	28
6 or 8	20	32
10 or 12	24	36
14 or 16	30	42

Table 1-9.   Trench Widths for PVC Pipe

Pipe Diameter in.	Trench Width Minimum in.	Maximum in.
4	18	29
6	18	31
8	21	33
10 and above	1 ft greater than outside diameter of pipe	2 ft greater than outside diameter of pipe

The bottom of the trench must be dug to the specified depth while maintaining the specified grade. Both should be double-checked. The bottom of the trench should form a continuous, even support for the pipe. This will, most likely, require some hand work and possibly some special fill material, as discussed later in this module under bedding.

Excavation of the trench should not be extended too far ahead of pipe laying. This will minimize the possibility of cave-ins or flooding of the trench during rainy weather. An open trench not only presents a danger to the worker, but also to traffic and pedestrians. Danger is often greater after working hours than it is during construction hours. These dangers can be minimized by keeping open sections of trench as short as possible, and by using written warnings, proper barricades, signals, and flaggers. Local regulations may require that the trench be filled or protected in a specific way overnight. Long sections of open trench may also cause disruptions and inconvenience to local residents, certain municipal services, and emergency vehicles. The following paragraphs deal with problems that may be encountered during trenching.

Rock excavation.   The term rock applies to solid rock, ledge rock (that is, hardpan or shale), and loose boulders more than 8 in. (200 mm) in diameter. In any type of rock formation, the rock must be excavated to a level 6 to 9 in. (150 to 230 mm), depending on pipe size, below the grade line of the pipe bottom. Excavated rock should be hauled away and not used for backfill.

Some materials may require blasting, which is expensive and dangerous, and which should be avoided if possible. If blasting is unavoidable, every precaution should be taken to prevent damage and injuries. The operator should keep dated records of conditions prior to blasting and also of any damages or injuries incurred as a result of the blasting.

Bad soil.   Where soil conditions are bad (for example, coal mine debris, cinders, sulfide clays, mine tailings, factory waste, or garbage), the poor soil should be excavated well below the grade line, hauled away, and replaced with more suitable material.

Ground water.   Ground water will enter the trench when the trench bottom is below the water table. Attempting to lay or join pipe in water is not a good practice. Disposal of this water is not only expensive, but it can also be hazardous. Where high ground water is unavoidable, a system of WELL POINTS should be placed at intervals along the trench, as shown in Figure 1-25. The points should then be connected to a pipe manifold and pumping system. This system lowers the ground-water level and allows work to proceed. Be sure to check with local regulatory agencies to assure proper disposal of the pumped water.

Trench-wall failure.   There is a tendency to make trench walls too steep and trenches too narrow when digging in firm soil and to make trenches too wide when digging in wet silts or free-running soils. Soil types, water content, and slope stability are constantly changing factors, requiring modifications in trenching methods.

Soils are generally characterized as clays, TILLS, sands, and silts. Firm clay and tills with below-optimum moisture content can be excavated easily and safely,

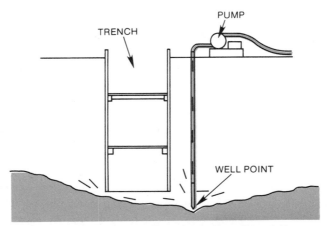

Figure 1-25.   Well Point Pumping to Keep Trench Dry

but require careful control during backfill. Dry silt, which is fairly uncommon, behaves in the same manner. Operating in dry sand requires special care during excavation, because sand can slip or run easily; backfilling with sand requires no special care. Wet silts, which are common, often require special treatment because of their unpredictability and potential hazard.

The possibility of trench-wall failure and cave-in is undoubtedly the greatest danger for workers in the trench. This failure may occur due to a variety of conditions, including:

- Water pressure in the soil
- External loads (for example, equipment being moved close to the trench)
- Excavated soils lying too close to the trench
- Trench walls that are too steep for the type of soil being excavated.

Failures occur more often in winter and early spring due to weather conditions and periods of high runoff. These should be considered particularly dangerous times for excavation. Failures give little warning and occur almost instantaneously. Some of the danger signs operators should look for are:

- Tension cracks in the ground surface parallel to the trench, often found a distance of ½ to ¾ of the trench depth from the edge
- Material crumbling off the walls
- Settling or slumping of the ground surrounding the trench.

Trench-wall failures can be prevented if problem areas and conditions are recognized before they occur. Techniques for preventing cave-ins include proper sloping of the trench walls, the use of shields or shoring, and limiting the loads applied to the soil near the excavation.

An important factor in preventing trench-wall failure is a steep wall slope. Narrow vertical-sided trenches should never be constructed except in rock. Slope-sided or U-shaped trenches, with sides having a 1:1 (45°) slope, are usually

Figure 1-26.   Sloped Trench Wall

Figure 1-27.   Examples of Shoring (left) and Shielding (right)

dug in normal soils. In very loose sand, the trench sides may require a 2:1 (30°) slope. Figure 1-26 illustrates a sloped trench wall.

Trenches should be shored and braced or movable shields should be used whenever soil, weather, or excavation conditions require. Local regulations and Occupational Safety and Health Administration (OSHA) regulations require shoring or shielding in certain situations. Examples of shoring and shielding are shown in Figure 1-27.

To limit the loads applied to the soil near the excavation, loaded trucks and other vehicles should be kept away from the excavation, and the soil should be placed as far away from the trench as practical. Trench safety is discussed in more detail in the Safety section at the end of this module.

Other utility lines.   The lines of other utilities (especially sewer pipes) must be avoided. This will usually require careful work with machines and by hand. Pipe must be routed around these lines or the lines themselves must be rerouted. Special fittings may be necessary for this. Pipe must not touch or rest on other lines or support another structure.

Potentially serious problems can result when sanitary sewers and water mains are buried close together. Sewer lines often leak, and before the operator is aware of it, a water main could be surrounded by sewage-contaminated soil. If a break occurs in that main, the operator is faced with a serious public health hazard. Whenever a water line approaches a sewer line, the two lines must be separated by a safe distance, both vertically and horizontally. Refer to local or regulatory agency rules for minimum separation distances and other restrictions, such as pipe materials, joints, and connections.

Pipe for potable water should not be laid in the same trench with a sewer line. Wherever a water main crosses a sewer line, the water main should be at least 3 ft (0.9 m) above the sewer line, and the sewer line should be made of ductile iron with no joints, except pressure-type joints, for 10 ft (3 m) on either side of the crossing. If a sewer line must cross over a water main, the requirements are usually even more restrictive. If laid horizontally, the distance between the two pipes should be at least 10 ft (3 m).

Bedding.   The trench bottom must be properly leveled and compacted so that the barrel of the pipe will have continuous, firm support along its full length. A leveling board should be used to ensure that there are no voids or high spots and that the grade is correct. Any high spots should be shaved off, and voids should be filled with well-tamped soil. Care must be taken to avoid damage to any pipe that may be located beneath the trench bottom.

The practice of laying pipe on blocks or earth pads to allow room and position for joining pipe is not recommended. There are many situations where special pipe BEDDING may be required. Bedding is usually considered to be the material in which the pipe is partially or completely embedded. The use and type of bedding is usually specified by the design engineer. Bedding is, or should be, used when trench bottoms do not provide a suitable pipe bed or when a greater load-bearing capacity of the pipe is desired. Trench bottoms that are unsuitable for pipe beds include areas that are soft, contain large rocks or cobbles, consist of unstable material, or cannot be graded uniformly. The bedding material should be a well-graded granular material up to 1 in. (25 mm) in size. It should contain no lumps or frozen ground and should not be a material like clay that can be sensitive to water. The bedding material should be spread over the trench bottom to the full width of the trench.

When natural trench bedding is unable to structurally support the pipe, it might be necessary to over-excavate 12 to 24 in. (300 to 600 mm) and backfill with a coarse gravel material in well-mixed sizes up to 3 in. (75 mm). This material should be compacted and more added, if necessary, until the trench bottom has been brought up to the proper grade. This is especially important in areas where there is an upward flow of ground water or in muddy or soft material. The trench bottom has to be stabilized before the pipe can be laid. In extreme cases, pilings or timber foundations may be required.

Bedding is also used to increase the load-bearing capacity of the pipe. If the pipe is supported only over a narrow width, as with a round pipe on a flat-bottom trench, the intensity of the load at the bottom of the pipe will be considerable and failure will be more likely. Distribution of the load over a wider area will reduce the load intensity beneath the pipe and, consequently, reduce the likelihood of failure. This is particularly critical in shale or rock formations. Using the example shown in Figure 1-28, if a pipe is bedded over half-way to the SPRING LINE of the pipe, then it can support 36 percent more weight than it could without any bedding. If bedding is extended to the spring line of the pipe, then it will support 114 percent more weight than the original unbedded pipe. See Figure 1-28 for other examples of the correlation between bedding and supporting strength. For comparison, Figure 1-29 illustrates several examples of poor bedding practice.

Required bedding conditions vary greatly depending on the pipe material, size, and anticipated loading over the filled trench. Appropriate AWWA standards and manufacturer's recommendations should be consulted for specific information.

Laying Pipe

Inspection and placement.   After the trench bottom has been prepared, the pipe may be set in place. The proper procedure varies somewhat with the type of pipe, but the following general directions for laying pipe apply to all types.

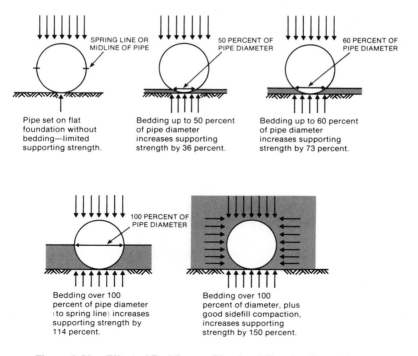

Figure 1-28.   Effect of Bedding on Pipe Load-Bearing Strength

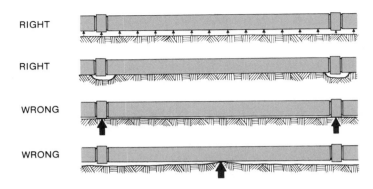

RIGHT

RIGHT

WRONG

WRONG

Courtesy of J-M Manufacturing Co., Inc.

Figure 1-29.   Good and Bad Pipe Bedding

Before the pipe is lowered into the trench, it should be inspected for damage and any unsatisfactory sections should be rejected. The inside of each pipe length should also be inspected and any dirt, oil, grease, animals, and other foreign matter removed. If the pipe has been tapped before being placed in the trench, the holes or corporation stops in the pipe should be covered to keep dirt out during pipe placement.

If mud and surface water have been permitted to stand or flow through strung-out pipe, the inside should be swabbed with a strong hypochlorite solution. This will save time and expense later when the pipe is disinfected. All gaskets should be kept clean and dry.

The pipe should be lowered into the trench by mechanical equipment, if possible. It should never be rolled into the trench from the top. Smaller diameter pipe may be lowered into the trench by two people using two ropes, one rope looped around each end of the pipe. A knot should be made in one end of each rope. While standing on the end of the rope with the knot, the rest of the rope should be let out hand-over-hand until the pipe rests on the bottom of the trench.

Larger pipe sizes are best handled with power equipment. When pipe is lowered by machinery, it is usually supported by a sling in the middle of the pipe length. The sling must be removed once the pipe is down. This requires that a certain amount of earth be taken out. The space under the pipe should not be longer than 18 in. (460 mm), and bedding or backfill must be replaced and tamped. When a lifting clamp is used, this excavation is not necessary.

Pipe that is joined by couplings may be laid in either direction. Belled-end pipe is normally laid with the bells facing in the direction in which the work progresses, except downhill, where the direction is reversed.

A conscious effort should be made to keep the inside of the pipe clean. When pipelaying is not in progress, the open ends of installed pipe should be plugged to prevent the entrance of animals, dirt, and trench water. Pipe lengths should never be deflected in the joint to any greater degree than that recommended by the manufacturer.

Jointing

Jointing pipe is an important part of pipe installation and must be done correctly to ensure long and satisfactory (watertight) service. The method for making up a joint depends on the type of pipe material and the type of joint. In all cases, sand, dust, tar, and other foreign material should be wiped carefully from the gasket recesses in the bells, otherwise the joint may leak. The spigots should be smooth and free from rough edges.

Joint holes must be dug by hand where pipes will be joined. This will allow room for the joint to be installed while the remainder of the pipe rests on the bed. Joints or couplings should not be allowed to support the pipe. The size of the hole will vary with the type of pipe and joint used. Joint holes that are too large result in undue stress on the pipe.

Specific directions provided by each manufacturer for making up joints should be followed. General directions for the more common joint types are given in this section.

1. Clean bell and end. Be sure no dirt can lodge between the ring and the bell or pipe end.

2. Set ring in groove with painted edge facing toward the end of the bell.

3. Lubricate pipe end with a light film of lubricant.

4. Push end in so that reference mark on spigot end is flush with end of bell.

Courtesy of J-M Manufacturing Co., Inc.

Figure 1-30.   Assembling PVC Pipe With Rubber Ring Push-On Joint

Push-on joints.   The PUSH-ON JOINT is probably the most common type of joint used today. It consists of a special bell that is part of the pipe. Inside the bell is a groove in which a heavy rubber gasket rests. The spigot of the bell is beveled. This joint is generally completed in three steps, shown in Figure 1-30 and described as follows:

1. Clean the inside of the bell and the outside of the spigot.

2. Turn the rubber gasket in the proper direction and slip it into the gasket recess in the bell socket. Note: There are several different shapes of push-on rubber gaskets used in the industry, and the inside of the bells are shaped differently for each kind of gasket. It is important to use the correct gasket for each joint and to make sure that the gasket faces in the right direction.

3. Apply a thin film of recommended lubricant to the inside of the gasket and to the beveled spigot. Never use oil or grease—oil ruins the rubber of the gaskets and grease causes bacteria growth.

4. Push the spigot end completely into the bell with a forked tool or a jack.

Mechanical joints.   MECHANICAL JOINTS may be used when it is important that the joint or the pipe remain firmly in place. A mechanical joint consists of four parts—a flange cast as part of the pipe bell, a rubber gasket that fits a groove in the bell, a gland (follower ring) to compress the gasket, and nuts and bolts to tighten the joint. The installation generally involves the following three steps (Figure 1-31):

1. Wipe the pipe ends clean; paint with a soap solution. Place the gland on the spigot end with the face toward the bell; then place the rubber gasket over the spigot end, with the thick edge facing the follower gland.

2. Push the pipe forward to set the spigot, and press the gasket into place within the bell.

3. Insert the bolts in the flange and the follower gland; screw the nuts on finger tight. Tighten the nuts at opposite positions with a torque wrench or ratchet wrench until all are tightened equally.

ACP couplings.   Asbestos–cement pipe is joined by a coupling made of asbestos–cement with rubber ring gaskets. It is easy to assemble and gives a tight seal. The steps in making up the joint are as follows:

1. Place the rubber gaskets inside the coupling.

2. Clean and lubricate the pipe ends and the coupling with the special lubricant supplied by the manufacturer. Do not use oil or grease.

3. Place the coupling on the installed pipe length (unless the coupling comes ready-made with the pipe).

4. Align the pipe ends.

5. Insert last length into coupling and push on the free end until the pipe is pushed home. This is done with a crowbar against a block of wood across the free end of the pipe (Figure 1-32) or a mechanical puller device for large pipe.

1. Clean pipe; install gland and rubber gasket; soap gasket and pipe.

2. Push pipe forward to set spigot; press gasket into place within bell.

3. Insert bolts; tighten nuts first with fingers, then with ratchet or torque wrench.

Courtesy of Pacific States Cast Iron Pipe Company

Figure 1-31.   Assembly of a Mechanical Joint

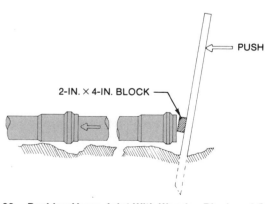

PUSH

2-IN. × 4-IN. BLOCK

Figure 1-32.   Pushing Home Joint With Wooden Block and Crowbar

For any type of pipe or joint an operator should:

- Closely follow the manufacturer's instructions for different kinds of joints and pipe.
- Keep the inside of the pipe, joints, gaskets, and associated grooves clean.
- If calcium hypochlorite tablets are to be used to disinfect the completed main, attach the tablets to the inside of each pipe section with gasket cement before the pipe sections are joined.
- Install polyethylene wrapping (if it is used for protection of metallic pipelines in corrosive soil) before joining the pipes.
- Keep the pipe horizontal when joining.
- Use a bar or a come-along when homing the pipe (Figure 1-32). The bucket of the backhoe can break or damage the pipe or coupling.
- When using a trench box, anchor the pipes so they will not move when the box is pulled ahead.

After the pipe is assembled in the trench, the joints should be checked to make sure that the ends abut each other in such a manner that there is no unevenness of any kind along the inside of the pipe. Also, the jointing material or gasket must be properly and solidly in place.

Fittings.   Crosses, tees, ells, hydrants, valves, and services are planned for well in advance, and it is best to install them as the pipeline is laid. They are usually connected with joints such as those previously discussed. For discussion of valves, hydrants, services, and meters, refer to Module 3, Valves; Module 4, Hydrants; and Module 5, Services and Meters.

Pipe Under Railroads, Streets, and Highways

When pipelines are installed under railroad tracks, certain streets and highways, and other obstructions, the installation must be such that the pipe is accessible for repair or replacement and that it is protected against superimposed extra dead and impact loads. One method used to provide access and protect the pipe is to install a casing pipe and then place the water main or carrier pipe inside of it. When casing pipe is used for highways or railroad crossings, the project should be completed in accordance with applicable federal, state, and local regulations.

In the case of railroad crossings, the project should comply with regulations established by the railroad company. General practice permits boring for casing diameters through 36 in. (920 mm) with a maximum length of about 175 ft (55 m), jacking for diameters 30 in. through 60 in. (760 mm through 1500 mm) with lengths of about 200 ft (60 m), and tunneling for pipes 48 in. (1200 mm) and larger for longer lengths. The casing pipe should be 6–8 in. (150–200 mm) larger than the outside diameter of the water main bell ends. Carrier pipe can then be pushed or pulled through the completed casing pipe. Chocks or skids should be placed on the carrier pipe to ensure approximate centering within the casing pipe and to prevent damage during installation. Care must be exercised in order to avoid metal-to-metal contact. In order to avoid the transfer of earth and possibly live animals or insects to the carrier pipe, the space between the carrier and casing pipes should not be filled completely.

Thrust Blocks and Anchors

Water under pressure and water in motion can exert tremendous forces inside a pipeline. One of these forces, THRUST, pushes against fittings, valves, and hydrants, and can cause couplings to leak or pull apart entirely. All tees, bends, caps, plugs, hydrants, and other fittings that either change the direction of flow or stop the flow completely should be restrained or blocked.

Although thrust is caused primarily by water pressure and water hammer, it can be caused by any factor that produces a pressure or force in the pipeline. Thrust almost always acts perpendicular (at a 90° angle) to the inside surface it pushes against, as shown in Figure 1-33. In the figure, the thrust acts horizontally outward, tending to push the fitting away from the pipeline. Uncontrolled, this thrust can cause movement in the fitting or pipeline that will cause leakage or complete separation at either coupling A or at any other nearby coupling upstream of the fitting.

In general, thrust can occur wherever the pipeline changes direction, such as at tees, bends, and crosses; wherever the pipeline changes diameter, such as at reducers; wherever the flow stops, such as at closed valves, blind flanges, and dead ends; and wherever the flow is controlled, such as at valves and hydrants.

The amount of thrust exerted at any particular fitting depends on four factors:

- Type of fitting
- Diameter of fitting
- Water pressure
- Water hammer.

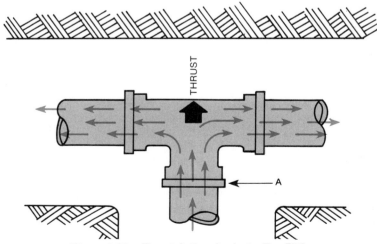

Figure 1-33.   Thrust Acting Against a Tee Fitting

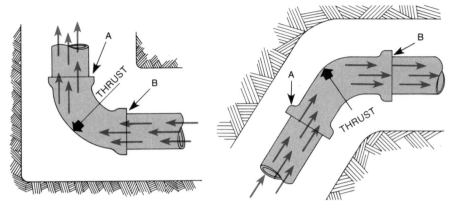

Figure 1-34. Thrust on 90° Fitting   **Figure 1-35. Thrust on 45° Fitting**

Detailed diagrams showing the direction of thrust exerted on several common pipe and fitting configurations are shown in Figures 1-33 through 1-36. Figure 1-33 shows a top view of a pipeline and tee fitting in a trench. Water flows into the fitting through the leg of the tee. In this case, the thrust is caused by the pressure of the water pushing against the back of the tee. The resulting thrust force, shown by the heavy arrow, acts perpendicular to, or at 90° to, the back of the tee. This thrust can be surprisingly large. If this example were a 12-in. (300-mm) pipeline and tee operating at 150 psig (1030 kPa [gauge]), the total thrust would amount to 30,760 lb (13,950 kg). Without some sort of blocking or anchorage at the top of the tee, a thrust of such size could easily push the tee off the pipeline at point A. Under such a force, unless the thrust-control device is properly constructed, the slightest movement could cause leakage, either at point A or at another coupling upstream of point A.

Both Figures 1-34 and 1-35 are top views of pipelines in trenches, showing how thrust acts on 90° and 45° fittings. Notice in both cases that the thrust acts perpendicular to the fitting at a point midway between the fitting ends. Again, without some sort of thrust-control device, the elbows could be pushed off the pipes at points A or B.

Thrust can also occur at valves, as shown by the sequence in Figure 1-36. The open valve (A) creates no thrust. As the valve is closed (B), the gate creates a drag or flow resistance in the moving stream of water and thrust begins to develop. When the valve is fully closed (C), it acts just like a tee or dead-end fitting. The upstream water pressure and the additional pressure due to water hammer create the thrust.

The examples shown so far demonstrate how thrust occurs in a horizontal run of pipe. Thrust will also develop whenever the direction of flow changes from horizontal to sloping or horizontal to vertical.

Thrust can be controlled by restraining all movement of the pipeline. There are several ways to do this. One way is to use strong couplings, such as ball, threaded, flanged, and locked joints, which will not leak or separate when pulled on by the thrust force. The push-on type of coupling used today does not have the strength

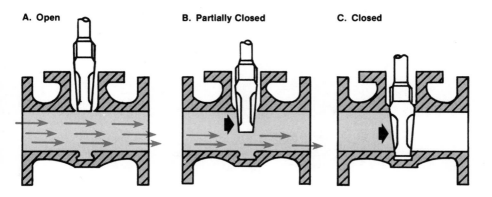

A. Open   **B. Partially Closed**   **C. Closed**

Figure 1-36.   Thrust on a Gate Valve

to resist the thrust force. Where push-on couplings are used, one of two other thrust-control devices can be used—the THRUST BLOCK or the THRUST ANCHOR.

Thrust blocks.   A thrust block is a mass of concrete that is cast in place between the fitting being restrained and the undisturbed soil at the side or bottom of the pipe trench.[5]

There are three important features to notice about thrust blocks. First, the block is centered on the thrust force. This ensures an even transfer of the force from the pipe to the block. Second, the block partially cradles the fitting. This further ensures even thrust transfer, yet leaves couplings accessible for inspection and maintenance. Finally, the bearing face is cast against undisturbed soil. This is extremely important. Undisturbed soils can be depended on to provide a known minimum amount of bearing support. Disturbed soils, such as backfill, cannot be relied on unless they have been adequately compacted and tested. This can be time-consuming and costly. (Once in place, excavation behind the thrust block can disturb the bearing soil and destroy the usefulness of the block. If the bearing soil is disturbed for any reason, the thrust block should be removed and a new one constructed against undisturbed soils.)

The bearing face of a thrust block should always be perpendicular to the direction of thrust. (This is not the case with thrust anchors, discussed in the next paragraph.) Figure 1-37 summarizes eight types of concrete thrust blocks commonly used in horizontal pipe runs, showing the direction of flow and the direction of the thrust.

Thrust anchors.   Another common type of thrust control, the thrust anchor, is shown in Figure 1-38. A thrust anchor is a massive cube of concrete that is cast in place below the fitting to be anchored. As shown in Figure 1-38, embedded steel shackle rods anchor the fitting to the concrete cube, effectively resisting upward thrusts. Notice that thrust anchors provide the same three important features as thrust blocks. However, instead of a concrete cradle, strap rods provide the rigid connection to the anchor block.

---

[5] *Basic Science Concepts and Applications*, Hydraulics Section, Thrust Control.

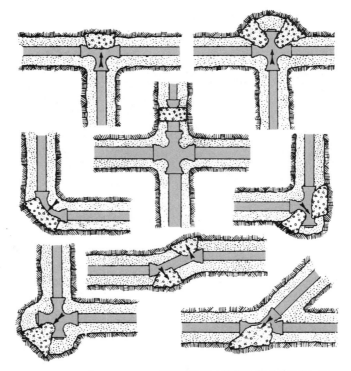

Courtesy of J-M Manufacturing Co., Inc.

Figure 1-37.   Common Concrete Thrust Blocks

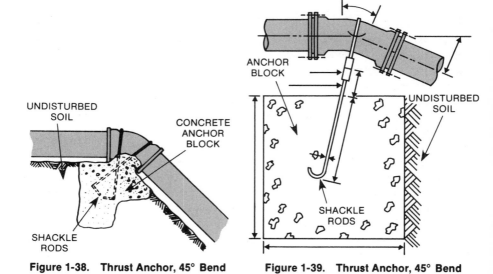

Figure 1-38.   Thrust Anchor, 45° Bend   **Figure 1-39.   Thrust Anchor, 45° Bend**

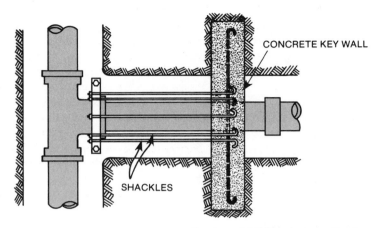

Courtesy of J-M Manufacturing Co., Inc.

Figure 1-40.   Key Wall and Shackles for Tee Fitting

A second type of thrust anchor for vertical bends is shown in Figure 1-39. For any thrust anchor, the amount of concrete needed and the imbedded length of shackle rod depends on the thrust, the type of fitting, the pipe diameter, and the water pressure.

There are many special conditions occurring in pipelines that may require an unusual type of thrust control. Where soils are unstable or the required soil bearing surface is not available, a reinforced concrete key wall is sometimes used. Figure 1-40 shows a key wall and shackles for a typical tee fitting.

Backfilling

The purpose of the backfill material, which is placed around and over the pipe, is to:

• Provide support for the pipe

• Provide lateral stability between the pipe and the trench walls

• Form a cushion over the pipe to prevent damage.

Backfilling is usually done mechanically. The general procedures and suggestions in the following paragraphs apply regardless of the type of pipe. Special considerations and problems are discussed later.

Only clean, washed sand or selected soil should be used for the first layer of backfill. This can be either suitable existing soil or imported soil. Backfill material should contain enough moisture to permit thorough compaction and support (free of voids) under and around the pipe. Large rocks, boulders, roots, construction debris, and frozen backfill material should not be used.

The first layer of backfill should be placed on both sides of the pipe, joints, valves, fittings, and so on simultaneously, up to the center line of the pipe, and then compacted. This is sometimes referred to as haunching.

Above the spring line, backfill practices vary considerably, depending on local conditions and practices. In general, an initial backfill should be placed around the upper half of the pipe and compacted, by hand or by approved mechanical equipment, to avoid damage to or movement of the pipe. The trench should be filled and compacted this way to a depth of 12 in. (300 mm) over 8-in. (200-mm) pipe and a depth of 18–24 in. (460–610 mm) over larger-sized pipe. All material close to the pipe should be selected to meet compaction specifications. This initial backfill will protect the pipe during the remainder of the backfilling process.

The remainder of the trench should be backfilled by placing the material in layers and compacting it thoroughly. This backfill does not need to be quite as carefully selected, placed, or compacted. The fill should, however, be uniformly dense. Unfilled space should be avoided. If trenches are in a road right-of-way or where there will be a sidewalk, completed backfill must meet the compaction requirements of the applicable agency. Backfill in other trenches need not be compacted to such a degree. Topping off backfill is an acceptable practice when compacting of backfill is otherwise not necessary.

Compacting.   The person in charge should insist that the backfilling be done gently and thoroughly. Generally, compacting of the backfill will be done in one of three ways: by hand, by water settling, or by mechanical compaction. All three types of compaction will allow the backfill material to provide the proper support for the intended surface.

Hand tamping may be done in the area adjacent to and immediately above the pipe. Curved tampers, shown in Figure 1-41 or other tamping devices may be used.

Where water is economically available and the soil is a free-draining type, water settling can be used. This is done by flooding the backfilled or partially backfilled trench or by jetting. Generally, this method will compact the backfill to within 5 percent of the maximum density.

Mechanical compaction is normally done when settling must be kept to a minimum and when the backfill must support the surface loads. The equipment used for this purpose is often air compressed or mechanically driven, and the backfill must be placed and compacted in 6- to 12-in. (150- to 300-mm) lifts.

When compacting in shored trenches, timbers and shoring should be withdrawn in stages equivalent to the layers of earth being placed. Shoring that extends below the spring line of the pipe may not be able to be withdrawn without disturbing the pipe bedding and should be cut off near the top of the pipe.

Where poor ground conditions exist from the bottom of the trench up to the surface, it may be necessary to leave shoring in place for the whole depth, except perhaps for the top 2 ft (0.6 mm), to maintain stability of the installation. Compaction methods must ensure that all voids (hollow spaces) caused by the removal of shoring are filled.

Equipment.   A balanced operation should be the goal in backfilling and tamping. A small bulldozer or a loader works well in combination with a self-propelled vibratory roller. It is poor practice to try to use the same machine for backfilling and as a crane for lowering pipe.

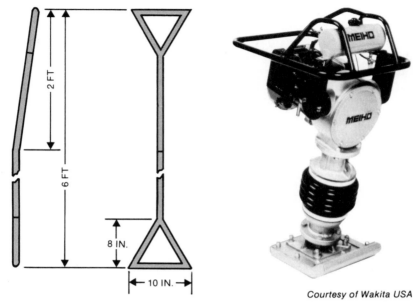

Courtesy of Wakita USA

Figure 1-41.   Hand-Tamping Device   **Figure 1-42.   Mechanical Plate Tamper**

Some types of mechanical compacting equipment and appropriate applications are as follows (This equipment is generally available in vibrating or static models.):

- Irregular drum tampers—clays, tills, silts
- Hand-controlled plate tampers (Figure 1-42)—sand in shallow lifts
- Boom-mounted plate tampers—clays or sands in deep, narrow trenches that preclude the operator's entry into the trench.

Even with the best equipment and the best intentions, acceptable results may still not be achievable. Manufacturer's recommendations should always be obtained and followed.

Pressure and Leak Testing

After the trench has been partially backfilled, the pipe, regardless of type, should be tested for amount of leakage and ability to hold pressure. This can be done in sections as the pipe is completed or it can be done after the entire pipe has been laid. In either case, it should be done before complete closure of the trench.

Leakage is defined as the volume of water that must be supplied to the pipeline to maintain a specified test pressure after all air has been expelled from the line. The test pressure and the allowable leakage for a given length and type of pipe and joint are given in AWWA standards and manuals. Much of the allowable leakage is due to settling in of the line under pressure, although a slight amount of actual leakage may be allowed with certain types of joints.

Leakage and pressure tests are usually made at the same time, observing the following procedure.

1. Check that all anchors and thrust blocks have been installed. Allow at least five days for the concrete to cure.

2. Install a pressure pump equipped with make-up reservoir and a method for measuring the amount of water pressure and water pumped in the system to be tested. The method used to measure water volume could be a calibrated make-up reservoir (preferably), a calibrated positive-displacement pump, or a very sensitive water meter. This setup should be installed at the end of the pipe, at a service connection, or at a hydrant, as shown in Figure 1-43. Alternately, a line with a pressure gauge, pump, and sensitive water meter could be connected from an existing live hydrant to the line being tested. A double-check valve should also be placed in the line to prevent any backflow into the supply line or tank.

3. Close appropriate valves, corporation stops, etc. Notify customers, if necessary.

4. Slowly fill the test section with water while expelling air through valves and hydrants.

5. Start applying partial pressure with the pump. Before bringing the pressure to the full test value, bleed all air out of the mains by venting through service connections and air-release valves. Corporation stops may have to be installed at high points in the pipe for this.

6. Once the lines are full and all air has been let off, leave on partial pressure and allow the pipe to stand for at least 24 hours.

7. For pressure testing, subject the test line to the hydrostatic pressure specified in the applicable AWWA standard. A pressure of 150 psi (1030 kPa) for a period of 30 min is usually the minimum.

8. Examine the trench for visible leaks or pipe movement. Any joints, valves, etc. that show leakage should be checked, adjusted, or repaired as needed. The test may need to be repeated after any adjustments or repairs.

9. After the test pressure has been maintained for at least two hours, conduct a leakage test by measuring, with the make-up reservoir, the amount of water that has to be pumped into the line in order to maintain the specified test pressure.

10. Compare the amount of leakage to the suggested maximum leakage given in the appropriate AWWA standards and manuals. A swift loss of pressure is likely due to a break in the line or a major valve opening. A slow loss of pressure may be due to a leaking valve or a curb stop not shutting off.

If the line has failed the leakage test, it is necessary to find where the excessive leakage is taking place. There are a few steps that can be taken to ensure that there is a pipe leak and not a leak at a fitting. First, leave the line under normal pressure. The next day, repeat the test. If the leakage measured the next day is greater than before, the leak probably is in a pipe joint or a damaged pipe. If the leakage is the same, it is probably in a valve or service connection.

To locate the leak, insert the key for the curb stop in each shutoff and listen at the top of the key. It may be possible to hear a leak, since the key acts as a stethoscope. If a leak is heard, open the shutoff and close it again. If there is no

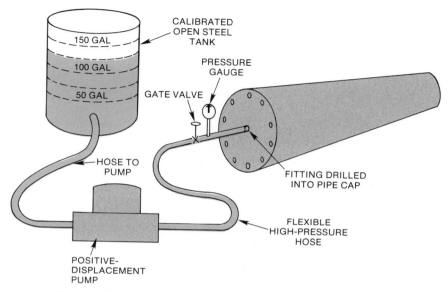

Figure 1-43.   Equipment for Leakage and Pressure Testing Pipeline

audible leak now, test the section again. If no leaking curb stops are found, crack the main valves at the ends of the test section several times and then close them. This should flush out any sand grains in the valve seats that may be preventing the valves from closing completely, thereby causing slight leakages.

If it is found that the leakage is not occurring at either of the above points, it is then necessary to try and find a leak through trial and error. Some sort of leak detector, such as a sensitive microphone with an amplifier and ear phones, is necessary.

The following is a useful checklist of causes for failure to pass a leak test:

• A hydrant valve is held open by a piece of rag, wood, or some other foreign object.

• There is dirt or foreign material under the coupling gasket.

• Fittings and valves in the test section are not sufficiently blocked (by thrust blocks).

• Corporation stops are not tightly closed.

• There is leakage through the valve at the end of the test section.

• The packing on valves is leaking.

• The test pump is leaking. This could be the check valve or the gate valve.

• The test includes too long a section of pipe.

• The saturation time was too short—it should be 24 hours.

• There is a break in the pipe—either a crack or a blowout.

- There is a broken coupling.
- There is some faulty accessory equipment—possibly a valve, fitting, hydrant, saddle, corporation stop, or relief valve.
- The test gauge is faulty.
- The test pump suction line is drawing air.

Any leaks in the line should be repaired, and the line retested until the measured leakage is less than the allowable leakage. If an operator is simply testing between valves, a pressure of 225 psi (1550 kPa) maintainable for 4 hours will indicate that the leak has been stopped.

Flushing and Disinfection

Any new or repaired water main must be thoroughly cleaned (flushed), disinfected, and tested for bacteriological quality before it can be put into service. This is necessary in order to:

- Protect consumers, since there is a good possibility that the new system will have become contaminated during the transit, storage, and laying of the pipe
- Prevent the growth of nuisance organisms in the distribution system, which create bad taste, odors, and discolored water
- Prevent contamination of the existing water distribution system by dirty new main
- Remove any dirt and debris in the piping that, if left in the main, will shield bacteria and result in incomplete disinfection
- Alleviate customer complaints.

Flushing. Mains should be flushed prior to disinfection in order to remove any foreign material that may interfere with disinfection or reduce water quality. This is usually (but not always) done following the pressure and leakage testing.

Flushing should be done through a convenient hydrant or other blowoff. It should be done at a velocity of 2.5 fps (0.8 m/s) to obtain proper flushing action. Table 1-10 provides some information on how much water must be used to adequately flush various pipe sizes.

Disinfection. Chlorine is the only chemical used as a disinfectant for pipelines. Calcium hypochlorite and sodium hypochlorite solution are the forms in which it is usually used.

Application point. The chlorine solution is usually injected through a corporation stop at one end of a pipe section at the time that water is bled from the opposite end of the section. Precautions must be taken to prevent dosed water from flowing into the potable water supply. All high points on sections being treated should be properly vented.

Chlorine dosage. The chlorine requirement depends on local and state requirements, the degree of contamination, contact time allowed, and the pH of

Table 1-10.   Required Flow and Openings to Flush Pipelines (40-psi Residual Pressure in Water Main)*

Pipe Diameter in.	Flow Required to Produce 2.5 fps (approx.) Velocity in Main gpm	Size of Tap in.			Number of 2 ½-in. Hydrant Outlets*
		1	1½	2	
		Number of Taps on Pipe†			
4	100	1	—	—	1
6	200	—	1	—	1
8	400	—	2	1	1
10	600	—	3	2	1
12	900	—	—	2	2
16	1600	—	—	4	2

*With a 40-psi pressure in the main with the hydrant flowing to atmosphere, a 2½-in. hydrant outlet will discharge approximately 1000 gpm and a 4½-in. hydrant nozzle will discharge approximately 2500 gpm.
†Number of taps on pipe based on no significant length of discharge piping. A 10-ft length of galvanized iron (GI) piping will reduce flow by approximately one third.

the water. However, rate of application should result in a uniform concentration of at least 25 mg/L at the end of the section being treated. Under certain conditions, higher chlorine dosages may be required.

Contact period.   The average retention period should be 24 hours. If unfavorable or unsanitary conditions exist, the period may have to be extended to 48 or 72 hours. If shorter retention periods must be used, the chlorine concentration should be increased to 50 or 100 mg/L. The chlorinated water should be flushed to a storm-sewer storage pond or flood-control channel until the chlorine residual is gone or approaches normal. Never discharge highly chlorinated water without checking with local or state regulatory agencies. The efficiency of disinfection must be checked through bacteriological tests before the line is placed in service.

Calculations.   The amount of chlorine and water needed for proper disinfection can be calculated by determining:

- The capacity of the pipeline (using pipe size and length)
- The desired chlorine dosage
- The concentration of the chlorine solution (usually 1 percent)
- The pumping rate of the chlorine-solution pump
- The pipe flow-through (or fill) rate.

Charts such as those in Tables 1-11 and 1-12 can be used to determine some of these numbers.[6]

Procedures.   Three of the most commonly used methods of disinfection are the continuous-feed method, the slug method, and the tablet method.

---

[6] *Basic Science Concepts and Applications*, Chemistry Section, Dosage Problems (Milligrams-per-Litre to Pounds-per-Day Conversions).

Table 1-11.   Hypochlorite Required for 50 ppm of Chlorine in Pipe

Pipe Size Inside Diameter in.	Contents in a 100-ft Section			Amount of Hypochlorite per 100-ft Length to Give 50 ppm Available Chlorine*		Length of Pipe in Which 1 oz of Hypochlorite Will Produce 50 ppm Avail. Cl_2
	cu ft	lb	gal	oz (approx.)	lb	
4	8.75	545	65.5	$5/8$	0.039	168
6	19.65	1 225	147	$1\,3/8$	0.087	71.9
8	34.90	2 180	261	$2\,1/2$	0.159	39.4
10	54.55	3 405	408	$3\,7/8$	0.244	25.6
12	78.55	4 905	587	$5\,5/8$	0.350	17.9
14	106.9	6 670	800	$7\,5/8$	0.476	13.2
16	139.6	8 725	1044	10	0.621	10.0
20	218.2	13 635	1632	$15\,1/2$	0.972	6.45
24	314.2	19 635	2350	$22\,3/8$	1.400	4.45
30	390.9	30 680	3672	35	2.185	2.95
36	706.9	44 180	5285	$50\,3/8$	3.150	2.0
42	962.1	60 130	7197	69	4.30	1.45
48	1256.6	78 535	9400	$89\,3/4$	5.610	1.06

*For 20-ft lengths of pipe, divide these values by five.
Source: *Hypochlorination of Water.* Olin Corporation, Stamford, Conn. (1971).

Table 1-12.   Number of 5-g Calcium Hypochlorite Tablets Required for Dose of 25 mg/L*

Pipe Diameter in.	Length of Pipe Section ft				
	13 or less	18	20	30	40
	Number of 5-g Calcium Hypochlorite Tablets				
4	1	1	1	1	1
6	1	1	1	2	2
8	1	2	2	3	4
10	2	3	3	4	5
12	3	4	4	6	7
16	4	6	7	10	13

*Based on 3.25 g available chlorine per tablet; any portion of tablet rounded to next higher number.

In the CONTINUOUS-FEED METHOD, water from the distribution system or other approved source along with the type of chlorine selected are fed at a constant rate into the new main at a concentration of at least 50 mg/L available chlorine. A properly adjusted hypochlorite solution injected into the main with a hypo-chlorinator or liquid chlorine injected into the main through a solution-feed chlorinator and booster pump can be used. The chlorine residual should be checked at intervals to ensure that the proper level is maintained. Chlorine application should continue until the entire main is filled. The water should then remain in the pipe for a minimum of 24 hours, during which time all valves and hydrants along the main must be operated to ensure that they are also properly disinfected.

In the SLUG METHOD, a continuous flow of water is fed with a constant dose of chlorine (as in the continuous-feed method), but with rates proportioned to give a chlorine concentration of at least 300 mg/L. The chlorine is applied continuously for a period of time to provide a column of chlorinated water that will contact all interior surfaces of the main for a period of at least three hours. As the slug passes tees, crosses, etc., proper valves must be operated to ensure their disinfection. This method is used primarily for large-diameter mains where continuous feed is impractical.

With the TABLET METHOD, calcium hypochlorite tablets are placed in each section of pipe, in hydrants, and in other appurtenances. The main is then slowly filled with water at a velocity of less than 1 fps (0.3 m/s) to prevent washing of the hypochlorite to the end of the main. The final solution should have a residual of at least 25 mg/L and should remain in contact for a minimum of 24 hours.

The tablet method does not allow preliminary flushing. Consequently, it is considered the least satisfactory method of disinfection. However, it is commonly used for small-diameter mains.

Regardless of the method used, it is necessary to make certain that backflow of the strong chlorine solution into the supplying line does not occur. Following the prescribed contact period, the chlorinated water should be flushed to waste until the remaining water has a chlorine residual approximating that throughout the rest of the system. Contact local regulatory agencies to assure proper disposal of the highly chlorinated water.

Bacteriological testing.   After the chlorinated water is flushed out of the pipe and the pipe is refilled with water from the rest of the system, bacteriological tests, as prescribed by the applicable regulatory agencies, should be taken. Safe samples are usually required before the new main can be placed into service.

If the results fail to meet minimum standards, the water will have to be tested again and shown to be free of bacteria before placing the main in service. Sometimes the entire disinfection procedure may have to be repeated. It should be kept in mind, however, that too much chlorine for an excessive time can damage brass and other fittings.

Final inspection.   Before putting the line into service, AS-BUILT PLANS should be completed and used as the basis for the final inspection. The plans should then be filed for future use. Detailed maps should be completed that identify the type and location of valve boxes on the new line, as well as the location of hydrants and other appurtenances.

All valves on the line should be operated and left in the full-open position. The number of turns needed to close and open each valve should be counted and recorded, along with the direction of opening. Each fire hydrant should also be tested to see if it is in good operating condition, and a record should be made of the direction of valve operation.

Restoration.   Good restoration requires common sense. It should take whatever form local conditions require. Since restoration is performed at the end of the project, there is often a tendency to give it low priority. However, operators must remember that the community will judge the water utility by what they see after the work is completed. "Before" and "after" photographs or video tapes of the job site are useful in evaluating the quality of restoration. The following observations on restoring job sites to their original condition are generally applicable.

Backfilling trenches.   Properly compacted backfill in trench cuts will reduce settling, which is one of the major nuisances faced by street maintenance departments.

Asphalt or hard surfacing.   Because some settling will occur, many municipalities request temporary repairs with cold patch and a delay of at least six months before performing a permanent repair.

Sodding or spraying of grass.   A good grade of sod or grass should be used. This should be placed over a layer of topsoil. Sod is usually used for lawn and boulevard restoration in urban areas. Sod must be constantly watered until there has been a reasonable growth, usually after about a month. Grass seed is sprayed on as a mulch in more rural areas, such as fields and ditches. Sprayed seed does not require routine watering, but is slower than sod to grow and may require respraying of spotty areas at a later date. Mulch should always be used to prevent erosion.

Ditches and culverts.   Ditches and culverts in the project area should be checked for proper drainage. Any excessive silt or debris should be removed. A plugged ditch in a heavy storm can flood the surrounding area. Culverts should be checked for any damage and arrangements made for repair.

Trees and shrubs.   Damaged roots on a tree can result in a dead tree up to five years after completion of the project. A qualified expert should be consulted if any trees have been damaged.

Utilities.   Other underground services uncovered during construction must be supported during the work phase and properly restored at the conclusion of the project.

Curbs, gutters, and sidewalks.   When reconstructed, new concrete work should match the old work as nearly as possible, including the texture finish.

Machinery and construction sheds.   All construction machinery and structures should be removed, the ground graded, topsoil added, and the land brought back to its original grade and condition.

Watercourses and slopes.   If a stream or wetland must be disturbed during construction, check with local, state, or federal regulatory officials regarding the need for a permit. Slopes should be structurally restored, possibly with sod and riprap. Stream beds should be inspected for excessive mud and debris for a considerable distance downstream from the actual construction site. Cleanup operations should be conducted promptly to limit the extension of silt and debris.

Roadway cleanup.   Any type of construction will track some dirt and dust over the roads being used. During wet weather, arrangements should be made to properly wash the mud from the wheels of the trucks at the site, before it is tracked throughout the municipality. During dry weather, dust forms on the roads. This must be swept or shoveled from hard-surface roads and graded to the shoulder or treated with calcium on gravel roads. Calcium should not be placed on the dust on hard-surface roads, since it may contribute to forming a muddy slick that is dangerous during braking and turning.

Traffic.   When construction is complete, restoration may involve removing signs and filling holes, removing any temporary roads, and putting the area back into its original shape by grading and sodding.

Miscellaneous.   All construction debris must be removed; private driveways, walks, fences, lawns, and private appurtenances must be returned to their original condition and the area around them thoroughly cleaned.

1-5.   Operation and Maintenance

The performance of a distribution system depends on the ability of the pipe to resist unfavorable conditions and to operate at or near the capacity and efficiency that existed when the pipe was laid. This performance can be checked in several ways: measurement of flow, fire-flow tests, loss of head tests, pressure tests, simultaneous flow and pressure tests, tests for leakage, and chemical and bacteriological water tests. These tests are important to maintenance and should be scheduled as part of the regular operation of a pipeline.

Pipeline maintenance, as discussed in this section, includes routine sampling and inspection, leak repairs, maintaining carrying capacity, pipeline cleaning, lining mains in place, main-break repair, and thawing of frozen pipelines.

Routine Water Quality Sampling

Sampling of water in the distribution system should provide evidence that the water being delivered to the customers is safe to drink and desirable to use. Tests to ensure that water treatment and chemical-addition processes are functioning properly and producing the desired result in a distribution system must be included. Testing may also give an indication of the STABILITY of the water and the resulting condition of the distribution system—whether it is corroding or accumulating scale.

Sampling the distribution system for bacteriological quality is the major sampling responsibility for many operators. A guide for bacteriological evaluation of public water supplies is presented in Appendix A. Testing and sampling procedures for bacteriological and other water quality parameters are also discussed at length in Volume 4, *Water Quality Analysis*. The water quality characteristics affecting or indicating the stability of water are also covered in Volume 2, *Water Treatment*, Module 9, Stabilization.

Pressure and Flow Checks

A water system is designed and installed to provide a designated amount of water at a designated pressure. The purpose of the distribution system is to consistently supply the desired flow and pressure to the customer. Part of an operator's job is to operate and check the system to ensure and document the continued supply of acceptable water flow and pressure.

Pressure and flow should be checked because most regulatory agencies require maintenance of a normal system pressure somewhere between 100 psi (590 kPa) maximum and 35 psi (240 kPa) minimum, or 20 psi (140 kPa) minimum under fire-flow conditions. This is mainly for the protection and convenience of the customers being served. Fire insurance companies and their representatives are also very interested in system water pressure and flow rate.

Even though pressure and flow were adequate when the system was installed, many changes can occur that reduce (or occasionally increase) pressure and flow, often to unacceptable levels. These include:

- System expansion or changes in configuration
- New water lines installed at undesirably high or low elevations
- Additional customer services
- Closed or partially closed valves
- Broken water mains
- Leaking or broken service lines
- Water meters, service lines, and plumbing in vacant homes or buildings that have been damaged by freezing
- A change in elevated-storage water level or operation
- Reduction of pipe capacity due to corrosion, pitting, tuberculation, sediment deposits, or slime growth.

Minimum pressures must be maintained to ensure adequate customer service when and where it is needed, even during peak flow periods, and to ensure sufficient fire protection. Adequate positive pressure in the mains also helps protect against backflow or backsiphonage from cross connections in the distribution system or the customer's system.

Maximum allowable pressures should not be exceeded. This will prevent customers from having their plumbing damaged or their water-heater safety valves "blow" due to excessive system pressure, perhaps in conjunction with water hammer.

Pressure checks.   A test of system pressure can be made using a fire hydrant pressure gauge. This pressure gauge, threaded into a hydrant cap, can be purchased or homemade (Figure 1-44). The cap and gauge are screwed onto a hydrant outlet nozzle, the hydrant is opened, the air is allowed to escape, the cap is tightened down, and the pressure is read in pounds per square inch (psi) or kilopascals (kPa).

Loss of head can be determined by isolating a section of the system from all branch lines and services, so that no water can be withdrawn or added except from the source, and then measuring and comparing the pressure at separate points along the main. Pressure in customer plumbing systems may be measured with a gauge attached to a fitting that can be screwed onto a threaded hose bibb.

Flow checks.   Flow readings are usually taken at hydrants, using a device called a PITOT GAUGE (also called Pitot tube, Figure 1-45). This is essentially a hollow tube, open at one end and closed at the other, connected to a pressure gauge. A Pitot gauge can be constructed as described in Appendix B.

A Pitot gauge may be used on any size opening. It is held in the flow stream, as shown in Figure 1-46. The gauge reading, in psi or kPa, is converted to a flow

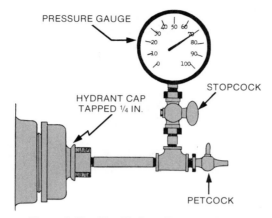

Figure 1-44.   Fire-Hydrant Pressure Gauge

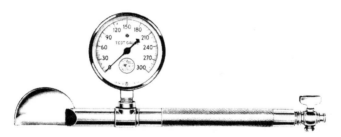

Figure 1-45.   Pitot Tube and Gauge

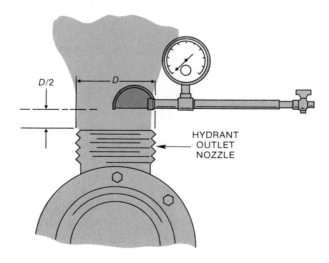

Figure 1-46.   Positioning Pitot Gauge to Take Reading at Hydrant

reading, in gpm or L/m, using tables and a hydrant coefficient, such as the ones found in Appendix C. Best results with any type Pitot gauge are obtained from Pitot pressure readings between 10 and 30 psi (70 and 210 kPa).

Loss of carrying capacity, caused by pipe interior roughness or size reduction, can also be measured using a Pitot gauge. This is called a flow-coefficient test or C value test. The C value is a number indicating the carrying capacity of the pipe.

To determine the C value, flow readings must be obtained by isolating a section of the system from all branch lines and services, inserting a Pitot gauge at each end of the section, and recording the pressure values over a period of time. The C value is then calculated, using the velocity of flow obtained indirectly from the Pitot gauge readings, the slope of the pipeline, and the diameter of the pipe. The value obtained should be compared with the value for the pipe when it was new or from a previous observation to determine if there has been a loss of carrying capacity.

Combination flow and pressure checks. Additional information about a system can be obtained by using flow and pressure measurements together. The procedures for determining the amount of water that can be allowed to flow from the system at a certain location (for example, in case of a fire), while maintaining a 20-psi (140-kPa) system pressure, are discussed in Appendix C.

Routine Inspection

Operators, meter readers, and street-crew personnel should observe and take appropriate action on the following situations: unapproved encroachment, unapproved crossings, damage, vandalism, leaks, cross connections, and improper or unauthorized hookups.

Flushing and Cleaning

Several procedures may be used to maintain or improve the system carrying capacity and water quality. Of these, the most common practice is flushing.

Flushing. Sand, corrosion products, and other solids have a tendency to settle in the pipeline, especially in dead ends or areas of low water consumption. These deposits can reduce the carrying capacity of the pipe and are often a source of color, odor, and taste. Slime growths may also be a problem. Flushing at high velocities, at least twice a year in problem areas, will normally remove most of the settled substances and discolored or stale water.

Flushing procedure. Flushing is accomplished by fully opening a hydrant located near the problem area. The hydrant should be kept open as long as it takes to flush the sediment out. Only through on-the-job experience will an operator be able to tell how often or how long certain areas should be flushed.

It is not normal practice (especially in large cities) to flush the entire system. Operators must, however, respond to customer complaints of dirty water and flush problem areas. If flushing is made a part of routine scheduled maintenance and if it is done properly, it will eliminate many customer complaints, meter repairs, and service-line blockages. Flushing also provides a good opportunity for operators to perform scheduled hydrant inspections.

Flushing programs. The following points should be considered when developing a program or performing flushing. A more complete discussion of hydrants and their operation can be found in Module 4, Fire Hydrants.

A map of the system and past experience should be used to plan the flushing schedule, paying particular attention to potential problem areas and areas where there have been a high incidence of customer complaints. The flushing procedures should be performed first in the area of the well or treatment plant and should proceed outward.

Flushing the system late at night will achieve greater flows through the line and cause fewer customer complaints. Customers also will not be able to see how bad the water sometimes looks coming out of the system. If possible, media announcements should be made to explain about the flushing schedule in advance of the work, alerting customers that there may be a temporary condition of discolored water. Customers should be advised that, after flushing is completed, they should not use water from their service until they have run enough water to thoroughly flush their service lines.

The work of flushing crews should be coordinated to avoid flushing too many hydrants at once, which could cause a negative pressure to be created, increasing the chances of backflow through any existing cross connections. Before flushing, the area in which the flushed water will drain should be inspected to ensure that the water will not flow into basements, excavations, or buildings.

Higher pumping pressures and/or shutting off branch lines will increase the effectiveness of the operation. A pipe flow of at least double the velocity normally recommended (2.5 fps or 220 gpm in a 6-in. main [0.76 m/s or 830 L/min in a 150-mm main]) is usually required for effective flushing.

The hydrants must be opened fully—the hydrant valve is not a throttling valve. The hydrant valves should be opened and closed slowly, so as to prevent water hammer. A diffuser, screen, length of fire hose, or other means of breaking up the force of the water stream is recommended, especially in unpaved areas (Figure 1-47).

Courtesy of Leo J. Wilwerding (Designer), Metropolitan Utilities District, Omaha, Neb.

Figure 1-47. Flow Diffuser Used in Flushing Hydrants

Flowing hydrants should not be left unattended, and flushing should be stopped if the water is damaging a roadway or parkway. If the street or parkway is accidentally damaged, it should be marked with a lighted barricade and the location recorded so that the damage can be repaired as soon as possible. Flushing should be continued no longer than necessary to remove sediment and ensure that the hydrant is operating properly.

When flushing of a hydrant is completed, the hydrant should be checked to ensure that it drains when it is shut off. Check either by feeling by hand for a slight vacuum at the open nozzle, by listening for air being drawn in with the outlet cap on loosely, or by running a weight on a line down inside the hydrant. Plugged hydrants must be pumped out if freezing is a possibility. Nozzle caps must be tightened so they cannot be removed by unauthorized persons.

Records should be kept for each hydrant flushed, length of time flushed, water condition at the start and end of flushing, and other special notations. Defective hydrants should also be noted, flagged as inoperative, and reported for immediate repair.

Chlorine treatment.   When loss of carrying capacity is caused by slime growths in pipelines, chlorine may be effectively used to solve the problem. The system, or section of the system affected, is usually given a "slug" dose of chlorine to kill the bacteria causing the problem. This should be followed by thorough flushing to get the slime and the chlorine out of the system. All of the precautions discussed previously concerning flushing are even more important in this situation, especially those concerning informing customers to thoroughly flush their service lines before using the water.

A second dose of chlorine may be needed to complete the job, and a free chlorine residual should be maintained throughout the system following the treatment to prevent regrowth of the bacterial slime.

Cleaning.   Mechanical cleaning may be necessary in areas where tuberculation and deposits on older cast-iron pipes exist or where iron bacteria and slime growth are a severe problem. Initial cleaning of a main usually involves flushing, but when this has proven inadequate, it may become necessary to use either air purging or cleaning devices such as swabs or pigs. In addition to the removal of objectionable material from a main, the cleaning operation can increase the flow rate through the main. This reduces operating pressure losses, which can result in a reduction in pumping time and lower energy costs.

Removal of encrustations may not be a permanent solution to dirty-water problems. In some cases, removal of encrustations or tuberculation may cause leaks if during removal the pipe wall is damaged. Without lining the pipe or treating the corrosivity of the water, the problem may reoccur rapidly. However, experience has shown that in many cases leaving just a few mils of iron oxide on the smooth interior of the pipe wall delays the occurrence of red water and the regrowth of encrustations.

Initial cleaning procedures.   Thorough planning should precede actual cleaning. The section of main or system to be cleaned should be mapped (Figure 1-48). The order of work, source of water, entry and exit points, and disposal of

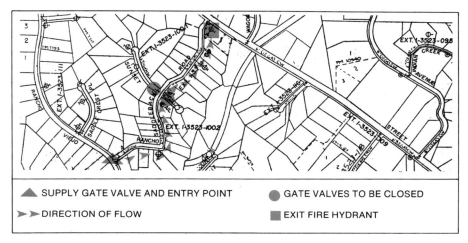

SUPPLY GATE VALVE AND ENTRY POINT   ● GATE VALVES TO BE CLOSED

DIRECTION OF FLOW   ■ EXIT FIRE HYDRANT

Figure 1-48.   Main Cleaning Map

the flushed water should all be determined. The vehicles to be used, size of crew, and equipment and materials that will be required can then be listed and made available.

Before cleaning, valves and hydrants should be checked to ensure they are operable. Customers should be notified concerning the date and time the system will be out of service, and temporary water service should be arranged for any customers who must have water for medical reasons.

Other utilities and agencies that will be affected by the planned operation should be notified, including police and fire departments. Regulatory agencies should be consulted concerning any special requirements, and any necessary safety procedures should be planned.

Before flushing or cleaning any main, provision should be made to control pressure surges. A sudden stopping of flowing water can occur if a line valve is operated rapidly or if a pig or swab suddenly slows or stops moving. Such surges can raise system pressure 20 to 60 psi (450 to 1350 kPa) for each foot per second (metre per second) of velocity change. These surges can destroy water mains and appurtenances.

Air purging.   In the AIR PURGING process, air mixed with water is used to clean small mains up to 4 in. (100 mm) in diameter. Before performing the procedure, all services must be shut off. Air from a compressor is then forced into the upstream end of a main after the blow-off is opened at the downstream end. Spurts of the air–water mixture will remove all but the toughest scale.

Swabbing.   The SWABS used in pipe cleaning operations are polyurethane foam plugs, somewhat larger than the inside diameter of the pipe to be cleaned, that are forced through the pipe by water pressure. Swabs can remove slime, soft scales, and loose sediment. They wear quickly in heavily encrusted mains, and they will not significantly affect hardened tuberculation.

Swabs can be purchased commercially either in specific sizes or in bulk, which can then be cut to size. The swab material is available in soft and hard grades. Soft swabs are typically used in exploring mains of uncertain cross section or condition, in mains where a reduction in diameter of 50 percent or more is expected, in mains with severe encrustations where flexibility is needed, and as a seal for swabbing operations. Hard swabs are commonly used in new mains, in mains with minor reductions in diameter, and in mains where deposits need continuous hard-swab pressure during the operation.

An experienced crew can swab up to several thousand feet of main per day if the operation is planned properly. Swabbing procedures vary with each job. A typical procedure is as follows:

1. Be sure initial procedures, such as shut off and notification of the public, have been completed.

2. Install equipment necessary at entry and exit points for launching and retrieving swabs.

3. Isolate water main or portion of system to be cleaned; make sure all valves in the main are open.

4. Open valve on upstream water supply to launch swab and control speed.

5. Run swab at 2 to 5 fps (0.6 to 1.5 m/s). If swabs travel too fast, they remove less material and wear out more rapidly.

6. Estimate flow rate with a Pitot gauge at exit or meter on inlet supply.

7. Note entry time and estimate time of exit. If travel time is too long, reverse flow and calculate location of blockage.

8. Make sufficient swab runs so flushing water clears within 1 min.

9. Account for all swabs. A typical cleaning operation may take from 10 to 20 swabs.

10. Run final flush until water is clear of swab particles.

Pigging. Pipe-cleaning PIGS are stiff, bullet-shaped foam plugs that are forced through a main by water pressure. They are similar to swabs, but are harder, less flexible, and more durable, which allows them to remove harder encrustations. However, their more limited flexibility somewhat reduces their capability to change direction at fittings and points where there are significant changes in pipe cross section.

Pigs are purchased commercially in various sizes, densities, and grades of flexibility and external roughness. A number of different types of pigs are available for use in system cleaning, and special pigs can be made for most situations. The ones most commonly used are classified as

- Bare pigs
- Cleaning pigs
- Scraping pigs.

Bare pigs of high-density foam are sent through a tuberculated main to determine the inside diameter of the pipe (including deposit) to be cleaned. Cleaning pigs have a tough coat of polyurethane synthetic rubber applied in a criss-cross pattern. When sent through a main, cleaning pigs remove most types of encrustation and growths. A bare pig of low-density foam or a swab can be sent behind an undersized cleaning pig to maintain the seal. Scraping pigs have spirals of silicon carbide or flame-hardened steel-wire brushes. These pigs are used to remove the harder encrustations and tuberculation. Cleaning or scraping pigs of increasing size may be sent through the main to gradually remove layers of encrustation.

Launching pigs using permanent or portable launchers are shown in Figure 1-49. An external source of water, either a 2½-in. (64-mm) connection to another hydrant not in the isolated section or a small high-pressure pump with an independent water supply, is used to force the swab or pig down the hydrant into the main. After this is done, the hydrant branch can be isolated, and the swab or pig will be pushed along the main by upstream water from the distribution system. It may be possible to place the pig into larger mains by removing the works of an in-line valve and using this as a point of entry.

Pigging procedures vary with the type of problem in the pipeline, the location of the problem in the system, and the type of pig to be used. An independent review should be made of the procedure to be used at each location, especially for

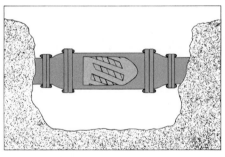

B. Over-Sized Spool Inserted in Line

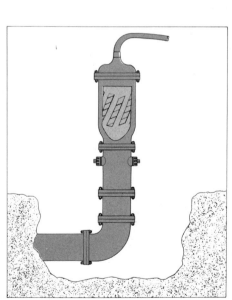

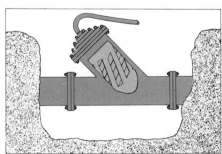

C. Y-Section Inserted in Line

A. Swage Reducer Attached to Fire Hydrant With Valve Removed

Courtesy of Girard Industries

Figure 1-49.   Launching Methods for Pipe-Cleaning Pigs

larger pipe sizes. The first time pigs are used, the utility's crew should work with a person experienced with the use of the equipment and the cleaning operation. Assistance is available from pipeline-cleaning firms and manufacturers of cleaning devices. A general procedure for cleaning a line with pigs is as follows:

1. Be sure initial procedures have been completed.

2. Install the equipment necessary at entry and exit points for launching and retrieving of pigs. In many cases, it will be necessary to cut into the main to install the equipment.

3. Isolate water main to be cleaned; be sure all gate valves are open.

4. Make provisions to control surges.

5. Open upstream water supply to launch pig.

6. Time passage of the pig in order to properly gauge the valve setting required to achieve the desired speed.

7. Control speed of pig with downstream hydrant or blowoff valve. Typical speeds are 1 to 5 fps (0.3 to 1.5 m/s). If pigs travel too fast, they remove less material and wear out more rapidly.

8. Take care to avoid sudden changes in speed or a stoppage in pig movement, which will cause destructive surges in water pressure.

9. Run final flush until water turns clear.

Final cleaning procedures.   After cleaning a water main with swabs or pigs, flush the main until the water runs clear. Test the quality of the water in the main, and chlorinate if necessary before returning the line to service. Check that all valves are in proper operating position and that all services have been reactivated.

A flow test should be conducted on the main before and after swabbing or pigging is performed. After cleaning has been carried out, another test should be made to determine if further cleaning is necessary. Conditions before and after pigging are illustrated in Figure 1-6.

Lining.   Cleaning can restore the interior surface of pipe to a condition close to that of a newly laid main. However, experience has shown that cleaning mains without lining them is only a temporary solution. Cleaning does not remove the causes of pipe deterioration. Tuberculation happens much faster after cleaning, and the flow coefficient declines back to its previous level. For this reason, cleaning alone is an expensive and relatively inefficient way to maintain carrying capacity.

After cleaning, pipe can be lined in place with a thin layer of cement mortar. This not only prevents recurrence of interior surface deterioration, but it also prevents red water and stops leakage. Cleaning and lining pipe will result in increased water quality, volume, and pressure to the customer. It will also decrease pumping, operations, repair, and replacement costs to the community.

The cost of mortar lining in place depends on pipe diameter and length, condition of the pipe, layout and profile of the line, number of bends, location and type of valves, bypass needs, type and depth of soil cover, accessibility, and traffic conditions. Valves, tees, and services require special attention to ensure they are functional after lining. The longer the length of pipe that can be lined in

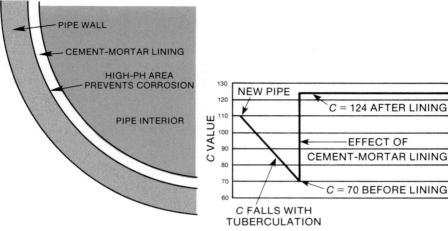

Figure 1-50.   Pipe With Lining   **Figure 1-51.   *C* Values Change With Lining**
(Riveted Steel Pipe)

one operation, the greater the production rate and, therefore, the lower the cost per foot. However, it is still an expensive process. Selection of a protected pipe for initial installation and chemical treatment of the water where necessary are more desirable alternatives. Figure 1-50 illustrates a pipe with lining and Figure 1-51 is a graph of increased *C* values.

Another method of lining existing water mains that are prone to tuberculation is slip lining with a high-density polyethylene (HDPE) pipe. Plastic pipe is lightweight and flexible and can be either pulled or pushed into the existing main from access points cut into the system. Valves, tees, and services must be recut, the same as with the cement-mortar lining.

Leak Detection and Repair

Distribution system leaks and the resulting water loss fall into two categories:

- Emergency leaks
- Nonemergency leaks.

Emergency leaks require immediate attention. Nonemergency leaks include known leaks that are repaired when time permits, as well as unknown leaks that are located and repaired during a leak-detection program. Although the repair procedures may be the same for both leak conditions, the method of detection and sense of urgency involved are often quite different.

Nonemergency/unknown leaks.   Water and revenues are lost whenever the amount of water pumped into the distribution system is greater than the quantity billed. These amounts are measured by the master flow meter at the well, treatment plant, or pumping station, as compared to the customers' meters, with the difference being referred to as UNACCOUNTED-FOR WATER. The most obvious cause of this unaccounted-for water is leakage in the distribution system.

However, before surveying the system to find leaks, several other factors should be considered, including the percentage of unaccounted-for water, water losses caused by inaccurate metering, and unmetered water use.

Percentage of unaccounted-for water.   The point at which the cost of lost water becomes high enough to make it profitable to control is different for each community. It depends on the availability of water, the cost of treatment, and other factors. A 15-percent loss has generally been considered acceptable for large water utilities. Greater percentage losses often occur in smaller utilities. It is probably profitable to control any loss above 10 percent. Most authorities recommend implementing a water-loss control program where losses exceed 20 percent or more.

Water unaccounted for because of inaccurate metering.   Not all unaccounted-for water is physically lost. Either the utility's master meter or some of the customers' meters may be inaccurate. In cases of inaccurate customer meters, the water is "lost" for purposes of billing and accounting. Any meter inaccuracies make it difficult to assess the amount of water actually lost through leaks. Therefore, it is important for a utility to reduce all metering errors.

Customer meters generally tend to underregister. Sometimes customer meters do not work at all. Meters that underregister or do not function may account for significant water and revenue losses. An ongoing customer-meter testing and replacement program is essential to alleviate this problem. A large midwestern utility with a 9-percent water loss calculated that 6 percent of the loss was due to customer meter underregistration in spite of a comprehensive and efficient meter testing program. Some error is also caused by overlaps, gaps, or inconsistencies in the meter-reading schedule.

Master meters, including totalizers, transducers, and recorders, can produce significant over- and underregistration errors. Master meters should be tested every two years with a calibrated meter, such as a venturi, ultrasonic, or replacement propeller meter.

Unmetered water use.   Authorized and unauthorized unmetered flows, such as from hydrants, may account for significant water loss, especially in the summer. A procedure should be developed for metering or otherwise measuring, reporting, and occasionally policing water used for fire training, fire fighting, system flushing, flooding ice rinks, filling pools, street sweeping, sewer cleaning, and construction.

Methods for locating leakage.   Any water loss not accounted for because of inaccurate metering or unmetered use must be due to system leakage. Table 1-13 relates water loss to leak size. Underground leakage control can be accomplished only by a painstaking survey of the entire system. The two basic methods are listening surveys and a combination of listening surveys and flow-rate measurements sometimes referred to as a water audit.

Listening.   Listening involves the systematic use of sound-intensifying equipment to locate leaks. In all leaks, the water escaping through a pipe wall looses energy to the wall and the surrounding area, and the energy released appears, in part, as sound waves. These sound waves can be picked up by

Table 1-13. Water Loss Versus Pipe Leak Size

Pipe Leak Size*	Water Loss	
	Per Day gal	Per Month* gal
●	360	11 160
●	3 096	95 976
●	8 424	261 144
●	14 952	463 512

*Based on approximately 60-psi pressure.

sensitive instruments and amplified so that an operator can hear them. In the hands of an experienced operator, these instruments can help locate a leak with remarkable accuracy.

The most commonly used sound-intensifying or amplifying equipment is either mechanical or electronic. The two most widely used mechanical devices are the aquaphone and the geophone. The aquaphone resembles an old-fashioned telephone receiver with a metal spike protruding where the telephone wire should go. The listening end of the geophone looks like a medical stethoscope, but the listening tubes are connected to two diaphragms rather than one. This gives the operator a desirable stereo effect that aids in determining the direction of the leak.

Although electronic instruments are delicate, they are capable of filtering out unwanted background noise. A power source and more care in handling and storing are required than are usual for mechanical instruments. All the electronic devices can be easily carried by the operator. At least one can be connected to an electronic console and housed in a truck to form a mobile laboratory. Electronic leak detection equipment is shown in Figure 1-52.

The procedure for a listening survey can vary from "skimming" (just listening on selected hydrants) to completely covering a system by listening on all hydrants, valve stems, and services. The survey is usually conducted at night, when background noise is at a minimum. When a leak is found it can usually be heard on adjacent valves or hydrants. Listening over the connecting main will indicate where the sound is the loudest, and the leak will usually be found below that point.

Dead ends, crosses, tees, and partially closed valves may make location and verification of the leak more difficult. Such configurations have fooled many experienced observers. Another problem with the listening method is that not all leaks may be detected by listening on surface appurtenances. Some extremely

Courtesy of Heath Consultants Inc.

Figure 1-52. Electronic Leak-Detection Equipment

large leaks have been found that made no noise at all until a metal bar was driven to the pipe wall and sounded.

A successful listening or sound survey includes the following steps:

1. Surveying the system

2. Checking storm sewers

3. Evaluating hydrants

4. Listening at valves

5. Checking the service line

6. Finding the leak location

7. Pinpointing the leak.

The survey of the system is performed at night, when the system is relatively quiet and at low flow levels. All hydrants are inspected using a system map. Those that are noisy or have water leakage sounds and those that have normal system sounds are appropriately identified on the map. Personnel from distribution and sewage departments are consulted for any information that might account for the noisy hydrants. All remarks concerning abandoned service, previous repairs, and construction are recorded and filed for future reference.

Storm sewers are checked for possible flow from the potable system. If flow is found, it is sampled for chlorine and other chemicals added to the treated water. A finding of these chemicals helps to confirm the existence of a leak somewhere in the area near where the samples were taken.

After listening has identified hydrants that may have leakage nearby, the survey crew takes additional steps to locate the leak more precisely. The valves on suspect hydrants are exercised several times; if the noise stops or changes pitch as the valve is adjusted, then the problem is with the hydrant valve itself. Where the hydrant is not found to be a problem, the valve isolating each hydrant from the main is checked, together with the intersection valves on the mains, the surrounding curb stops, and the customers' outside sill cocks or faucets. Usually one of those locations will have a significantly louder leakage noise, indicating that it is near the leak.

Any noise that appears to be on the customer service line is checked by valving off the location and opening an outside faucet to relieve the pressure. If the noise disappears, then the leak is presumed to be on the customer side of the service, and an effort is made to locate the problem there. If the noise is still present after the service is valved off, then the leak must be on the utility side of the service, most probably on the main near the customer corporation tap. In this case, the road surface over the line is sounded with a ground microphone. Where the loudest sound is located, a mark is placed on the road to indicate the possible leak. The outside limits where the noise is observed to change or disappear are also indicated on the road surface. In this way, a working area can be identified.

To further pinpoint the leak, with as little damage to the road surface as possible, the following procedure is used. First, a 3-in. (0.75-mm) hole is drilled through the pavement over the main. Then, a compressed air supply is used to blow a hole down through the subsoil until the main is reached. If the leaking area has been located, water will usually come to the surface through the probe hole. In areas where probe holes cannot be made, metal rods are driven down to contact the main, and listening devices are applied to the rods.

Factors affecting leak detection. Several factors affect the success of leak detection. Better system continuity increases the chance for successful leak detection. Where main sections have gasket joints that isolate each pipe section, the leakage noise will be restricted to the section that is leaking. Type and size of the water main or piping are other considerations. Copper transmits sound best, followed in order by steel, cast iron, ductile iron, plastic, asbestos–cement, and concrete. Smaller pipes transmit sound well, but as the diameter increases, the quality of the sound diminishes. Tees, elbows, and other fittings consistently amplify sounds and cause difficulty in determining the location of the leak.

Leak size and system pressure also affect the sound of leaks. Systems running at 50–70 psi (350–480 kPa) will transmit sound over some distance, but leaks in low-pressure networks may be hard to detect even when close to the source. Small leaks under higher pressures will tend to cause shrill sounds. Leaks from beam breaks under the same pressures may cause quiet moving-water noises. Other factors that affect the sound of leaks include soils and road construction over the pipeline. Dry, sandy soils produce the best noise transmittance, whereas sounds are deadened somewhat in loamy soils and reduced even further in clay soils. Gravel roads and lawn areas are poor transmitters of sound compared to paved surfaces. The noise contributed by other buried utilities can also cause difficulty in pinpointing a leak.

The major advantages of listening are that no excavations are needed for tapping the mains, distribution crews are not required to operate valves and hydrants, and the work can be performed easily by other utility personnel. Listening can be effectively used where a small section of a system is suspected of having major leakage.

Water audits. A WATER AUDIT is a combination of flow measurements and listening, in an attempt to give a reasonably accurate accounting of all the water entering and leaving a system. To ensure that the measurements used in the audit are accurate, the master meters (which measure the total water entering the system) must be checked for accuracy; industrial meters must be tested in place for accuracy, without interrupting service, if feasible; unauthorized use of water through fire lines or other means must be identified; and underground leakage must be identified through systematic measurements and soundings.

The first step in performing a water audit is usually a 24-hour measurement of the water entering the system. A portion of the unaccounted-for water is sometimes found at this stage, if master meters are determined to be over-registering. The water system is then divided into districts that are measured independently for a 24-hour period. It is unusual for a system to show a uniform distribution of underground leaks, and it is wasteful to spend time and money performing a leak survey in areas with a low potential for leakage. Decisions as to where leakage is likely to be occurring can be made by analyzing the results of measurements within each district.

The ratio of minimum night rates of consumption to total consumption is a key to analyzing the results of measurements within the districts. When the ratio is 0.35 to 0.40 in a residential district, little leakage is likely to be found through subsequent investigations. However, where the ratio is greater than 0.40 and cannot be attributed to industrial night use or to other large consumers, districts should be subdivided into smaller areas for further investigation.

The flows into each of the smaller areas should be measured during night hours, usually between 10:00 p.m. and 6:00 a.m. Any areas showing unusually large flow rates should be sounded. This procedure will usually locate the leak and identify the most probable type of leak and its size.

After the leak is reported and repaired, flow into the area can be measured again to determine if any potential for additional leakage remains or if the potential for unauthorized use through fire lines exists, and at what rate. When the survey is complete, all leakage is tabulated along with meter under-registration and unauthorized use. Estimates are made of unmetered public use and unavoidable leakage that would be too costly to repair. Water that remains unaccounted for is attributed to domestic meter underregistration. If the amount is sizable, the need for more frequent meter testing and rotation is indicated.

An example of a summary from a water audit is shown in Figure 1-53. The principle advantage of having a water audit is that it produces an accounting of all water flowing through the distribution system. Additional benefits of an audit include a partial valve-and-hydrant maintenance program and an update of distribution maps.

	ML/day	mgd	Percent
Water sold through domestic meters	9.4	2.5	50
Water sold through industrial meters	4.7	1.25	25
Underground leakage located	1.8	0.5	10
Underregistration of industrial meters	0.38	0.1	2
Unauthorized consumption	0.37	0.1	2
Total accounted-for water	16.65	4.45	89
Unmetered use (sewer flushing, fire fighting, street washing)	0.18	0.05	1
Loss through unavoidable leakage at 200 gal/day per mile of main	0.07	0.2	4
Unmetered use and underregistration of domestic meters	1.1	0.3	6
Total unaccounted-for water	1.35	0.55	11
Total water produced	18	5	100

Figure 1-53.   Typical Water Audit Results

A disadvantage of the audit method is the need for night work. The method also requires that gauging points be excavated, where necessary, and taps installed. Once the mains are exposed, many utilities install street connections as permanent gauging points, so these expenses need only be incurred once.

Some utilities do not worry about water losses because they have an abundant source of water available. However, they pay increased chemical and energy costs to produce the extra water. Leakage also directly influences the ability to provide adequate fire protection and adequate service, hastening the time when capital expenditures will be necessary to improve the pipeline system, expand filtration plant capacity, construct new pumping stations, and locate new water sources. Leakage also increases the possibility of cross connections and backflow of nonpotable fluids into the water system.

Emergency leaks.   An emergency leak is usually a broken main or a severe service leak. An action plan to deal with main breaks should be established and coordinated with police, fire, and street department personnel as necessary. Trained personnel, records, maps, and repair parts should be available at all times in case a major leak occurs.

If a major leak is reported, it should be investigated at once to determine the severity of the problem and to see whether the repair is the responsibility of the utility or of the owner of the service line. If the leak is serious or in an area where it could quickly do extensive property damage, then it should be repaired immediately.

The first consideration in dealing with the leak should be to get the water loss under control by complete or partial shutdown of the line. Where damage to property is occurring or is likely to occur, water should be shut off immediately. Otherwise the flow should be reduced by closing all valves except one. Customers who will be without water, including buildings with sprinkler systems, water-cooled refrigeration units, or other industrial water users, should be notified not to draw water during the period of time that it will take to repair the break. If buildings are unoccupied, then the owner, manager, or agent should be notified, if possible. If the leak is at the bottom of a hill, customers should be notified to shut off their inlet valves to prevent siphoning of the hot-water tanks or softeners. If the occupants are not at home, the curb stops should be shut off. After customers have been notified, the final valve can be shut down. An up-to-date system map with accurate valve locations and main sizes is valuable at this time, and the results of a regular valve locating/exercising program will be appreciated.

The valves that are operated and their operating conditions should be recorded. The section of the system that is out of service should be noted, and the location of all hydrants affected should be reported to the fire department.

Locating the leak.   When water is flowing up out of the ground, it may seem obvious where to locate the leak. Time and money can be wasted, however, by digging in the wrong spot. The leak is often not directly below where the water surfaces. Sometimes the surface flow may be a long way from the actual source of the leak. The leak should be pinpointed using geophones, drilling over the main, or any other method previously discussed. Some water pressure will have to be left on or temporarily restored to the affected area during the leak-location phase.

Several problems may occur at a main break. Roads, walks, railways, or other utilities may have been undermined by the water flow. Basements may have been flooded, and slippery driving conditions may result from flooding or freezing water. Problems will usually be worse where sewers are plugged or restricted. It may be necessary to arrange for police assistance for traffic control, and salting or sanding of roadways may be necessary.

Excavation.   Before excavating a leaking line for repair, other utilities should be contacted to determine where their underground lines are located. In addition to excavating equipment, the following items should be on hand:

- A pump for dewatering the excavation
- Traffic control equipment, including barricades, flashers, and cones
- Pipe cutters and/or saws
- Proper-size clamps, couplings, pipe, and sleeves
- An air hammer and, in winter, a propane torch
- A generator, lights, and flashlights, if the work will be done at night
- Safety lines, if personnel must work where sudden flooding or cave-in is a possibility.

Safety procedures should not be ignored in the haste to handle the emergency—accidental injuries are especially likely in unplanned situations.

Leak repair.   If the break is severe, the damaged section may have to be cut away with a saw or pipe cutter (Figure 1-54) and a new section installed. Often the leak can be fixed with a flexible clamp or coupling of the correct size and material for the pipe being repaired. Various types of repair devices are shown in Figure 1-55. If electrical continuity needs to be maintained, use a ductile-iron repair

Courtesy of Wheeler Manufacturing Corp.

Figure 1-54. Pipe Cutting Equipment

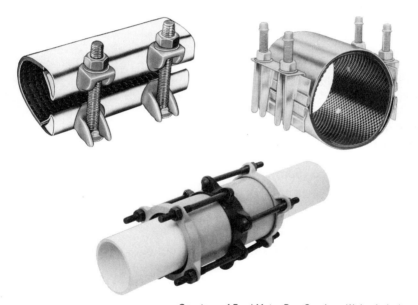

Courtesy of Ford Meter Box Co., Inc., Wabash, Ind.

Figure 1-55. Pipe Repair Devices

sleeve or a stainless-steel repair sleeve with metal inserts. Follow the manufacturer's recommendations when installing the repair device.

Several preliminary procedures should be observed to help ensure a good repair. The diameter of pipe should be checked to make certain the correct-size clamp is being installed. If a stainless-steel clamp is used, the bolts and nuts must be stainless steel or cadmium plated, to minimize galvanic corrosion. After the pipe is uncovered, the area where the clamp will be installed should be scraped to remove as much dirt and corrosion as possible, making the surface smooth.

In applying the clamp, care should be taken that no foreign material sticks to the gasket as the bottom half is brought around the pipe or becomes lodged between the gasket and pipe as nuts are tightened. Loose-fitting wrenches and wrenches that are too short to achieve proper torque should not be used, and the threads of the clamp bolts should be kept free of foreign material to facilitate tightening.

After the clamp is installed, it should be tar coated for rust prevention, and the line should be pressure tested for leaks before backfilling the excavation. When the repair has been completed and tested, the main should be flushed to remove any air or dirt that may have entered while it was under repair. The entire line should be chlorinated to reduce the danger of contamination from soil, materials, and backsiphonage.

Restoration.   All unsuitable excavated material should be hauled away and sand or crushed rock should be used for backfilling with careful lift compaction. This should allow permanent street repairs to be made at once. The area should be checked to make sure no equipment or traffic hazards, such as ice or erosion, remain. All valves should be in the proper position, and customers and police should be notified of completion of the repair, as appropriate.

Reporting requirements.   It is helpful to determine the type and cause of breaks (for example, shear, split, blow out, joint). Information should be recorded on a water-main-failure report similar to the one shown in Figure 1-56.

The importance of good distribution system records cannot be overstated from the standpoint of operating and maintaining the system. Scheduled inspections, maintenance, and repairs cannot be made to the system in a timely, efficient manner without records that are complete, current, accurate, and accessible. The importance of mapping and other record-keeping requirements related to pipelines and appurtenance is discussed in detail in Module 10, Maps, Drawings, and Records.

1-6. Safety

Specific areas where the need for safety precautions must be recognized and observed when dealing with pipe installation and maintenance include:

- Material handling
- Trenches
- Barricades, warning signs, and traffic control
- Personal-protection equipment
- Chemicals
- Hand tools
- Portable power tools.

Material Handling

Lifting an object by hand can be done safely (and easily) if people use common sense and follow a few basic guidelines. Do not lift or shove sharp, heavy, bulky

FIELD DATA FOR MAIN BREAK EVALUATION

DATE OF BREAK _____ TIME _____ A.M. _____ P.M.

TYPE OF MAIN _____ SIZE _____ JOINT _____ COVER _____ FT _____ IN.

    THICKNESS AT POINT OF FAILURE _____ IN.

NATURE OF BREAK: Circumferential ☐   Longitudinal ☐   Circumferential & Longitudinal ☐
Blowout ☐ Joint ☐ Split at Corporation ☐ Sleeve ☐ Miscellaneous _____ ☐
                                                          (describe)

APPARENT CAUSE OF BREAK: Water Hammer (surge) ☐   Defective Pipe ☐   Corrosion ☐
Deterioration ☐   Improper Bedding ☐   Excessive Operating Pressure ☐

Differential Settlement ☐ Temp. Change ☐ Contractor ☐ Misc. _____ ☐
                                                          (describe)

STREET SURFACE: Paved ☐ Unpaved ☐   TRAFFIC: Heavy ☐ Medium ☐ Light ☐

TYPE OF STREET SURFACE _____ SIDE OF STREET: Sunny ☐ Shady ☐

TYPE OF SOIL _____ RESISTIVITY _____ ohm/cm

ELECTROLYSIS INDICATED: Yes ☐ No ☐   CORROSION: Outside ☐ Inside ☐

CONDITIONS FOUND: Rocks ☐ Voids ☐ PROXIMITY TO OTHER UTILITIES _____

    DEPTH OF FROST _____ IN.   DEPTH OF SNOW _____ IN.

OFFICE DATA FOR MAIN BREAK EVALUATION

WEATHER CONDITIONS PREVIOUS TWO WEEKS _____

SUDDEN CHANGE IN AIR TEMP.? Yes ☐ No ☐   TEMP. ____ °F RISE ____ °F FALL ____ °F

WATER TEMP. SUDDEN CHANGE: Yes ☐ No ☐ TEMP. _____ °F RISE _____ °F FALL _____ °F

SPEC. OF MAIN _____ CLASS OR THICKNESS _____ LAYING LENGTH _____ FT

                               OPERATING   PREVIOUS BREAK
DATE LAID _____ PRESSURE _____ PSI REPORTED _____

INITIAL INSTALLATION DATA
TRENCH PREPARATION: Native Material _____ ☐ Sand Bedding ☐ Gravel Bedding ☐
                                      (describe type)

BACKFILL: Native Material ☐ DESCRIBE _____ Bank Run Sand & Gravel ☐

    Gravel ☐ Sand ☐ Crushed Rock ☐ OTHER _____

SETTLEMENT: Natural ☐ Water ☐ Compactors ☐ Vibrators ☐ OTHER _____
                                                    (describe)

ADDITIONAL DATA FOR LOCAL UTILITY USE

LOCATION OF BREAK _____ MAP NO. _____

REPORTED BY _____

DAMAGE TO PAVING AND/OR PRIVATE PROPERTY _____

REPAIR MADE (Materials, Labor, Equipment) _____

REPAIR DIFFICULTIES (If any) _____

INSTALLING CONTRACTOR _____

Figure 1-56. Report of Water-Main Failure

objects without help or tools. The main guideline to proper lifting is to bend at the knees to grasp and lift the object; do not lift with your back. Everyone should know the proper way to lift heavy objects by hand. The correct procedure is:

1. Get a good footing with feet about shoulder-width apart.
2. Bend at the knees to grasp the weight, keeping the back straight.
3. Get a firm hold.
4. Keeping the back as straight and upright as possible, lift slowly by straightening the legs.

To change direction, turn the whole body, including the feet, rather than just twisting the back. Never lift a load that is too heavy or too large to lift comfortably. Use a mechanical assist when possible to help prevent injuries. Even though pipes are big and look tough, handle them carefully. Pipes should be carefully lowered from the truck to the ground—not dropped. If several people are trying to move or place pipe, they should work together. Only one person should give directions and signals. Where a crane or other machine is handling pipe, one person should direct the machine operator.

If pipe is to be lowered by skids, make sure the skids can hold the weight and are firmly secured. Snubbing ropes should be right for the job and in good condition. Never let anyone stand between the skids and the pipe when it is being lowered. Individuals working with ropes should wear gloves to prevent rope burns. It is often a good idea to control the pipe by using wood chocks.

For large valves, fork lifts or slings around the valve body or under the skids should be used for loading. Only hoists and slings with adequate load capacity should be used. Do not hook hoists onto or fasten chains around bypasses, yokes, gearing, motors, cylinders, or handwheels.

Equipment for transporting objects (wheelbarrows, hand trucks, or carts) should be properly maintained and not overloaded. When such equipment is used, clearances should be carefully judged and horseplay should be prohibited.

Trenches

Trenches can be made safe if proper shoring rules are followed and proper equipment is used. Everyone in the trench must wear a hard hat and individuals working around jackhammers should wear proper eye protection and foot gear.

Proper trench shoring cannot be reduced to a standard formula. Each job is an individual problem and must be considered under its own conditions.

Under any soil conditions, cave-in protection is required for trenches or excavations 5 ft (1.5 m) deep or more. Where soil is unstable, protection may be advisable even in more shallow trenches.

These are three basic means for cave-in prevention and protection:

- Sloping
- Shielding
- Shoring.

SLOPING involves excavating the walls of the trench at an angle, so the downward forces on the soil are never allowed to exceed the soil's cohesive strength (Figure 1-57).

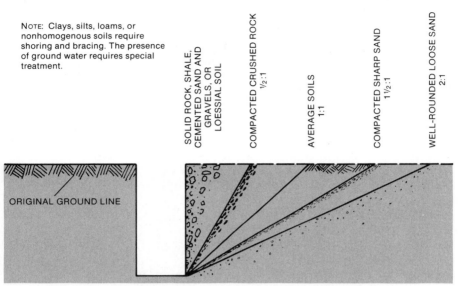

NOTE: Clays, silts, loams, or nonhomogenous soils require shoring and bracing. The presence of ground water requires special treatment.

SOLID ROCK, SHALE, CEMENTED SAND AND GRAVELS, OR LOESSIAL SOIL

COMPACTED CRUSHED ROCK ½:1

AVERAGE SOILS 1:1

COMPACTED SHARP SAND 1½:1

WELL-ROUNDED LOOSE SAND 2:1

ORIGINAL GROUND LINE

Figure 1-57. Approximate Angle of Repose for Sloping of Sides of Excavations

For any section of an excavation, there will be a certain angle, called the ANGLE OF REPOSE, where the surrounding earth will not slide or cave back into the trench. The angle of repose varies with the type of soil, the amount of moisture it contains, and with surrounding conditions, especially vibration from machinery.

If the excavation is done in solid rock, shale, cemented sand and gravel, or LOESSIAL soil, then the wall may be vertical—straight up and down. If the excavation is in compacted, crushed rock, then the bank must be cut back 6 in. for each foot of depth (0.5 m for each metre). For soils of average cohesion, the angle is one-to-one; that is, the bank is cut back 1 ft for each foot of depth (1 m for each metre). If the soil is mostly sharp, compacted sand, the angle is 1½ ft back for each foot down (1½ m for each metre). If it is loose sand, the angle of repose is 2 ft back for each foot down (2 m for each metre).

The second method of cave-in protection, SHIELDING, involves the use of a steel box, open at the top, bottom, and ends. The box is placed into the ditch so workers can work inside it (Figure 1-58). As the work progresses, the protective box is moved or towed to provide a continuing shield from any caving in of the walls. This open-ended box is called a trench shield, portable trench box, sand box, or drag shield.

Shielding does not prevent a cave-in. The shield cannot fit tightly enough in the trench to hold up the trench walls. However, if a cave-in does occur, the worker within the shield is protected.

The shield is constructed of steel plates and bracing, welded or bolted together. It is important that the shield extend above ground level or that the trench walls above the top of the shield be properly sloped.

A major disadvantage of the shield is that workers have a tendency to leave its protection in order to check completed work, or to help adjust pipe placement, or

Figure 1-58. Trench Shield

just to get out of the way of the job in progress. The shield *only* protects those workers actually within it. Because the shield offers protection in only a limited area, the trench should be backfilled, if possible, as soon as work in an area is completed and the shield moved forward.

The third method of protecting workers in trenches is SHORING. If properly installed, shoring will actually prevent the caving in of excavation and trench walls.

Basically, shoring is a framework support system of wood, metal, or a combination of both. It is important that the type and condition of the soil be checked before excavating, so that the correct shoring can be selected for the conditions encountered. The type and condition of the soil also needs to be checked during the operation. Soil conditions can vary greatly within a few feet, and conditions may change if they are affected by weather or vibration.

A shoring assembly has three main parts (Figure 1-59):

- UPRIGHTS—the vertically placed boards that are in direct contact with the earthen faces of the trench; spacing between the uprights will vary, depending on soil stability. When the uprights are tight against each other and form a solid barrier against the faces of excavations and trenches, they are usually called sheeting, sheet piles, or close sheeting.

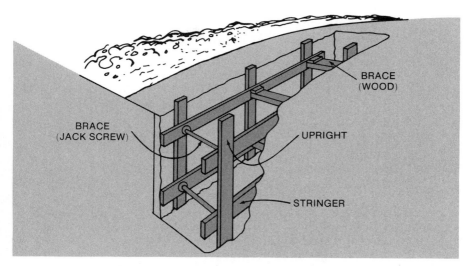

Figure 1-59. Construction of Trench Shoring

- STRINGERS—the horizontal members of the shoring system to which the braces are attached. The stringers are also known as whalers, because they are similar to the reinforcements in the hulls of wooden ships.
- TRENCH BRACES—the horizontal members of the system that run across the excavation.

To prevent movement and failure of the shoring, there should not be any space between the shoring and the excavation sides. After the shoring has been installed, all open spaces should be filled in and the backfill material compacted.

A basic principle of shoring operations is that shoring and bracing should always be done from the top down, then removed from the bottom up after the operation has been completed. This will give maximum protection to workers constructing or dismantling the shoring.

When a job in a shored excavation is finished, continue to use shoring while the trench is backfilled and compacted. The recommended procedure is to raise the shoring a few feet, then place and compact a layer of backfill. When the first layer is in place, raise the shoring a few feet more and place more backfill.

In some installations, uprights can be left in place until all or most of the backfill has been placed, then removed with power lifting equipment. The sheeting itself should be raised to clear the backfill being placed and compacted, but should not be removed until trench depth has been reduced to a point where workers no longer need protection. As backfill reaches the bottom of jacks or braces, they should be removed.

When all work has been completed and no compacting of backfill is needed, ropes should be used to pull jacks, braces, and other shoring parts out of the trench, so workers do not have to enter the trench after the protection has been removed.

Barricades, Warning Signs, and Traffic Control

Barricades, traffic cones, warning signs, and flashing lights are usually used to inform workers and the public of when and where work is going on. These devices should be placed so there is plenty of space between the work and the warning and so that they are easy to see. If necessary, a flag person should be used to slow traffic or direct it. Everyone involved in the work should wear bright reflective vests.

All barricades and signs should meet OSHA (Occupational Safety and Health Administration) and local standards for color and placement. Examples of recommended barricade placement are shown in Figures 1-60 and 1-61.

Personal-Protection Equipment

Most utilities now provide a broad range of personal-protection equipment, which helps to prevent serious injuries. Hard hats, safety goggles, and steel-toed shoes are probably the most widely used safety equipment.

Face shields and goggles to protect the face and eyes are required wherever there is a chance of injury that could be prevented by their use. Steel-toed safety shoes and boots are helpful in preventing damage to feet, caused by dropped objects. Where protection is needed from very heavy impacts, such as when operating tamping equipment, thick-flanged and corrugated-sheet-metal foot guards are recommended.

Gloves are necessary for protection from rough, sharp, or hot materials. Special long-length gloves are available to provide wrist and forearm protection. Rubber gloves should be worn when handling oils, solvents, and other chemicals. However, gloves should not be used around revolving machinery. Specially designed insulating rubber gloves should be worn wherever there is an electric shock hazard (ordinary rubber gloves provide little or no protection against electrical shock).

Hard hats, which are necessary whenever an operator is working in a trench or near electrical equipment, have been very successful in reducing serious injuries or deaths due to head injuries. However, metal hard hats should never be used where there is an electrical hazard. Hard hats are required whenever an operator is working in a trench or could be injured by falling objects.

Chemicals

The main chemical used in distribution system operations is probably chlorine, which is used for disinfecting mains. It can be purchased and stored as a liquid, gas, or solid. Chlorine must be treated, stored, and used properly to prevent accidents. Even with proper use, accidents can still happen, and the operator should know how to react to them. Depending on the concentration and the length of exposure, chlorine can cause lung irritation, skin irritation, burns, and a burning feeling in the eyes and nose. If chlorine gas is used, supplied-air masks (air packs) should be available, and operators should be trained in their use. Anyone caught around a chlorine gas leak without a mask should leave the area immediately, keeping head high, mouth closed, and

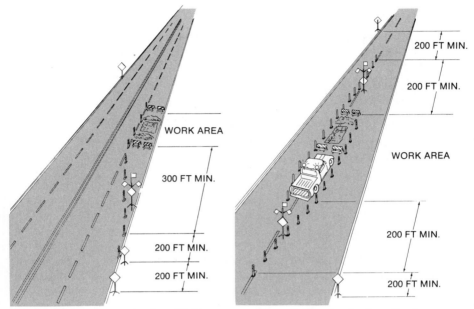

**Figure 1-60. Barricades—Right
Lane Closed**

**Figure 1-61. Barricades—Center
of Road**

avoiding coughing and deep breathing. Detailed safety precautions for chlorine gas are covered in Volume 2, *Introduction to Water Treatment*, Module 10, Disinfection.

In the distribution system, chlorine is most often used in the form of calcium hypochlorite or sodium hypochlorite. Calcium hypochlorite is available as a dry powder, crystals, or tablets. It is corrosive in small amounts of water and can support combustion. If stored or used improperly near or with organic material, it will require the same safety precautions as chlorine gas, except that it should be handled with gloves. Sodium hypochlorite is a strong acid and can cause similar problems. Proper protective equipment, including eye goggles, gloves, and a coat, should be worn.

Hand Tools

Basic rules for hand-tool use are as follows:

- Always select the appropriate tool for the job, never a makeshift.
- Check the condition of a tool frequently and repair or replace it if it is defective.
- Avoid using tools on moving machinery; stop the machine before making adjustments.
- Check clearance at the work place to make sure there is sufficient space to recover a tool if it should slip.

- Have good support underfoot so that there is no hazard of slipping, stumbling, or falling.

- Where appropriate, protect hands by wearing gloves; rings should not be worn.

- Carry sharp or pointed tools in covers; be sure they are pointed away from the body in case of a fall; do not carry sharp-edged tools in trouser pockets.

- Wear eye protection when using impact tools on hard, brittle material.

- After using tools, wipe them clean and put them away in a safe place; keep the work place orderly.

- Do not lay tools on top of stepladders or on other places from which they could fall on someone below.

- Learn and use the right way to work all hand tools.

Portable Power Tools

Electric power tools can cause shocks. These tools should always be grounded, and a ground-fault interrupter (GFI) circuit should be used whenever the tool is used outside or near water. The cord should be in excellent condition and never used to lift the tool or as a line. The cord or tool should never be left in the way to cause tripping.

Air tools can be dangerous if the hoses and connections are not correctly maintained. Do not point air tools at anyone, and do not clean off any part of the body or clothing with compressed air.

Selected Supplementary Readings

Beck, R.A.; Buttle, J.L.; & Wolfe, T.A. Water-Jet Technique Used to Clean Encrusted Pipe. *OpFlow*, 9:5:1 (May 1983).

Brainard, F.S. Jr. Leakage Problems and the Benefits of Leak Detection Programs. *Jour. AWWA*, 71:2:64 (Feb. 1979).

Brown, T.G. Basic Leak Detection is Necessary for Any System. *OpFlow*, 11:10:1 (Oct. 1985).

Causes and Control of Water Hammer. *OpFlow*, 9:9:3 (Sept. 1983).

Cleaning Cement-Mortar Lining Relieves Red Water Problem in Belmont, MA. *OpFlow*, 3:3:4 (Mar. 1977).

Coal, E.S. Methods of Leak Detection: An Overview. *Jour. AWWA*, 71:2:73 (Feb. 1979).

Cole, G.B. Leak Detection: Two Methods That Work—Part I. *OpFlow*, 6:5:3 (May 1980).

Cole, G.B. Leak Detection: Two Methods That Work—Part II. *OpFlow*, 6:6:3 (June 1980).

Cole, G.B. Leak Detection Program Reduces Unaccounted-for Water. *OpFlow*, 9:4:1 (Apr. 1983).

Considine, Edward Jr. Cost Data Report of Main Rehabilitation. *OpFlow*, 5:4:3 (Apr. 1979).

Cooper, R.E. & Knowles, W.L. Cleaning Improves Capacity of Large Diameter Pipeline. *OpFlow*, 2:5:1 (May 1976).

Disinfecting Water Mains. AWWA Standard C651-86. AWWA, Denver, Colo. (1986).

What Do You Know About The Nature of Water?—Part VIII. *OpFlow*, 1:10:1 (Oct. 1975).

Heim, P.M. Conducting a Leak Detection Search. *Jour. AWWA*, 71:2:66 (Feb. 1979).

Hudson, W.D. Improving Water System Efficiency Through Control of Unaccounted-for Water. *Jour. AWWA*, 70:7:362 (July 1978).

Introduction to Water Sources and Transmission. AWWA, Denver, Colo. (1979).

Is Your Main-Flushing Program Ready to be Activated? *OpFlow*, 4:9:1 (Sept. 1978).

Jackson, Rodney. Corrosion Control for Water Transmission and Distribution Systems. Proc. AWWA DSS, Los Angeles, Calif. (Feb. 1980).

Jones, L.V. Effective Cleaning of Water Mains. *OpFlow*, 11:3:6 (Mar. 1985).

Kerr, M.K. Cleaning Water Mains Using Polyurethane Pigs. Proc. AWWA DSS, Birmingham, Ala. (Sept. 1983).

Kingston, W.L. A Do-It-Yourself Leak Survey Benefit Cost Study. *Jour. AWWA*, 71:2:70 (Feb. 1979).

Kroon, J.R.; Stoner, M.A.; & Hunt, W.A. Waterhammer—Causes, Effects, and Solutions. Proc. AWWA DSS, Birmingham, Ala. (Sept. 1983).

Landes, Mike. An Effective Method for Cleaning Pipelines. *OpFlow*, 7:1:3 (Jan. 1981).

Laverty, G.L. Leak Detection: Modern Methods, Costs, and Benefits. *Jour. AWWA*, 71:2:61 (Feb. 1979).

Laverty. G.L. A Perspective on Leak Detection—Part I. *OpFlow*, 6:12:1 (Dec. 1980).

Laverty, G.L. A Perspective on Leak Detection—Part II. *OpFlow*, 7:1:1 (Jan. 1981).

Lohmiller, P.A. & Curtiss, J.F. Leak Detection Equipment—What's Right for You? Proc. AWWA DSS, Syracuse, N.Y. (Sept. 1984).

Maintaining Distribution-System Water Quality. AWWA, Denver, Colo. (1986).

Power, J.A. Sanitary Considerations in Constructing Water Mains. Proc. AWWA DSS, Birmingham, Ala. (Sept. 1983).

Safety Practice for Water Utilities. AWWA Manual M3. AWWA, Denver, Colo. (1983).

Steel Pipe—A Guide for Design and Installation. AWWA Manual M11. AWWA, Denver, Colo. (1985).

Stevens, R.L. & Dice, J.C. Disinfecting Mains and Storage Tanks—Part I. *OpFlow*, 3:9:1 (Sept. 1977).

Stevens, R.L. & Dice, J.C. Disinfecting Mains and Storage Tanks—Part II. *OpFlow*, 3:10:1 (Oct. 1977).

Stevens, R.L. & Dice, J.C. Chlorination—A Proven Means of Disinfecting Mains. *OpFlow*, 7:10:1 (Oct. 1981).

Stevens, R.L. & Dice, J.C. Chlorination—Pills for Pipe and Techniques for Tanks. *OpFlow*, 7:11:1, (Nov. 1981).

Vest. R.L. Traffic Controls Ensure Work is Done Safely. *OpFlow*, 9:9:1 (Sept. 1983).

Wallaceburg, Ont. Flushes Main With Foam Swab Technique. *OpFlow*, 1:6:1 (June 1975).

Glossary Terms Introduced in Module 1

(Terms are defined in the Glossary at the back of the book.)

Air purging
Anaerobic
Angle of repose
Appurtenance
Arterial-loop system
As-built plan
Backfill
Bedding
Bimetallic
Cathodic protection
Continuous-feed method
Corrosion
Crushing strength
C value
Distribution main
External load
Flexural strength
Flow coefficient
Galvanic cell
Galvanic corrosion
Grid system
Hydrostatic pressure
In-plant piping system
Internal load
Loessial
Mechanical joint
Pig
Pitot gauge
Push-on joint
Regular operating delivery pressure
Roughness coefficient

Service
Service connection
Shielding
Shoring
Sloping
Slug method
Smoothness coefficient
Spring line
Stability
Stringer
Superimposed load
Surge pressure
Swab
Tablet method
Tensile strength
Thickness class
Thrust
Thrust anchor
Thrust block
Till
Traffic load
Transmission line
Tree system
Trench brace
Tuberculation
Tubercule
Unaccounted-for water
Upright
Water audit
Water hammer
Well point

Review Questions

(Answers to the Review Questions are given at the back of the book.)

1. What type of distribution system configuration is not recommended and why?

2. Explain what determines the selection of main sizes.

3. What is water hammer and why is it important in distribution system operations?

4. Explain what is meant by the *C* value.

5. List four types of corrosion that can occur in a distribution system and summarize the causes of each.

6. Explain what causes tuberculation and what effect it has on the distribution system.

7. Identify those factors that can cause corrosion cells on the outside of a pipe.

8. List four methods used to prevent external corrosion.

9. Identify three types of piping material commonly used in distribution systems.

10. List advantages and disadvantages of ductile-iron, asbestos–cement, and PVC pipe.

11. List four common joints used with DIP and describe the conditions under which each may be used.

12. Identify a common disadvantage shared by plastic and AC pipe and describe how the condition can be overcome.

13. Why should pipe be inspected before or during unloading?

14. How should pipe be unloaded?

15. List at least five important points to be considered when stringing pipe at a jobsite.

16. What factors usually determine the depth and width of a trench?

17. Why should very wide trenches be avoided?

18. What is the angle of repose and why is it important?

19. List some of the typical causes of trench cave-ins.

20 What are some of the danger signs that indicate potential trench failure?

21. What recommended methods or practices should be followed to prevent trench failure?

22. What is a major disadvantage of shielding when working in a trench?

23. Explain why the proper placement of bedding material is important.

24. Why must joint materials and pipe ends be kept clean?

25. What type of lubricant should be used for rubber gaskets on pipe joints and why?

26. Explain why thrust blocks and anchors are important and indicate where they should be used.

27. Is it true that thrust blocks should only be placed against undisturbed soil?

28. List three methods that can be used to compact backfill material.

29. Why is proper compaction of backfill material important?

30. Explain what steps are to be followed after pipe is installed but before the new line is placed into service.

31. List and describe three methods of disinfecting water mains.

32. Stipulate the chlorine residual and contact time required for disinfecting new mains using the methods described in question 31.

33. Explain the importance of as-built plans (conforming to construction records).

34. Explain why routine water-quality monitoring of a distribution system is important.

35. Indicate why a routine hydrant-flushing program is considered to be good operating practice.

36. What are pigs? Explain how they are used in distribution system operations.

37. Explain the term unaccounted-for water and its importance to management and operations.

38. What are the two basic methods for conducting leak surveys?

39. What precautions are necessary when installing distribution mains in traffic areas?

40. List recommended personal-protection gear that should be used during the installation of a pipeline.

Study Problems and Exercises

1. Your utility serves a population of 10,000 people. To meet increased demand in a new service area, a main-extension program is being considered. The soil in the area ranges from hardpan to cobbles. Prepare statements supporting your recommendations for (a) selection of pipe materials and (b) minimum size main to be installed.

2. You have just been employed to supervise the installation of a mile of new 8-in. PVC water main that will service a newly annexed subdivision. Outline the steps you would take from delivery of the pipe to final inspection to assure proper and safe installation of the pipeline.

3. You receive a telephone call at 3:00 a.m. Sunday morning from a police dispatcher who informs you that a patrol car has reported large quantities of water running down Front Street. Prepare recommended guidelines for dealing with this situation.

Module 2

Tapping

As a distribution system expands, lines of various sizes are added to connect the existing lines to new points of use. These new lines may be LATERALS (usually larger than 2-in. [50-mm] diameter) needed to furnish water to new service areas or they may be SERVICE LINES to individual customers.

Lateral-to-main connections can be made under line pressure by the use of specialized fittings and machines, or by shutting down the line and installing a conventional tee. Large service-to-main connections, 3 in. (75 mm) and larger, are usually installed using the same procedures as lateral-to-main connections.

Small service-to-main connections, 2-in. (50-mm) diameter and less, are normally installed under line pressure. A drilling-and-tapping machine can be used to install a valve (CORPORATION STOP) directly into the wall of the main, or a connecting device to hold the valve may be installed first and a hole drilled through the wall of the pipe with a drilling machine.

TAPPING is the procedure of cutting threads on the inside of a drilled or cut hole. However, connecting any type of line to an existing main is also referred to as tapping, whether or not a threaded connector is used.

The distribution operator should be familiar with the procedures for making lateral connections for both large and small lines, whether dry or under pressure. This module covers those subjects; the service lines themselves are discussed in Module 5, Services and Meters.

After completing this module you should be able to

- Recognize the different types and sizes of connections (taps) used to connect lateral and service lines to mains.

- Understand the use of corporation stops and identify important factors in selecting the type of corporation stop required for a given installation.

93

- Recognize the equipment used to install small and large taps.
- Explain the general procedures and precautions used when tapping small and large connections.
- Recommend appropriate records used in connection with tapping operations.
- Explain the maintenance required for tapping machines.
- Recommend safety precautions related to tapping operations.

2-1. General Procedures and Materials

The tapping procedures and materials required in a given situation depend primarily on the size of the main, the size of the lateral to be installed, and the inconvenience that would be involved in shutting down the main. Additional factors affect each individual situation, and detailed planning is necessary before beginning any tapping operation. The operator must be sure that the correct size and type of connection has been selected and that all necessary materials are on hand.

Wet and Dry Taps

Connections can be made either when the main is empty (DRY TAPS) or when the pipe is under pressure (WET TAPS). Dry taps are usually made only during the installation of a main. Most tapping done today, whether on new or existing mains, is wet tapping. This method is preferred when adding a service connection to an existing main because it allows the connection to be made without turning off the water and interrupting service to existing customers. There is also less chance of contamination, since the pressure in the main tends to expel any foreign matter.

Small Connections

Connections for lateral service lines 2 in. (50 mm) and smaller in diameter are considered small connections. These connections are usually made for service lines to homes or small businesses. Most utilities make some small taps each year and own the equipment needed to make taps for one or more of the smaller, more common sizes.

Corporation stops.   The fitting used to connect small-diameter service lines to a water main is called a CORPORATION STOP. It is generally a brass shut-off valve, threaded on both ends (Figure 2-1). The corporation stop is also referred to as the corporation cock, corporation tap, corp stop, corporation, or simply corp or stop.

The corporation stop is usually screwed directly into a threaded hole in the pipe wall. Drilling-and-tapping machines allow this operation to be performed without shutting down the main. However, where the pipe wall is too thin or too soft, or when for other reasons it is not practical to directly tap the main, a SERVICE CLAMP, or SADDLE, must be used (Figure 2-2). This device is attached

*Courtesy of A.Y. McDonald Manufacturing
Company, Dubuque, Iowa*
Figure 2-1   Corporation Stop

*Courtesy of A.Y. McDonald Manufacturing
Company, Dubuque, Iowa*
Figure 2-2   Brass Saddle

directly around the main, and the corporation stop is then threaded into the service clamp, rather than into the pipe wall. A drilling machine is used to activate the service.

Corporation stops are available in many different sizes and styles, including several combinations of threads for the inlet and outlet ends of the stop. The standard AWWA corporation-stop thread is the one most commonly used on the inlet end of the corporation stop. This thread is commonly known as the Mueller thread. It is also sometimes referred to as corporation cock (CC) or corporation-stop (CS) thread. The other thread in general usage is the iron-pipe thread. Both threads have approximately the same number of threads per inch, but the corporation-stop thread has a larger diameter and a steeper taper. This gives it greater strength and exerts less wedging pressure on the pipe into which it is installed. Figure 2-3 illustrates the standard AWWA corporation-stop thread.

Corporation-stop outlet ends are available for a variety of service-line materials. These include flared copper service connections, iron-pipe threads, increasing iron-pipe threads, lead flange connections, and compression couplings for various materials. Some corporation-stop outlets also include an internal driving thread used for the attachment of an installation tool.

In selecting the correct type of corporation stop for a given installation, the following should be considered:

- Service-line diameter
- Service-line material
- Use of a service clamp or direct insertion
- Type of connection to service line
- Size and material of main being tapped.

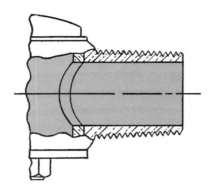

Figure 2-3   Standard AWWA Corporation-Stop Thread

Large Service/Lateral Connections

Large connections are generally defined as those 3 in. (75 mm) or larger in diameter. These include large service connections and branch mains. Large connections may be made by emptying the main and installing a tee fitting, or they may be tapped into the main without shutting off water service.

Tee connections.   In the past it was practical to shut down a main during a low-use period and install a tee (Figure 2-4) for any large lateral connection. The job was often performed at night or very early in the morning, so that few customers were aware that they were without water. Advanced planning was required to locate and prepare the proper valves in order to shut down the main where the connection would be made.

In most systems today, the characteristics of users are such that temporary shutdowns are no longer acceptable unless absolutely necessary. Every community has some customers who must have water at all times. Shutting down service involves the danger of losing part of the fire-protection system during the period of shutdown. It also creates water loss, premium charges for night labor, and increases the likelihood of contamination and complaints of dirty water. As a result, tee connections should only be installed when the lateral to be connected is too large to be made by tapping, or when the main will be shut down for some other reason.

Tapped connections.   The use of drilling machines to install large connections reduces the problems associated with the interruption of water service. The equipment used to tap large connections is, however, larger and more expensive than that used for small services.

The opening cut into the main for large taps is not threaded. Connection to the main is made with a TAPPING SLEEVE or SPLIT TEE FITTING (Figure 2-5), and the appropriate size shut-off valve (TAPPING VALVE) is used instead of a corporation stop. Large lateral connections require proper restraint and blocking (refer to Module 1, Pipe Installation, for information on thrust restraint).

**Figure 2-4   Tee Used for Large Lateral
Connections**

Courtesy of US Pipe and Foundry Company

Figure 2-5   Tapping Sleeve

Tap and Service Size

The size of the hole to be made and the connecting fitting to be used are usually determined by the size of the service pipe required, which is in turn determined by the water pressure and flow requirements of the customer to be served. Service sizes may be determined by local plumbing codes or insurance company requirements, or they may be calculated using guidelines from such codes and from AWWA Manual M22, *Sizing Water Service Lines and Meters*. A ¾-in. (19-mm) connection is generally the standard requirement for residential service. Sizing services is discussed in greater detail in Module 5.

Tapping size is also somewhat dependent on the main. The minimum recommended size of cast-iron main required for each tap size in the ½-in. to 2½-in. (13-mm to 64-mm) range is given in Tables 2-1 and 2-2.

In some instances it is impossible to make the service-to-main connection as large as required, either because of limited capacity of the available equipment or because the hole size would be too great for the main. In such cases two or more small connections may be made and joined together, using a Y-branch connection, to supply the larger service line.

Standardization

Using standard procedures and materials for common tap sizes will help to ensure fast, competent installation of taps. Standardization of large connections is especially important for a small utility, since maintaining a varied stock of large fittings for repairs may be too costly. Even for smaller connections, standardization is beneficial in simplifying stock ordering and maintenance.

Table 2-1. Pipe Thicknesses Required for Different Tap Sizes as per ANSI B1.20.1* for Standard Taper Pipe Threads With Two, Three, and Four Full Threads

Pipe Size in.	No. of Threads	Tap Size—in.							
		1/2	3/4	1	1 1/4	1 1/2	2	2 1/2	3
		Pipe Thickness—in.							
3	2	0.18	0.21	0.28					
3	3	0.26	0.29	0.37					
3	4	0.33	0.36	0.46					
4	2	0.17	0.19	0.26	0.31				
4	3	0.25	0.27	0.35	0.40				
4	4	0.32	0.34	0.44	0.49				
6	2	0.17	0.18	0.23	0.27	0.30			
6	3	0.25	0.26	0.32	0.36	0.39			
6	4	0.32	0.33	0.41	0.45	0.48			
8	2	0.16	0.17	0.22	0.24	0.27	0.33		
8	3	0.24	0.25	0.31	0.33	0.36	0.42		
8	4	0.31	0.32	0.40	0.42	0.45	0.51		
10	2	0.15	0.17	0.21	0.23	0.25	0.30	0.44	
10	3	0.23	0.25	0.30	0.32	0.34	0.39	0.56	
10	4	0.30	0.32	0.39	0.41	0.43	0.48	0.69	
12	2	0.15	0.16	0.20	0.22	0.24	0.28	0.40	0.48
12	3	0.23	0.24	0.29	0.31	0.33	0.37	0.52	0.60
12	4	0.30	0.31	0.38	0.40	0.42	0.46	0.65	0.73

*Pipe Threads, General Purpose (Inch). ANSI/ASME Standard B1.20.1-83. ANSI, New York (1983).

Table 2-2. Pipe Thicknesses Required for Different Tap Sizes as per AWWA C800* for Standard Corporation Stop Threads With Two, Three, and Four Full Threads

Pipe Size in.	No. of Threads	Tap Size—in.						
		1/2	5/8	3/4	1	1 1/4	1 1/2	2
		Pipe Thickness—in.						
3	2	0.21	0.24	0.25	0.33			
3	3	0.29	0.32	0.33	0.41			
3	4	0.36	0.39	0.40	0.49			
4	2	0.19	0.22	0.23	0.30	0.36		
4	3	0.27	0.30	0.31	0.38	0.45		
4	4	0.34	0.37	0.38	0.46	0.54		
6	2	0.18	0.20	0.20	0.26	0.30	0.35	
6	3	0.26	0.28	0.28	0.34	0.39	0.44	
6	4	0.33	0.35	0.35	0.42	0.48	0.53	
8	2	0.17	0.18	0.19	0.24	0.27	0.31	0.39
8	3	0.25	0.26	0.27	0.32	0.36	0.40	0.48
8	4	0.32	0.33	0.34	0.40	0.45	0.49	0.57
10	2	0.17	0.17	0.18	0.23	0.25	0.28	0.35
10	3	0.25	0.25	0.26	0.31	0.34	0.37	0.44
10	4	0.32	0.32	0.33	0.39	0.43	0.46	0.53
12	2	0.16	0.17	0.17	0.22	0.24	0.26	0.32
12	3	0.24	0.25	0.25	0.30	0.33	0.35	0.41
12	4	0.31	0.32	0.32	0.38	0.42	0.44	0.50

*Standard for Underground Service Line Valves and Fittings. AWWA Standard C800-84. AWWA, Denver, Colo. (1984).

2-2. Equipment

Although the construction of tapping equipment varies with the manufacturer, all the available machines perform similar functions. The machines discussed in this section are representative of those commonly available.

Small tapping machines. Tapping a water main and inserting a corporation stop directly into the pipe wall requires a tapping machine, more properly called a drilling-and-tapping machine. This machine actually performs three operations—drilling, tapping (threading), and inserting the corporation stop.

Small tapping machines (Figure 2-6) are used for direct installation of ½-in. through 2-in. (13-mm through 50-mm) corporation stops into a main.

Small drilling machines. Small drilling machines are used to install lines 2 in. (50 mm) and smaller, using a service clamp and corporation stop. A manual drilling machine is shown in Figure 2-7.

Large drilling machines. An example of a drilling machine for 3-in. (75-mm) diameter and larger taps is shown in Figure 2-8. Supplementary pieces of equipment, some supplied with the machines and some that must be ordered separately, include drill bits, shell cutters, and adapters. Large drilling machines are either manually operated or powered by electric, hydraulic, or pneumatic (air) motors.

A large drilling machine with cutters, adapters, and other components is a fairly large investment for a small community. As an alternative, rental equipment, often with an equipment operator provided, may be available from service companies.

Courtesy of Mueller Company

Figure 2-6 Manual Tapping Machine

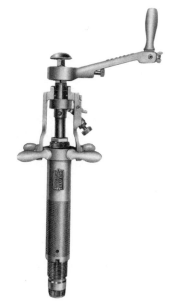

Courtesy of Mueller Company

Figure 2-7 Manual Drilling Machine

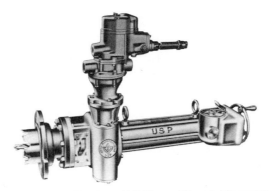

Courtesy of US Pipe and Foundry Company

Figure 2-8 Drilling Machine for Large Taps

2-3. Tapping Procedures

After the method of tapping has been selected and the necessary tools and materials have been identified and obtained, the line can be excavated and tapped. The work crew should be prepared to perform the tap as quickly and efficiently as possible, to avoid leaving the excavation open any longer than necessary.

An operator should always consult the instructions supplied with the tapping equipment or seek the advice of a more experienced operator to be sure of what steps to follow and what to expect at each step. Smooth, trouble-free tapping and good quality taps depend on the operator's training, experience, and preparation, as well as proper equipment in good condition.

Excavation

Whenever connections are to be made on an existing main, the location of the main must be determined. This can usually be done by referring to the utility's maps and records. However, if adequate maps and records are not available, it may be necessary to locate the main with instruments (Figure 2-9).

Excavations must be protected from cave-in as discussed in Module 1, and must be of sufficient size and shape to safely accommodate the machine and the operator. After the tap is completed, proper backfill and compaction procedures should be followed, especially if the area is in or near a roadway. Utilities should ensure that excavation, installation, and area restoration are done in a manner such that the work will not inconvenience the public any more than necessary. (Refer to Module 1 for more complete information on excavation, backfilling, restoration, and related safety procedures.)

Procedures for Small Connections

Direct-insertion corporation stops.   Direct-insertion corporation stops are screwed directly into the pipe wall. The first step in this procedure is to clean the excavated pipe. The drilling-and-tapping machine is then clamped onto the pipe.

Courtesy of Metrotech Corporation

Figure 2-9   Locating a Main Using a Metal Detector

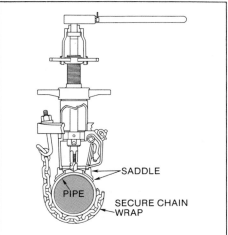

A. With the drilling-and-tapping machine, first drill a hole into the main.

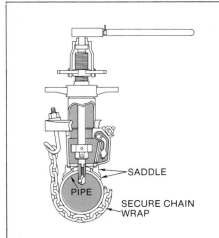

B. After the hole is drilled, it is threaded.

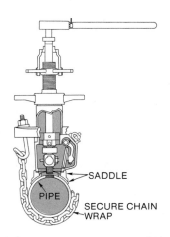

C. Again, with the drilling-and-tapping machine, insert the threaded inlet corporation stop. The service can then be activated.

Courtesy of Mueller Company

Figure 2-10   Installation of a Direct-Insertion Corporation Stop

As shown in Figure 2-10, the first operation with the machine is to bore a hole into the pipe wall. The water pressure is contained within the sealed body of the machine. As the boring bar is advanced further into the pipe, it cuts threads for the corporation stop. The boring bar is then removed, and the drill-and-tap bit on the end of the bar is replaced with a corporation stop. The bar carrying the corporation stop is reinserted into the machine and the corporation stop is screwed into the threaded hole. Finally, the machine is removed and the corporation stop is ready to have a service line attached.

A. With the service clamp attached to the main, the corporation stop is threaded into the clamp. The machine is then mounted on the corporation stop, and the stop is opened.

B. The drill penetrates the main without water escaping.

C. The drill bit is retracted, and the corporation stop closed. The stop now controls the water.

D. The machine is removed, the service line connected, and the corporation stop reopened to activate water service.

Courtesy of Mueller Company

Figure 2-11   Installation of a Corporation Stop Using a Service Clamp

Service clamps.   Where the corporation stop will be screwed into a service clamp instead of the pipe wall, the first step is to clean the pipe all the way around, then install the service clamp. As shown in Figure 2-11, the corporation stop is screwed into the service clamp, then the drilling machine is attached to the corporation stop. The drill bit of the drilling machine is extended through the open corporation stop and penetrates the wall of the main. The drill bit is then backed out of the corporation stop, the stop is closed, and the drilling machine is removed. The service line can then be attached to the corporation stop.

Precautions.   The following precautions should be followed when making taps for small connections:

- Perform a test tap in the shop with each machine before actual taps are made at the job site. Test procedures will help in determining the amount of pressure to use on the drilling bar and the rate of penetration of the drill bit.

- Use the correct saddle to fit the machine to the main.
- Use correct gaskets between the saddle and the main and between the saddle and the machine.
- Use the drilling-and-tapping bit intended for use with the material that the main is made of.
- Be sure the bit is clean and sharp, and do not force the drill through the pipe wall.
- Lubricate the cutting tool, if necessary, according to the manufacturer's instructions.
- Be sure the size and type of threads on the corporation stop match the threads on the tap.
- Be sure the corporation stop is in the *off* position before installation.
- Leave one to three threads showing on the corporation stop.

Tapping Large Connections

Procedures. As with small connections, procedures vary somewhat depending on the equipment being used. In general, however, basic procedures are described in the following paragraphs.

When excavation is complete, the outer surface of the main is cleaned and the appropriate tapping sleeve is bolted into place (Figure 2-12A). Next, the drilling machine with the proper cutting equipment and adapter is attached to the valve outlet flange, and support blocks are positioned (Figure 2-12B).

A. Installing tapping sleeve. **B. Setting up drilling machine.**

Courtesy of Mueller Company

Figure 2-12   Preparing to Tap Large Connection

The actual tapping procedure is detailed in Figure 2-13. The first step is to open the tapping valve and test the watertightness of the installation. Next, the cutter is advanced and a hole is drilled into the pipe inside the sleeve. The cutter is retracted and the tapping valve is closed. The drilling machine and adapter are removed from the sleeve and the new line or lateral is attached. The tapping valve can then be opened, placing the new lateral into service.

Close examination of the COUPON (the section of the main cut out by the drilling machine, Figure 2-14) can give a good indication of the condition of the water main—particularly the interior.

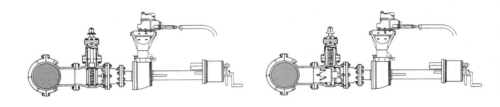

A. The tapping sleeve and valve are first attached to the main. Then, the drilling machine, with a shell cutter fastened to its boring bar, is attached to the tapping sleeve and valve using an adapter. The assembly should be pressure tested prior to making the cut.

B. With the tapping valve open, the shell cutter and boring bar advance to cut the main.

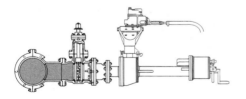
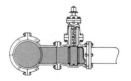

C. The boring bar is retracted and the tapping valve closed to control the water pressure.

D. With the machine removed, the lateral is connected and the tapping valve opened to pressurize the lateral and place it in service.

Courtesy of Mueller Company

Figure 2-13   Installation of Large Lateral Using Tapping Sleeve

Courtesy of T.D. Williamson, Inc.
Figure 2-14   Coupon Cut From Main

Precautions.   The following precautions should be followed when making taps for small connections:

- Follow manufacturer's instructions closely regarding lubrication of the cutting equipment.
- Make sure the tapping valve is open before advancing the cutter.
- Make sure that the cutting equipment is within $1/16$ in. (2 mm) of the pipe wall before starting cut.
- Make sure the cut is complete before withdrawing the cutting equipment.
- Be sure the coupon is recovered.

Flushing

After installation of the service-to-main connection and the connection of the service line, but prior to its use as a potable water source, it should be thoroughly flushed. Detailed procedures for flushing lines are covered in Module 1.

2-4. Record Keeping and Machine Maintenance

Tap installation records for small services should include the following:

- Name of the foreman, plumber, or contractor in charge of the job
- Date installed
- Exact location of tap and street address
- Tap size
- Main size.

For large connections, the records required are essentially the same, but additional information concerning the tapping valve should be included. (Refer to Module 3, Valves, for a more complete discussion of records necessary for large valves.)

Maintenance of Tapping Machines

Maintenance depends on the machine and its use. Basically, little maintenance is required beyond routine cleaning and lubricating. Procedures on maintenance are included with each machine. Necessary maintenance for a typical drilling-and-tapping machine for small services is discussed in the following paragraphs.

Maintenance before use.   Clean and lubricate all bearing and thread surfaces. Check and clean all tools, especially the shank end. If there is wear, check to see if lubrication would minimize the problem. Remove burrs or scale from tools, especially the boring bar and related parts, since proper sealing could be prevented by such materials.

Maintenance after use.   Clean all surfaces thoroughly and lubricate machined areas. Remove all chips from the inside of machine body. Do not bump the machine on any hard surface to dislodge particles. Store the machine and tools in a machine chest. If leakage occurs, change O-rings in the boring bar packing. Check the sharpness of the bit, and have it resharpened or replaced if necessary.

2-5. Safety

Specific areas where the need for safety precautions must be recognized and observed when dealing with tapping include:

- Material handling (refer to Module 1)
- Trenches (refer to Module 1)
- Barricades, warning signs, and traffic control (refer to Module 1)
- Personal-protection equipment (refer to Module 1)
- Hand tools (refer to Module 1)
- Portable power tools (refer to Module 1).

Precautions to be observed with tapping machines are similar to those used with any power or hand tools. With electric tapping machines, proper grounding is essential. An additional area of danger involves the pressure within a main being tapped. This pressure can dislodge an improperly seated tapping machine or blow out a corporation stop with considerable force. Tapping plastic pipe requires special care, and pipe manufacturer's recommendations should be strictly observed to avoid splitting the pipe.

A blowout of a corporation stop or breakage of the pipe can quickly flood an excavation. When making a wet tap, two precautions are important:

- A second worker or a supervisor should be nearby.
- A ladder or some other means of quick exit should be in place in the work area.

Selected Supplementary Readings

Fischrupp, J.C. Providing Uninterrupted Service to Customers. Proc. AWWA Ann. Conf., Minneapolis, Minn. (June 1975).

Franklin, B.W. Cutting, Repairing, and Tapping Cast- and Ductile-Iron Pipe. Proc. AWWA DSS, Milwaukee, Wis. (Mar. 1982).

Miller, Rick, et al. Wet Tapping, Line Plugging, and Inserting Valves. Proc. AWWA DSS, Los Angeles, Calif. (Feb. 1980).

Morrow, Darrel, et al. Lateral Connections to Pressurized Water Mains (Wet Tapping) Part I. *OpFlow*, 4:4:1 (Apr. 1978).

Morrow, Darrel, et al. Lateral Connections to Water Mains (Dry Tapping) Part II. *OpFlow*, 4:5:4 (May 1978).

Morrow, Darrel, et al. Large Lateral Connections to Pressurized Water Mains (Wet Tapping) Part III. *OpFlow*, 4:6:1 (June 1978).

Nesbeitt, W.D. Direct Tapping of PVC Pressure Pipe. Proc. AWWA DSS, Milwaukee, Wis. (Mar. 1982).

Reef, J.S. Wet Tapping, Line Plugging, and Inserting Valves. Proc. AWWA DSS, Milwaukee, Wis. (Mar. 1982).

Glossary Terms Introduced in Module 2

(Terms are defined in the Glossary at the back of the book.)

Corporation stop	Service line
Coupon	Split tee fitting
Dry tap	Tapping
Lateral	Tapping sleeve
Saddle	Tapping valve
Service clamp	Wet tap

Review Questions

(Answers to Review Questions are given at the back of the book.)

1. What is the purpose of tapping?

2. What is the difference between a wet and dry tap?

3. What are two advantages of wet taps?

4. What is a corporation stop?

5. When is a service clamp used?

6. What is the most common type of inlet thread on a corporation stop, and what is the major advantage of this type of thread?

7. List three factors that determine the type of corporation stop that should be used.

8. Name two methods of connecting laterals to mains.

9. What is the most common connection size for residential service lines?

10. List three important precautions to take when using a tapping machine.

11. What kind of information should be recorded regarding service taps?

12. What safety hazard exists when making a wet tap?

Study Problems and Exercises

1. Part of your system has copper service lines attached to ductile-iron mains. The system has just expanded into a new service area that has PVC mains and high-molecular-weight polyethylene service lines. All mains have a 6-in. diameter. Discuss the type of corporation stop and tapping procedures you would recommend for each situation.

Module 3

Valves

VALVES are mechanical devices that are used to stop, start, or regulate water flow in a water distribution system. This module describes the different types of valves, conditions under which different valves are used, and the proper installation, operation, and maintenance of the more common types of valves.

After completing this module you should be able to

- Explain under what conditions different types of valves are installed in the distribution system.
- Identify common valves and components from illustrations.
- Explain the purpose of a valve exercise program and make recommendations for establishing a scheduled program based on the age of the system and other local conditions.
- Describe what recommended inspection checks and maintenance should be routinely performed on valves.
- List items of information that should be recorded in conjunction with the installation, inspection, and maintenance of valves.
- Prepare guidelines on safety precautions that should be observed when installing, inspecting, and performing valve maintenance.

3-1. Purpose of Valves

In a water distribution system, valves can be used for such diverse purposes as isolating a section of a water main, draining the water line, throttling the flow, regulating water storage levels, controlling water hammer, bleeding off air, or preventing backflow.

Isolating a section of a water main.   Figure 3-1 shows several types of shut-off valves that can be used to stop the flow of water so that if a water main breaks, the area of the break can be isolated while the repair work is being done. For example, if the pipe broke at point A in Figure 3-1, valve B and another valve on the other side of the break (not shown) could be closed. The rest of the system could then continue to provide water service while the break was being repaired. Valves can also be used to isolate different parts of the system to help in finding large water leaks or difficult-to-locate main breaks. Figure 3-1 also shows a fire hydrant with an auxiliary valve that can be used to shut off the water if the hydrant must be repaired or replaced.

ISOLATION VALVES in the distribution system should be located less than 500 ft (150 m) apart in business districts and less than 800 ft (240 m) apart in other parts of the system. It is good practice to have valves located at the end of each block so that only one block will be without water during repair work.

SERVICE VALVES are used to shut off service lines to individual homes or businesses. Specific types of service valves include the CORPORATION STOP, which is tapped into the main, and the CURB STOP, which is located near the property line. Access to the curb stop is through a CURB BOX. The corporation stop is normally kept in the open position after the service has been installed. If a leak develops between the main and the curb stop, the corporation stop could be dug up and used to cut off the water. If a leak develops between the curb box and the building, a long-stemmed valve wrench (or VALVE KEY) can be used to easily close the curb-stop valve.

Draining the water line.   In hilly areas, as shown in Figure 3-2, drain or BLOWOFF VALVES can be installed at the low points to drain sediment from the main or to drain the entire main.

Courtesy of Mueller Company, Decatur, Ill.

Figure 3-1.   Shut-off Valves in the Distribution System

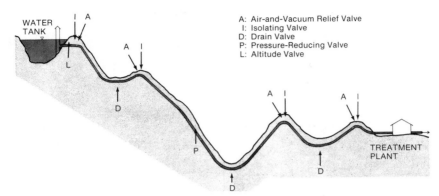

Figure 3-2.   Valve Locations in a Distribution System

Throttling the flow and regulating water storage levels.   In addition to on–off control of flow, valves can be used to regulate or throttle the flow. This can be done manually although it is usually done automatically. The ALTITUDE-CONTROL VALVE in Figure 3-2 regulates the flow to the water tank in order to keep the proper level in the tank without overflowing, in much the same way the ball-float valve in a toilet tank keeps it from overflowing. The PRESSURE-REDUCING VALVE, also shown in Figure 3-2, is used to change the high pressure from the reservoir at the top of the hill to a specified lower pressure for the customers in the valley.

Controlling water hammer.   The slam you sometimes hear or feel when a valve in a washing machine or dishwasher closes too quickly is called WATER HAMMER. Water hammer is caused by a rapid increase in pressure as the entire moving column of water in the pipe suddenly stops. The pressure wave that is created moves rapidly down the pipe and can damage valves, burst pipes, or blow pipe joints apart. In a water distribution system, opening or closing a valve too fast can also cause water hammer. PRESSURE-RELIEF VALVES are used to help control water hammer by releasing some of the energy that is created by a sudden stop in flow.

Bleeding off air.   Air that tends to collect in water lines can cause several problems. Under the pressure of the distribution system, air dissolves and can reappear in a customer's drinking glass as microscopic air bubbles, which give the water a cloudy appearance. This "white water" is safe to drink but is unappetizing to many people.

A more common operating problem occurs when air collects in high places in the distribution system producing air pockets. Air pockets cut down the area of pipe that water can move through, causing an effect known as AIR BINDING. The result is pressure loss and increased pumping costs.

AIR-RELIEF VALVES can be installed to eliminate these problems in pumping stations where air can enter the system and at high points where it can collect. Air-relief valves solve the problem by automatically venting any air that accumulates. An air-relief valve installed in a pump discharge line is shown in Figure 3-3.

Courtesy of Henry Pratt Company

Figure 3-3. Air-Relief Valve

Courtesy of M&H Valve Company

Figure 3-4. Check Valve

Preventing backflow. In most parts of the water system, it is not necessary to be concerned about the direction the water moves. However, there are several areas where the water should only move in one direction. These include customer service lines, water treatment plants, and pump discharge lines.

BACKFLOW, or reversed flow, in a customer line or in the treatment plant could result in contaminated or polluted water entering the public water system. The hazards involved and the valves that can be used to prevent backflow are covered in Module 6, Cross-Connection Control.

In a pump discharge line, reversed flow is not a health hazard, but it can damage equipment. When a pump shuts off, the water tends to drain back through the pump, causing the rotor to turn backwards, damaging both pump and motor. To prevent this, a CHECK VALVE is normally installed on the pump discharge line. This valve allows flow only in one direction—away from the pump. Typical check valves in a pump station are shown in Figure 3-4.

3-2. Types of Valves

Valves commonly used in water distribution systems (Table 3-1) include gate valves, bypass valves, butterfly valves, plug valves, ball and cone valves, altitude-control valves, air-and-vacuum relief valves, pressure-relief valves, globe valves, check valves, and sluice gates. Some of these terms describe the construction of the valve; for example, the gate valve is constructed with a moving gate that shuts off the water. Other terms refer to the use of the valve, for example, an air-and-vacuum relief valve releases air or vacuum from a pipeline.

Table 3-1. Valves Commonly Used in Water Distribution Systems

Type	Use	Advantages	Disadvantages
Gate	Isolation	Low cost in small sizes Low head loss when open Easy to install Good service life	High cost in large sizes Large sizes very heavy Not good for frequent operation Large sizes difficult to operate by hand
Butterfly	Isolation Check Throttling	Low cost in larger sizes Some styles have short lengths Easy to operate	Higher head loss than gate valves Must be removed for maintenance Difficult to clean or reline pipe
Globe	Isolation (small sizes) Altitude, flow, or pressure control (large sizes)	Simple construction Dependable Good for throttling Sediment or other matter unlikely to prevent complete closing	Very high head loss Very heavy Expensive in large sizes
Plug (also cone)	Isolation (curb stop and corporation stop) Throttling	Dependable Low friction loss Easy to operate Minimizes closing surge	Expensive Very heavy
Ball	Isolation (curb stop and corporation stop) Throttling	Dependable Low friction loss Easy to operate Minimizes closing surge	Expensive Very heavy

Gate Valves

The most common type of valve found in a water distribution system is the GATE VALVE. Gate valves are mechanical devices used to isolate specific areas of the system during repair work or to reroute water flow throughout the distribution system. An open gate valve allows water to flow through in a straight line (Figure 3-5). The valve may be closed during an emergency, such as a water main break, or during routine maintenance. However, gate valves are not designed to regulate or throttle flow. Sizes range from ½ in. (13 mm) in diameter (used to isolate water meters on a service line) to 72 in. (1830 mm) in diameter.

Gate valves are easily identified from the outside by the high bonnet into which the disk rises when the valve is open. Other parts of the gate valve (Figure 3-6) include the BODY, DISK, VALVE STEM, SEAT, and OPERATING NUT. The stem is sealed with O-rings (Figure 3-6) or with PACKING, PACKING GLAND, and STUFFING BOX. The gate-like disk, moved by a stem screw, moves up and down at right angles to the path of flow. When closed, the gate fits against two seat faces to stop flow. Gate valves offer little resistance to flow and less friction and pressure drop than

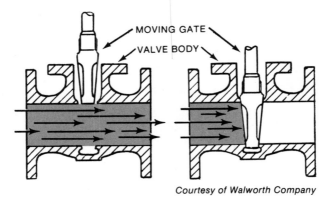

Courtesy of Walworth Company

Figure 3-5.   Operation of a Gate Valve

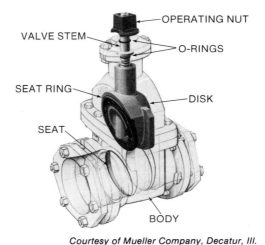

Courtesy of Mueller Company, Decatur, Ill.

Figure 3-6.   Parts of a Resilient-Seated Gate Valve

butterfly valves, provided the valve disk is kept fully opened. The wear in a gate valve will be found on the downstream faces of the seat and disk, because of line pressure forces.

Nonrising-stem valves.   The most common type of gate valve found in a distribution system is the iron-body bronze-mounted nonrising-stem double-disk gate valve. The NONRISING-STEM VALVE is commonly referred to as IBBM or NRS. The body is made of cast or ductile iron, and the sealing and operating parts are made of bronze. In closing, the rotation of the stem allows the disks to drop into the body. When they have reached the bottom of the valve body, wedging mechanisms between the two disks push them outward to seal against the seat rings. The valve stem itself rotates, but does not move up and down.

Because of the bronze-to-bronze sealing surface, double-disk gate valves will allow some leakage even when new. Since the double disks are loose in the valve body until they are fully closed, this valve should not be used for throttling purposes; a partially closed double-disk gate valve will vibrate and chatter causing damage to the seating surfaces.

Buried gate valves are usually nonrising-stem valves. In situations where an operator will need to know from inspection whether a valve is open or closed, a rising-stem valve with an outside screw and yoke (OS&Y) is often used (Figure 3-7A). To prevent water from coming past the stem, it is sealed with either O-rings (Figure 3-7B) or conventional packing (Figure 3-7C). The operation of a nonrising-stem valve is shown in Figure 3-8.

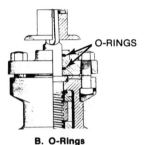

B. O-Rings

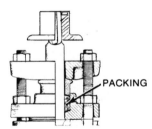

A. Outside Screw and Yoke
Courtesy of Mueller Company, Decatur, Ill.

C. Conventional Packing (Flax)
Courtesy of Terminal City Iron Works Ltd.

Figure 3-7.   Rising-Stem Gate Valve

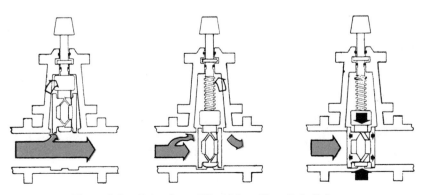

Figure 3-8.   Operation of Nonrising-Stem Gate Valve

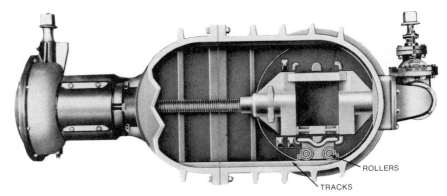

Courtesy of Mueller Company, Decatur, Ill.

Figure 3-9.   Horizontal Gate Valve

Courtesy of M&H Valve Company

Figure 3-10.   Gate Valve With Bypass Valve (Without Proper Support)

Gate valves of the double-disk type, in sizes 16 in. (400 mm) or larger, are often designed to lie horizontally so the operator does not have to lift the gate to open the valve. Also, since these valves do not rise as high above the pipe centerline, they allow better coverage of earth over the valve. Horizontal valves should be equipped with stainless-steel tracks fastened into the valve body and bonnet (Figure 3-9). The weight of the gate is carried by rollers on the disks, riding in the tracks throughout the entire length of travel. Bronze scrapers are also provided to move ahead of the rollers in both directions to remove any foreign matter from the tracks. In rolling-disk valves, the disks serve as rollers.

Bypass valves.   Where larger diameter gate valves (16 in. [400 mm] and greater) are used, the use of BYPASS VALVES (Figure 3-10) is recommended. Bypass valves are smaller diameter valves that will allow bypassing of the larger valve to equalize pressure, making the main valve easier to open and close. For example, if the water pressure in a main is 80 psi (550 kPa), the gate of a 16-in. (400-mm) gate valve could have 16 000 lb (7300 kg) of force pushing against it. Bypass

Table 3-2. Size Requirements for Bypasses

Valve Diameter in. (mm)	Bypass Diameter in. (mm)
16–20 (400–500)	3 (75)
24–30 (600–750)	4 (100)
36–42 (900–1050)	6 (150)
48 (1200)	8 (200)

valves also provide for low-volume flow without opening the main valve. Bypass valves can be installed on gate valves in which the main-valve stem is either horizontal or vertical. Bypass sizes for various size gate valves are shown in Table 3-2.

Tapping valves. A TAPPING VALVE (see Figure 3-11) is a gate valve that has one end machined to attach to a tapping sleeve for making lateral connections to an existing main under pressure. Tapping valves have a slightly larger inside diameter to allow the tapping machine cutter to pass through with no damage. (See Module 2, Tapping, for details.)

Cut-in valves. CUT-IN VALVES have oversized end connections that allow them to slip over oversized cast-iron water mains. They are used when a valve is needed in an area that previously did not have a working valve. A cut-in sleeve is used in conjunction with this operation to make it quick and easy. Water must be shut off during this installation.

Inserting valves. If a valve is needed and water flow cannot be stopped, an INSERTING VALVE is used. This type of valve can be installed under full line pressure when there is no alternative (see Figure 3-12).

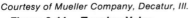

Courtesy of Mueller Company, Decatur, Ill.

Figure 3-11. Tapping Valve

Courtesy of Mueller Company, Decatur, Ill.

Figure 3-12. Inserting Valve

Courtesy of Kennedy Valve

Figure 3-13.   Resilient-Seated Gate Valve

Courtesy of Henry Pratt Company

Figure 3-14.   Butterfly Valve

Resilient-seated gate valves.   The RESILIENT-SEATED GATE VALVE consists of a single disk to which a resilient (rubber or synthetic) member is attached. It is either bonded to the disk or mechanically fastened. These valves offer a leak-tight shutoff of up to 200 psi (140 kPa) (see Figure 3-13).

Butterfly Valves

The BUTTERFLY VALVE (Figure 3-14) consists of a body in which a disk rotates on a shaft as it opens or closes. In the full open position, the disk is parallel to the axis of the pipe. Butterfly valves are easy to identify because of their very short body. A 16-in. (400-mm) diameter flanged-end butterfly valve is 8 in. (200 mm) long (flange to flange), whereas a 16-in. (400-mm) diameter gate valve is 17 in. (430 mm) long.

The parts of the butterfly valve (Figure 3-15) include the BODY, DISK, RUBBER SEAT, SHAFT, SHAFT BEARING, and THRUST BEARING. Because the disk stays in the water path even in the open position, the butterfly valve creates a higher head loss than the gate valve. The position of the disk also makes it very difficult to clean scale from a pipeline because the pigs or swabs are blocked by the valve disk. However, butterfly valves open easily, since the water pressure pushing on one half of the upstream side of the disk tends to force it open, balancing the pressure on the other half, which tends to force it closed (Figure 3-16). The cost of large butterfly valves is less than the cost of gate valves, since large gate valves usually require reduction gears, a bypass valve, rollers, tracks, and scrapers.

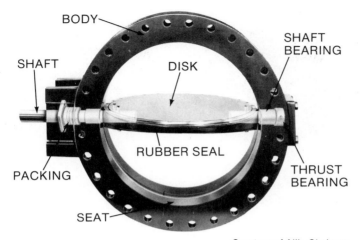

Courtesy of Allis-Chalmers

Figure 3-15.   Parts of a Butterfly Valve

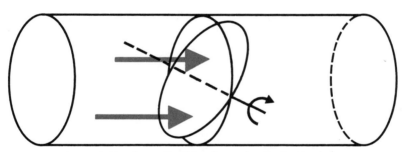

Figure 3-16.   Butterfly-Valve Operation Under Pressure

Although isolation is the main purpose of butterfly valves, they can be used in some situations for throttling under low-flow and low-pressure conditions.

Plug Valves

In a PLUG VALVE (Figure 3-17), the movable element is a cylinder-, ball-, or cone-shaped plug rather than a flat disk. The plug rotates and has a passageway or port through it. It requires one-quarter turn to move from fully open to fully closed. Small plug valves are used as curb stops and corporation stops. Large plug valves may be used for throttling.

BALL VALVES and CONE VALVES are types of plug valves whose plugs are balls or truncated (cut off) cones (Figure 3-18).

Altitude-Control Valves

ALTITUDE-CONTROL VALVES are a type of diaphragm valve used to control the level of water in a tank supplied from a pressure system (Figure 3-19). There are two general types of altitude valves—single acting and double acting.

PORT   CYLINDER

Courtesy of Xomox Corporation

Figure 3-17.   Plug Valve

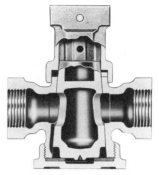

Courtesy of Mueller Company, Decatur, Ill.

Figure 3-18.   Cone Valve

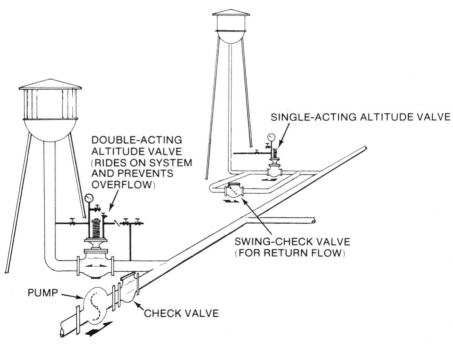

SINGLE-ACTING ALTITUDE VALVE

DOUBLE-ACTING
ALTITUDE VALVE
(RIDES ON SYSTEM
AND PREVENTS
OVERFLOW)

SWING-CHECK VALVE
(FOR RETURN FLOW)

PUMP

CHECK VALVE

Courtesy of GA Industries, Inc.

Figure 3-19.   Altitude-Control Valves

A single-acting altitude valve is used only for filling the tank. The tank discharges through a separate line or through a check valve in a bypass line around the altitude valve. Figure 3-20 shows a single-acting altitude valve. A bypass line with check valves around the altitude valve is needed to permit backflow out of the tank and into the distribution system when the inlet pressure is lower than the tank pressure.

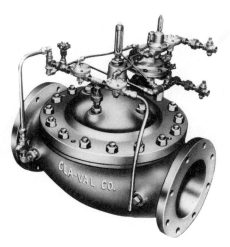

Courtesy of Cla-Val Co., Newport Beach, Calif.   *Courtesy of Cla-Val Co., Newport Beach, Calif.*

Figure 3-20.   Single-Acting Check Valve   **Figure 3-21.   Double-Acting Check Valve**

A double-acting altitude valve allows water to flow both to and from the tank. When the tank becomes full, the valve closes to prevent overflow. When the distribution pressure drops below the pressure exerted by the full tank, the valve opens to discharge water into the distribution system. A double-acting altitude valve is illustrated in Figure 3-21. Note that the essential difference between the single- and double-acting altitude valves shown is the external control mechanism, not the valve itself.

Air-and-Vacuum Relief Valves

In long pipelines, air can accumulate at the high points in the line. This accumulation causes air binding, the partial blockage of flow by the entrapped air. AIR-AND-VACUUM RELIEF VALVES reduce the problem by automatically venting the unwanted air.

Periodically, pipelines must be drained for routine maintenance or repair. As the water flows out through drain valves, a vacuum can be created inside the pipeline. If the vacuum is great enough, it can completely collapse the pipeline. Air-and-vacuum relief valves automatically allow air into the pipe to occupy the volume that was filled by water, so no vacuum is created. This prevents pipe collapse and also shortens draining times. The diagrams in Figure 3-22 show a typical air-and-vacuum relief valve.

Pressure Relief Valves

PRESSURE-RELIEF VALVES are designed to relieve excessive pressure from water lines. They are fitted with an adjustable spring to set the maximum pressure of the line. When line pressure becomes greater than the spring pressure, the valve opens until the pressure is equalized.

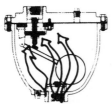

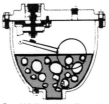
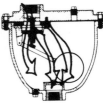

A. During the filling of the line, air entering the valve body will be exhausted to atmosphere. When the air is expelled, and water enters the valve, the float will rise and cause the orifices to be closed.

B. The large and small orifices of the air and vacuum valve are normally held closed by the buoyant force of the float.

C. While the line is working under pressure small amounts of trapped or entrained air are exhausted to atmosphere through the small orifice.

D. Air is permitted to enter the valve and replace the water while the line is being emptied.

Courtesy of GA Industries, Inc.

Figure 3-22.   Operation of an Air-and-Vacuum Relief Valve

Globe Valves

The GLOBE VALVE (Figure 3-23) is commonly used for ordinary household water faucets. It has a circular disk that moves downward into the valve port to shut off flow. Because of the turns the water must make moving through the valve (Figure 3-24), globe valves produce high head losses when fully open and, therefore, are not suited for distribution mains where head loss is critical. Where rapid draining is not important, globe valves may be used to drain lines. Though quicker to operate and less costly to repair than gate valves, 3-in. (75-mm) and larger globe valves are more expensive.

Courtesy of Walworth Company
Figure 3-23.   Globe Valve

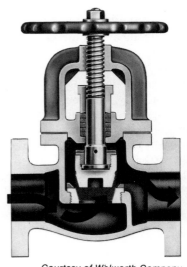

Courtesy of Walworth Company
Figure 3-24.   Flow Through Globe Valve

DOUBLE-DOOR CHECK

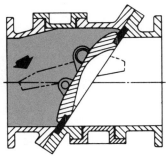

SLANTING-DISC CHECK

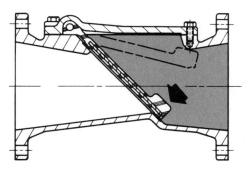

RUBBER-FLAPPER CHECK

SWING CHECK

DOUBLE-DOOR CHECK

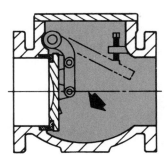

SLANTING-DISC CHECK

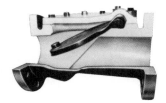

RUBBER-FLAPPER CHECK

SWING CHECK

Figure 3-25.   Four Types of Check Valves

Courtesy of Henry Pratt Company
Figure 3-26. Automatic Motor-Operated Check Valve

Check Valves

CHECK VALVES (Figures 3-25 and 3-26) are automatic valves used to prevent backflow. When a pump shuts down, the discharge line contains water that has just been pumped. Without a check valve this water would drain back through the pump, turning it backwards and damaging both pump and motor. Check valves prevent the damage by preventing the backflow. They also keep water in the discharge line, thereby providing pressure for the pump to work against when it is restarted. This helps prevent pump motor burnout under no-load conditions. Preventing backflow also saves the time and energy that would be needed to repump water that had drained back out of the discharge line.

Allowed to close unrestrained, a check valve may slam shut suddenly, creating water hammer that can damage pipes and valves. A variety of devices are normally installed on check valves to minimize slamming, including simple external weights, restraining springs, and automatic slow-closing motorized drives (Figure 3-26). Check valves are not adequate to prevent backflow of contaminated water into the potable system.

Sluice Gates

SLUICE GATES are used in open conditions, such as canals, reservoirs, filtration systems, and holding ponds. They consist of a frame that is mounted into the waterway with a single gate that can be lifted by a lever or rising or nonrising screw stem. They can be rectangular, square, or round in design, and they operate only under low head.

3-3. Valve Operation

Valves can be operated by one of several methods—manually (Figure 3-27), electrically (Figure 3-28), hydraulically (Figure 3-29), or pneumatically. The method of operation depends on the use of the valve and its function in the water system. It also depends on the source of energy available.

Courtesy of M&H Valve Company
Figure 3-27.   Handwheel Valve Actuator

Courtesy of M&H Valve Company
Figure 3-29.   Hydraulic Valve Operator

Courtesy of Kennedy Valve
Figure 3-28.   Electric Valve Actuator

Manual operation of a valve requires that the person operating the valve turn a valve-operating wrench (valve key) or handwheel, which opens and closes the valve. This is probably the most common method of operating a valve. Manually operated valves are generally used in the distribution system where valve operation is limited to on–off situations and occasional operation.

Buried distribution valves are usually manually operated by a valve wrench or valve key (Figure 3-30). These valves are equipped with a 2-in. (50-mm) square operating nut. Some are designed to open counterclockwise (to the left); others open clockwise (to the right). Service-line valves are equipped with tee heads and use slotted keys.

Valves in pumping stations or vaults can be operated by a handwheel, chainwheel, or FLOORSTAND (Figure 3-31). Gears are often used to make it easier to operate large valves (Figure 3-32). Large valves that are operated frequently are often power actuated.

Power Actuators

Electrical actuators.   The simplest type of power ACTUATOR is an electric motor that is geared to the valve stem so that when the motor is running, the valve disk or plug is moving or changing position. This motor continues to run until the

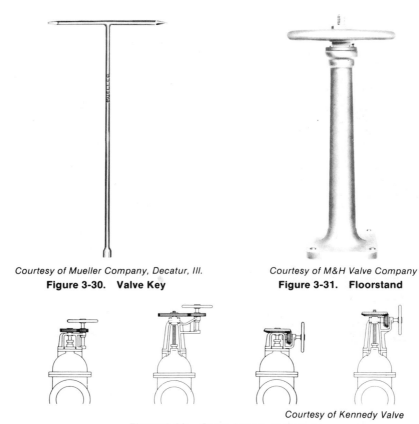

Courtesy of Mueller Company, Decatur, Ill.
Figure 3-30.   Valve Key

Courtesy of M&H Valve Company
Figure 3-31.   Floorstand

Courtesy of Kennedy Valve
Figure 3-32.   Geared Valve Actuators

valve disk has reached its desired or end position, at which time a limiting device turns the motor off. This type of valve and actuator can be used with many sensing and control devices to achieve the desired control. The electric-motor operator can be either AC or DC and is ideally suited for either local or remote operation. It can be started manually with a push button, or it can be automatically interlocked with pump or computer operators. This type of actuator is often used in complex control situations.

Cylinder actuators.   The hydraulic or pneumatic principle is used in cylinder actuators. In a simplified form, a piston is connected to the valve stem. The piston moves when oil, water, or other pressure is applied against it. Hydraulically actuated valves generally use water pressure as the actuating force. The actuator uses a piston to translate water pressure into valve-stem motion. Pneumatic actuators use air as the operating force on the valve stem. This type of installation requires a source of compressed air and can be either locally or remotely controlled. These actuators can be used on butterfly, cone, ball, plug, or globe valves. Although they can also be used on isolation valves, they are more common on control valves.

Accessories

The need for various accessories for valve actuators depends largely on the use and function of the valve. Buried valves in the distribution system are generally manually operated and have accessories that limit travel of the valve disk or stem, with no indicator other than the increase in resistance to turning as the valve reaches its limit. In any valve, the travel of either the valve disk or stem must be limited so that it does not go beyond either the open or closed position. Electric switches can be used on valve actuators or stems to restrict the travel of either the valve actuator or the valve disk, or to actuate some other device. Valve-position indicators, discussed in more detail in a later section, are useful in certain applications. Other accessories sometimes used are bypasses and test ports. These are useful in determining tightness of valve closure during valve operation and maintenance.

Valve Controls

A valve with an actuator can be made to function automatically when proper controls are furnished. The type of control device will vary depending on the function of the valve and the type of actuator used. A few of the many types of control schemes are discussed in the following paragraphs.

Two position. The two-position valve is fully opened or fully closed as determined by level, pressure, or pump-flow conditions. Sensing devices can be electrodes in a tank, pressure switches, or relays that cause the actuator to position the valve in one of its two extreme positions.

Speed control. Controlling the speed of a valve position change can be extremely important in preventing unnecessary water hammer. The method of speed control varies with the type of actuator involved. For electric actuators, it is accomplished with gearing coordinated with motor speed; for cylinder actuators, it is done with needle valves in the cylinder fluid-control circuit. At times it can become necessary to have a different speed for valve closing than for valve opening.

Timing and interlocking scheme. Frequently, the valve function is only part of an overall pump, motor, or auxiliary-device timing sequence. In this case, the valve actuator can be very complex, with many auxiliary relays actuating other devices at certain positions of the valve. The valve, its position, and its controller then become part of the system-control scheme, and system performance depends on proper application and operation of each individual device.

Emergency-closure situations. The normal valve-closing scheme, although preferable for prevention of water hammer, may not be adequate during emergency conditions. A valve may normally require as much as 5–10 min to travel from wide open to fully closed in order to accomplish equipment sequencing properly and provide smooth load changes. This length of time, however, may be too long in the event of some emergency condition, making an emergency closing speed necessary. At emergency closing speeds complete closure may be achieved in as little as 30 sec.

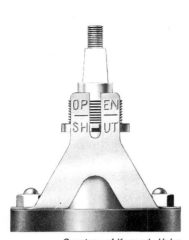

Courtesy of Kennedy Valve

Figure 3-33. Position Indicator

Courtesy of M&H Valve Company

**Figure 3-34. Position Indicator on
Floorstand**

Position Indicators

It is often important for an operator to know if a certain valve is open or closed without operating it. The simplest position indicator is the stem of a rising-stem valve. A needle-and-slot position indicator (Figures 3-33 and 3-34) or a barrel indicator is available for nonrising-stem gate valves. Open–close indicators are generally necessary on fire-protection system valves. One type shows the valve position by the word "OPEN," the shape of the post, and a colored band. A disk-position indicator—built into the actuator—is available on butterfly valves for installation in exposed locations. For buried distribution valves, a ground-line indicator is available for installation in the valve box.

3-4.   Storage and Installation

In order for valves to operate efficiently over long periods of time without the need for repair or replacement, proper storage and installation procedures must be followed.

Storage of Valves

Valves should be stored in the fully closed position to prevent the entry of foreign matter that could damage the seating surfaces. Whenever possible, valves should be stored indoors. If outside storage is necessary, protection from the weather should be provided for the gears, power actuators, cylinders, valve ports, and flanges. If valves are stored at subfreezing temperatures, it is absolutely essential to remove water from the inside of the valve. Valves stored outside in cold climates should have their disks in a vertical position. If the disks are in a horizontal position, rainwater can collect on top of the disk, seep into the valve body, freeze, and crack the casting.

BELL ENDS   MECHANICAL JOINT

FLANGED JOINT   ASBESTOS–CEMENT JOINT
Courtesy of Kennedy Valve

Figure 3-35.   Joints Used to Install Valves

Types of Valve Fittings

The joints used to install valves (Figure 3-35) are the same types used throughout the distribution system and will only be discussed briefly here. A detailed discussion is contained in Module 1, Pipe Installation and Maintenance.

The most common valve ends are flanged joints, mechanical joints, and push-on joints. Flanged joints are rigid connections that are used in accessible areas, such as in a vault or in a treatment plant. A series of bolts with a flat gasket between the flanges makes the joint. Mechanical joints are intended for buried service and consist of bolts, a rubber gasket, and a gland. By tightening the bolts the gland compresses the rubber into a recess where it compresses around the pipe to make the seal. Mechanical joints are more flexible than flanged joints and allow for some deflection of the pipe. Push-on valve ends are the easiest to install, and joints are made by pushing the valve onto the pipe end. A recess in the valve retains the rubber sealing ring, and water pressure from the back side forces it to seal around the pipe. As discussed in Module 1, mechanical and push-on ends must be installed with restraints to prevent movement in some situations.

Threaded valves are conveniently used on 2-in. (50-mm) and smaller valves. The valve end is generally a female iron-pipe thread and the pipe a male iron-pipe thread. The interference of the two threads creates a leak-tight joint.

Grooved-and-shouldered type end connections consist of a groove in both pieces of pipe to be connected. A full-circle ring with a gasket is bolted around the grooved ends to make a leak-tight seal. Grooved-and-shouldered end connections can be used on pipe ranging in size from ¾ in. (20 mm) to 30 in. (760 mm).

All bolts should be protected against corrosion with a suitable paint or a polyethylene wrapping. Where possible, valves should be located in unpaved areas to make access and maintenance easier. Valves should be installed in the closed position. This will prevent foreign materials from entering the valve and damaging the working parts or the seating surfaces.

Valve Boxes and Vaults

To ensure that isolation valves are accessible for turning on and off, VALVE BOXES (Figures 3-36 and 3-37) are generally used. Since valves are placed at various depths, valve boxes are made in two or more pieces that telescope to give

Courtesy of Mueller Company, Decatur, Ill.
Figure 3-36.   Arch-Type Valve Box

Courtesy of Mueller Company, Decatur, Ill.
Figure 3-37.   Minneapolis-Type Valve Box

Courtesy of Badger Meter, Inc.
Figure 3-38.   Valve Vault Under Construction

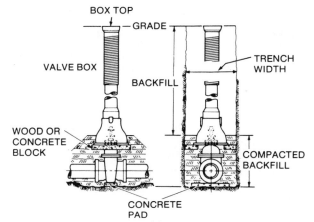

Courtesy of Ductile Iron Pipe Research Association

Figure 3-39.   Valve-Box Installation

adjustable lengths. The extension pieces are either the screw type (Figure 3-36) or the slide type (Figure 3-37). For large valves it is desirable to have the entire valve accessible for servicing; therefore, VAULTS with manholes (Figure 3-38) are used in place of valve boxes.

Proper installation of valve boxes helps to ensure against operating difficulties in the future. The soil should be firmly compacted around the bonnet of the valve to prevent settling. When cast-iron valve boxes are used, they should rest above the valve so that the weight of traffic passing over the street will not be transferred to the valve or the pipe (Figure 3-39). The bottom flared edge of the box may require extra support, such as a 2-in. × 6-in. × 18-in. (50-mm × 150-mm × 450-mm) timber on each side of the valve. Valve boxes must be installed vertically and centered around the operating nut of the valve to ensure that the valve wrench can engage the nut. Figures 3-36 and 3-37 show the arch and Minneapolis types of valve boxes, which are commonly placed over curb stops and referred to as curb boxes.

Valves in unusually deep trenches should have special provisions for operating the valve. A riser should be installed on the stem to allow a normal key to be used or else a note should be made on the valve record that a long key will be required.

When valves with exposed gearing or operating mechanisms are installed below ground, a vault must be installed. The operating nut should be accessible from the top opening of the vault with a valve key.

Access manholes should be large enough to allow removal of the valve if future replacement is necessary. If the valves are in a concrete or masonry vault, a space of at least 2 in. (50 mm) should exist between the concrete and the pipe to ensure that the weight of the vault will not rest on the water main. Valve vault drains should not be connected to storm drains or sanitary sewers because of possible contamination. Vaults should be drained to underground absorption pits or to the ground surface where they are not subject to flooding by runoff.

3-5.   Records and Maintenance

The following section discusses some basic valve-maintenance procedures and the importance of maintaining accurate records, not only on the specifics of the valve but also on the exact location of the valve.

Records

Before maintenance can be performed on any valve, accurate maps and records must be available providing all pertinent information. If the valve is buried, the exact location must be recorded using stationary landmarks. A sketch or drawing for referencing each valve on the valve record card (Figure 3-40) should be included, showing the location of stakes, buildings, or other reference points, where necessary. The master file of valve-maintenance record cards (Figure 3-41) should be kept in the valve book. A copy of the book should be made available to the maintenance crew.

To assist in locating valves, a dip needle or a metal detector can be used to locate valve boxes (Figure 3-42). After the valve has been located, errors in location tie-ins should be recorded on the service card and corrections made on the appropriate valve record card. Reference points or monuments that have been removed or relocated should be noted.

In addition to location, the type of valve installed should also be noted. If the valve is of a large diameter, it should be stated whether it is a vertical or horizontal installation, direct or gear operated, and if a bypass valve has been installed. If the valve is not manually operated, complete information should be recorded concerning the actuator's type, power source, and type of operation—local, remote, or automatic. Line pressure may also be noted as well as other information that would be of benefit to the operator.

When ordering repair parts, it is essential to know the manufacturer and date of manufacture. This information is cast onto the surface of the valve or noted on attached plates. The size and direction of opening must also be known before parts can be ordered. For maintenance purposes, the number of turns required for full opening and closing should be noted as well as whether the normal position of the valve is open, closed, or partially open.

All this information can be kept on file cards and/or in a loose-leaf notebook. Maps with numbered sections to correspond with valve records assist in locating buried valves. A master file should always be kept up-to-date, and copies of working documents should be made available for field use. Any discrepancies between the master file and records should be noted by the crew and updated on the master file.

Maintenance

Once the valve has been located, inspection and maintenance can begin. Remove the valve-box cover and use a flashlight to inspect the valve. Check the valve box for any damage or evidence of leakage in the area of the stuffing box or O-ring seal plate.

VALVE MAINTENANCE REPORT

XYZ WATERWORKS _____ VALVE NO. _____

LOCATION _____

MEASUREMENTS: CHECKED O.K. _____ MEASURED AS FOLLOWS:

_____ FT. _____ OF _____ P.L. OF _____

_____ FT. _____ OF _____ P.L. OF _____

VALVE TURNS _____ TO OPEN _____ NO. OF TURNS _____

FOUND _____ OR _____ TURNS CLOSED _____

PACKING: O.K. _____ TIGHTENED _____ REPLACED _____

STEM: O.K. _____ BENT OR BROKEN _____ REPLACED _____

NUT: O.K. _____ MISSING _____ REPLACED _____

GEARS: CONDITION _____ GREASED _____

BOX _____ OR VAULT _____ O.K. _____ REPLACED _____

BURIED _____ IN. PROTRUDING _____ IN.

TOO CLOSE TO STEM _____ RESET _____

COVER: MISSING _____ BROKEN _____ REPLACED _____

WEDGED IN _____ TARRED OR GROUTED IN _____

ANY OTHER DEFECTS _____

INSPECTED _____ BY _____

DEFECTS CORRECTED _____ BY _____

Figure 3–41.   Valve Maintenance Report

VALVE RECORD

XYZ WATERWORKS _____ VALVE NO. _____

LOCATION _____

_____ FT. _____ OF _____ PROP. LINE OF _____

AND _____ FT. _____ OF _____ PROP. LINE OF _____

SIZE _____ MAKE _____ TYPE _____

GEARING _____ BY-PASS _____

OPENS _____ TURNS TO OPERATE _____

SET IN _____ DEPTH TO NUT _____

REMARKS _____

(SKETCH ON BACK IF NECESSARY)

MAINTENANCE & INSPECTION RECORD

DATE	WORK DONE	O.K.	BY

(5" × 8" CARD)

Figure 3–40.   Valve Record Card

Courtesy of Metrotech Corp.
Figure 3-42.   Locating Valve Box With Metal Detector

Remove dirt, gravel, sand, and small rocks from valve boxes using a valve-box cleaner. Note whether the valve is found open or closed. An aquaphone or other similar device or method can be used to ensure that a valve is closed.

Where possible, check the operation by closing the valve completely and then opening it. Do not close valves completely on primary transmission mains or dead-end mains if service would be interrupted or severely curtailed. To inspect these, close the valve most of the way, then immediately reopen it.

Check the packing. Dry packing will cause difficult operation at all points of valve movement. If the packing leaks excessively, use a socket wrench with a long extension for tightening. Check that leakage drains from the valve box or vault. If necessary, replace the packing.

Gate valves.   Gate valves can be repacked without being removed from service. Before repacking, open the valve all the way. This prevents excessive leakage when the packing or entire stuffing box is removed by drawing the stem thrust collar tightly against the bonnet on a nonrising-stem valve.

Clean the stuffing box. Remove all the old packing from the inside of the stuffing box with a packing hook. Clean all adhering particles from the valve stem and polish it with fine emery cloth.

Insert new split-ring packing in the stuffing box and tamp it into place with the packing gland. Stagger ring splits. After the stuffing box is filled, place a few drops of oil on the stem, assemble the gland, and tighten it down on the packing.

On valves with O-ring seals, use the same procedure to replace the O-ring above the thrust collar. Check for a bent stem. Bent stems allow comparatively free operation when the valve is nearly open or closed but cause considerable binding in the middle portion of travel. Broken or stripped stems or stem nuts permit unlimited turning of the valve stem. If a damaged stem or stem nut is found, remove pressure from the main, open the valve, and repair.

Damage to valve stems usually results from using too much force when trying to close gate valves that are kept from seating by foreign matter, scale, or incrustation under the seat in the lower part of the body. To prevent this, open the valve a few turns after first closing it, and admit a good flow of water under the gate to wash out loose material. After repeating this operating procedure four or five times, a reasonably tight closure can usually be made.

Plug valves.   Adjust the gland. The adjustable gland holds the plug against its seat in the body and acts through compressible packing that functions as a thrust cushion. Keep the gland tight enough at all times to hold the plug in contact with its seat. If this is not done, the lubricant system cannot function properly and solid particles may enter between the body and plug, causing damage.

Lubricate all plug valves. Apply lubricant by removing the lubricant screw and inserting a stick of plug-valve lubricant appropriate for usual temperature conditions. Inject the lubricant into the valve by turning the screw down as needed to keep the valve in the proper operating condition. If lubrication has been neglected, several sticks of lubricant may be necessary before the system is refilled to proper operating condition. Be sure to lubricate valves that are not often used to ensure that they are always in operating condition. Leave the lubricant chamber nearly full so that an extra supply is available by simply turning down on the screw. Use lubricant regularly to increase valve efficiency and service, promote easy operation, reduce wear and corrosion, and seal the valve against internal leakage.

Cone valves.   Lubricate cone valves. Operate each cone valve manually or by changing the pilot control setting momentarily where the valve is used as a pilot valve. Grease or oil metal-to-metal contacts in pilot mechanism. Oil packing glands. Grease or oil all parts of the seating and rotating mechanism. If necessary, dismantle and service. Remove all corrosion on the plug or valve body. Clean with a wire brush, and paint inside and out with two coats of corrosion-resistant paint. Corrosion or tuberculation on the plug or inside valve body prevents tight seating.

Check valves.   Isolate and open the valve to observe the condition of the facing on swing-check valves equipped with leather or rubber seats on the disk. If the metal seat ring is scarred, dress it with a fine file and lap with fine emery paper wrapped around a flat tool. Check pin wear on balanced disk check valves, since the disk must be accurately positioned in the seat to prevent leakage.

Globe valves.   Lubricate packing and replace, if necessary. Operate globe valves to prevent sticking. Check the valve for leakage, adjust the packing nut, and replace the disk, if necessary.

Altitude and pressure-regulating valves.   Because many types of automatic valves and control equipment are used, specific maintenance instructions cannot be included here. They should be obtained from the manufacturer for each valve in service. The following is a typical maintenance procedure.

PILOT VALVES (Figure 3-43) on automatic valves should be tested and greased weekly. Turn the pilot control briefly to operate the main valve and then return

Courtesy of Cla-Val Co., Newport Beach, Calif.
Figure 3-43.   Pilot Valve

the control to its original setting. Grease the pilot settings with general-purpose grease. To close an open altitude valve, back off the nut on the diaphragm spring setting one or two turns. Return the nut to its exact original position after inspection.

Maintenance procedures after isolation of the pilot valve include the following:

- Dismantle the valve, remove the valve lid covering the piston chamber, withdraw the piston.

- Inspect the walls of the piston and cylinder liner for scoring, and smooth and polish with fine emery cloth.

- Inspect the piston face and seat; replace, if necessary.

- Check for leakage, keep port open at all times; leakage here indicates a worn piston face or a scored piston or liner surface; constant leakage from pilot-valve waste lines indicates a defective pressure valve, which must be reground or replaced; leakage through the main valve indicates worn seat rings.

Exercising valves.   Planned exercising of valves keeps them clean and operable, extends their life, and pinpoints problem valves allowing time for scheduling repair or replacement. Proper exercising of valves cannot be overemphasized. Any valve in a system can fail to operate at any time due to lack of proper exercising.

In general, valves should be exercised at least once a year and critical valves should be inspected more often. However, the age and water conditions of each system should be taken into consideration in determining the required frequency of valve operation. Most valve exercising programs are conducted in the spring or fall of the year when demand is not too great and the weather is still good. Each utility must determine the best time for a program.

Courtesy of E.H. Wachs Co.

Figure 3-44.   Power Valve-Actuating Equipment

The main obstacle to planned valve exercising programs has been that manual exercising of valves by "walking around the key" and keeping an accurate count of turns is a time-consuming process. Power equipment is now available that will reduce valve operating man-hours by at least 65 percent. The equipment selected can be portable, versatile, fast, and efficient (Figure 3-44). Most power equipment fits all standard valves from 6 in. to 60 in. (150 mm to 1500 mm). With power units it is also possible to get a valve operating that may have been considered inoperable. It may take an hour or more to do it, but that is minor when compared to the cost of replacing the valve.

When the valve is exercised check the operating nut—replace missing or badly chewed operating nuts. Check and lubricate gears—observe and correct any bad operating condition. Grease the gears. Check the bypass—note whether the bypass is open or closed and whether the small valve is operating satisfactorily. Check the condition of the box or vault—adjust the box or vault if buried or protruding. Reset if the box is too close to the stem. Check the cover and replace it if it is missing or broken.

3-6. Safety

Specific areas where the need for safety precautions must be recognized and observed when dealing with valves include:

- Material handling (refer to Module 1)
- Barricades, warning signs, and traffic control (refer to Module 1)
- Personal-protection equipment (refer to Module 1)
- Hand tools (refer to Module 1)
- Portable power tools (refer to Module 1)
- Confined spaces
- Manholes and vaults
- Hydraulic and pneumatic devices.

Confined Spaces

Personnel working in confined spaces face deadly risks from hazards that are not easily seen, smelled, heard, or felt. Primary dangers of confined spaces include atmospheric contamination, cave-in or collapsing scaffolding, and the possibility of injury from falling objects.

Atmospheric contamination is one of the most dangerous and most easily overlooked hazards. Anyone entering a confined space should remember a primary rule—never trust your senses. What looks like a harmless situation may be filled with danger. Strange odors that are noticeable when first entering a confined space can often impair the sense of smell and make the worker careless. In fact, some of the deadliest vapors and gases have no odor at all.

Three of the most common hazardous atmospheric conditions, which must be checked for before entering a confined space, are:

- Air containing less than 19.5 percent oxygen (normal air contains 20.9 percent oxygen)
- Air containing explosive gases and vapors, which can ignite or explode, such as methane or natural gas
- Air containing toxic gases and vapors, such as hydrogen sulfide or chlorine, which, even in low concentrations, can cause injury or death.

Precautions.   Several precautions should be followed when working in or around confined spaces.

At least three individuals should work together on every job that requires entering a confined space.

The absence of harmful or toxic gases in the confined space should be verified using approved instruments and methods before personnel are permitted to enter.

Adequate and continuous ventilation should be provided to ensure sufficient fresh air while personnel are within a confined area. When emergency conditions require working in unventilated spaces, blowers or self-contained breathing

Courtesy of Rose Mfg. Co.
Figure 3-45.   Safety Harness

apparatus should be used to ensure that all workers are supplied with adequate breathing air. Fresh air should be pumped into deep structures and any structure that does not have adequate ventilation. If a blower is used, the discharge end of the hose should be placed near the bottom of the structure, forcing the old air up and out.

Mechanical ventilation is usually needed to guarantee enough oxygen as well as to remove toxic fumes and dust produced during maintenance activities inside storage tanks. An air mask and air supply are necessary for the operator if concentrated chlorine solutions are being sprayed on the inside of storage tanks after cleaning and maintenance.

All personnel entering a confined area should always wear a safety harness (Figure 3-45), with lifelines attached so that any worker who collapses in the confined space can be removed quickly. Smoking should not be permitted in or near a confined space. Spark-proof tools should be used for all manual work performed in confined spaces, and a fire extinguisher should be readily available. Proper shoring and bracing should be used to prevent cave-ins while workers are in the confined space.

Tools should never be used or left in such a manner as to endanger the safety of personnel; subsurface workers should always wear hard hats in case something is dropped. Mechanical lifting aids should be used to raise, lower, or suspend heavy or bulky material to personnel working in subsurface confined spaces.

A ladder should always be used for entering or leaving a confined space over 4 ft (1 m) deep. The fixed metal-rung ladders installed in some confined spaces should be examined for damage due to corrosion before descending.

When using either portable powered equipment or truck-mounted powered equipment, the wheels of the equipment or truck should be properly blocked or chocked to prevent movement during operations.

Courtesy of Enmet Corp.

Figure 3-46.   Electronic Atmosphere Analyzer/Alarm

Sufficient lighting should be installed for working in the dark interior of tanks. Lighting provided should be explosion-proof flashlights or electric lights. Operators should wear safety goggles when painting or when wire brushing, chipping, or sandblasting paint. Ear protection may also be necessary if noise is excessive. This is likely to be the case inside tanks where noise may be intensified.

Confined-space entry program.   Every water utility should establish a confined-space entry program and provide training to any distribution personnel who will be working in confined spaces. The essential steps in such a program are listed in the following paragraphs:

The first step is to develop a policy statement specifying the hazards of confined space entry and the procedures to be followed to ensure the workers' safety.

A hazard assessment must be made of each confined space, both in terms of the atmospheric contaminants involved (Figure 3-46) and the specific hazards associated with the work to be done.

Entry-permit requirements and procedures should be developed that bring together all parts of the program and document the individual responsible for the various parts of the program, the existing conditions in the confined space, conditions to be met before work may begin, and the requirements that must be met when the work is in progress. An entry permit should only be valid for as long as the conditions in the confined space do not change, generally not more than eight hours.

All personnel working in confined spaces must be properly trained in safe entry and rescue procedures, since the best policies and procedures generally are of little value unless the employees are properly trained to carry them out.

Periodic evaluation of confined-space entry procedures is necessary to ensure that all the elements of the program are complete and are working as intended.

Manholes and Vaults

Care should be exercised in removing manhole covers. The proper use of mechanical assists will prevent any possible foot or back injuries.

Figure 3-47. Traffic Control Around Work Site

Guardrails should be placed around any open and unattended manholes. A toe board around the manhole should be used to prevent objects from falling into the structure. Individuals in the manhole should wear hard hats for additional protection.

Traffic should be carefully controlled when working in manholes in the street (Figure 3-47). If possible, a vehicle should be used as a protective wall between the structure and the traffic. All barricades, lights, cones, signs, and other traffic equipment must meet state regulations.

An individual ascending or descending a ladder should face the ladder. Only one person should be on a ladder at a time. The top of a stepladder should never be used as a platform on which to stand. Straight ladders should be placed so that the distance of the base of the ladder from the wall is 1 ft for every 4 ft (0.25 m for every 1.0 m) of the ladder. Everyone should make a habit of carefully inspecting any ladder before using it.

Hydraulic and Pneumatic Devices

Always regard hydraulic devices as power equipment capable of inflicting physical injury unexpectedly. Extra care is mandatory where hydraulic devices are incorporated into a control panel containing electrical instruments. Making disconnects on hydraulic lines can cause sudden electrical problems if a fitting fails. The operator should plan on the possibility and have precautionary measures set up.

Valve cylinders are usually trunnion-mounted so that the cylinder moves with the stroke, making it possible to get an arm trapped and crushed between the metal members. Workers should take special care to stay clear of such assemblies. A cylinder that has been repaired should never be tested until all the tie bolts have been secured. Temporary or partial securing can result in blowing off a cylinder head.

The same risks noted for hydraulic devices apply to pneumatic devices; however, it should be noted that pneumatic failure has a much greater range and speed of operation.

Selected Supplementary Readings

Antos, Lois. Application and Maintenance of Altitude Valves. *OpFlow*, 10:3:1 (Mar. 1984).

Ball Valves, Shaft- or Trunnion-Mounted—6 In. Through 48 In.—For Water Pressures Up to 300 psi. AWWA Standard C507-73. AWWA, Denver, Colo. (1973).

DiLorenzo, Ralph. Theory and Use of Air Valves. *OpFlow*, 9:12:4 (Dec. 1983).

Foland, D.L. Butterfly Valves in a Distribution System. *OpFlow*, 2:6:6 (June 1976).

Franklin, W.W. Maintaining Distribution System Valves and Fire Hydrants. *Jour. AWWA*, 74:11:576 (Nov. 1982).

Gate Valves, 3 through 48 In. NPS, For Water and Sewer Systems. AWWA Standard C500-80. AWWA, Denver, Colo. (1980).

Grant, J.F. City of East Lansing's Approach to Main Line Valve Maintenance. DSS Proc., AWWA, Birmingham, Ala. (Sept. 1983).

Handling an Installation of Double-Disk Gate Valves. *OpFlow*, 8:8:7 (Aug. 1982).

Haviland, B.F. Butterfly Valves for Distribution System Defense. *OpFlow*, 9:6:1 (June 1983).

Hopper, G.H. Reflective Valve-Markers Lead to System Upgrade. *OpFlow*, 6:8:1 (Aug. 1980).

Introduction to Water Sources and Transmission. AWWA, Denver, Colo. (1979).

Kimball, P. Valves—Some New Ideas on an Old Subject. *OpFlow*, 10:6:1 (June 1984).

Lescovich, J.E. Locating and Sizing Air-Release Valves. *Jour. AWWA*, 64:7:457 (July 1972).

Lindblom, M.H. Inspect and Maintain All Fire Service Lines. *OpFlow*, 7:1:1 (Jan. 1981).

Mogren, T.D. A Maintenance Program for Valves and Hydrants. *Jour. AWWA*, 65:7:507 (July 1973).

Resilient-Seated Gate Valves, 3 Through 12 NPS, For Water and Sewer Systems. AWWA Standard C509-80. AWWA, Denver, Colo. (1980).

Rubber-Seated Butterfly Valves. AWWA Standard C504-80. AWWA, Denver, Colo. (1980).

Safety Practice for Water Utilities. AWWA Manual M3, AWWA, Denver, Colo. (4th ed., 1983).

Seevers, Dick. Gate Valves—Miniature 'Dams' For Distribution Lines. *OpFlow*, 8:8:1 (Aug. 1982).

Skorcz, Dan. Maintaining Distribution System Valves. *Jour. AWWA*, 75:11:556 (Nov. 1983).

Staying Alive and Well When Working in Manholes. *OpFlow*, 8:8:6 (June 1982).

Swing-Check Valves for Waterworks Service, 2 in. Through 24 in. NPS. AWWA Standard C508-82. AWWA, Denver, Colo. (1982).

Valves—Care and Maintenance Part I. *OpFlow*, 3:11:1 (Nov. 1977).

Valves—Care and Maintenance Part II. *OpFlow*, 3:12:3 (Dec. 1977).

Valves—Care and Maintenance Part IV. *OpFlow*, 4:2:4 (Feb. 1978).

Valves—Care and Maintenance, Exercising Valves Part V. *OpFlow*, 4:3:1 (Mar. 1978).

Valves—The Selection of Butterfly Valves vs. Gate Valves Part III. *OpFlow*, 4:1:4 (Jan. 1978).

Glossary Terms Introduced in Module 3

(Terms are defined in the Glossary at the back of the book.)

Actuator
Air-and-vacuum relief valve
Air binding
Air-relief valve
Altitude-control valve
Backflow
Ball valve
Blowoff valve
Body
Butterfly valve
Bypass valve
Check valve
Cone valve
Corporation stop
Curb box
Curb stop
Cut-in valve
Disk
Floorstand
Gate valve
Globe valve
Inserting valve
Isolation valve
Nonrising-stem valve

Operating nut
Packing
Packing gland
Pilot valve
Plug valve
Pressure-reducing valve
Pressure-relief valve
Resilient-seated gate valve
Rubber seat
Seat
Service valve
Shaft
Shaft bearing
Sluice gate
Stuffing box
Tapping valve
Thrust bearing
Valve
Valve box
Valve key
Valve stem
Vault
Water hammer

Review Questions

(Answers to Review Questions are given at the back of the book.)

1. List six uses for valves in a water distribution system.

2. For each valve use listed in question 1, name one valve type suitable for that use.

3. List the three most common types of joints used to install valves.

4. What is the main purpose of valve boxes and vaults?

5. When is a vault preferred over a valve box?

6. What factor is important in providing drainage for valve vaults?

7. What precautions should be taken to protect workers doing inspections or repairs on valve vaults in streets?

8. List three types of power actuators for valves.

9. What is the primary purpose of a bypass valve?

10. What information can be learned from valve position indicators?

11. How often should distribution system isolation valves be operated or inspected?

12. What can happen if a valve is opened or closed too quickly?

13. List at least three items to check during routine inspection of a valve.

Study Problems and Exercises

1. The director of public works has asked you to recommend the types of valves to use for the following purposes in a 14-in. distribution main: (a) isolate the main, (b) throttle flow in the main, (c) prevent backflow into the main from a brewery. What are your recommendations and why?

2. You and a worker have been asked to shut off a valve in a vault. The worker enters the vault and collapses. What should you do?

3. Diagram or describe how you would pinpoint the location of a valve using landmarks.

4. You have just been employed as chief operator for a town serving 8500 people. You soon discover there are no valve records. Describe how you would locate the valves in the distribution system and the type of records you would establish.

Module 4

Fire Hydrants

A FIRE HYDRANT is a device, used primarily for fire fighting, that is connected to a water main and provided with outlet nozzles to which fire hoses can be attached for discharging water at a high rate. One or more valves are provided for the operation of the hydrant. The hydrant is one of the most important parts of the water distribution system and often one of the most ignored. Fire hydrants stand idle for long periods of time and are vulnerable to damage, yet are expected to work well in emergencies. To ensure their availability when needed, hydrants must be properly installed, operated, and maintained.

After completing this module you should be able to

- Explain the various uses of hydrants in a distribution system.

- Differentiate between dry-barrel and wet-barrel hydrants.

- Identify the main components of dry-barrel and wet-barrel hydrants.

- Explain the proper procedures for installing, operating, inspecting, and maintaining hydrants.

- Describe how a hydrant flow test should be properly performed.

- Recommend the type of information that should be recorded when fire hydrants are installed, inspected, and repaired.

- Explain what safety precautions should be observed when flushing and testing fire hydrants.

4-1. Purpose and Uses of Hydrants

The major purpose and function of a fire hydrant is public fire protection. Although the hydrant is the property and responsibility of the water utility, during emergencies it is operated by members of the fire department. The use of fire hydrants as a source of water for main flushing, street cleaning, construction projects, or any purpose other than fire fighting is outside the primary purpose for which the hydrant is installed. Such use should be rigidly controlled by the water utility so that the hydrants are kept in good working condition for emergency use.

The water utility, unless expressly relieved of its responsibility by the fire department in accordance with a written agreement, or by public ordinance, or by private ownership of the hydrant, should schedule frequent inspections of hydrants to ensure that they are in good working condition.

Problems can arise if the distribution system is not designed to handle the demands placed on it by hydrants during operation. For example, if the mains are not large enough to provide adequate FIRE FLOW, a fire-department pumper could create a negative pressure in the main. This negative pressure could cause any existing cross connections to siphon nonpotable water or hazardous chemicals into the distribution system; it could also cause hot water heaters to drain back through service lines, damaging customer meters. Standard practice is to install hydrants only on mains 6 in. (150 mm) or larger. Larger mains are often necessary to ensure that the pressure during fire flow remains above 20 psi (140 kPa) (residual pressure).

Whenever hydrants are operated, the increase in flow causes the water to move faster through the mains, scouring any sediment that may have accumulated. This material can produce discolored and cloudy water, and the decrease in water quality can result in numerous customer complaints if the public has not been notified of the problem beforehand. Suspended material can also cause customer meters to record inaccurately, which is a major cause of unaccounted-for water.

Fire hydrants and mains can be damaged by untrained operators or nonutility personnel using improper tools or procedures during the operation of hydrants. Correct operating procedures are discussed in detail later in this module.

4-2. Types and Description of Hydrants

Types of Hydrants

Depending on their construction and shape, hydrants may be classified under one or more of several headings, including:

- Dry-barrel hydrants
- Wet-barrel hydrants
- Warm-climate hydrants

- Post hydrants
- Flush hydrants.

Dry-barrel hydrants.   Figure 4-1 shows a DRY-BARREL HYDRANT. This type of hydrant is equipped with a main valve and a drain in its base. The BARREL (body) is filled with water only when the main valve is open. When the hydrant is not in use, the main valve is closed and the barrel drains. The dry-barrel hydrant can be used anywhere; it is especially appropriate where freezing weather occurs.

A DRY-TOP HYDRANT (Figure 4-2) is a dry-barrel hydrant in which the threaded end of the MAIN ROD and OPERATING NUT are sealed from water in the barrel when the main valve of the hydrant is open and the hydrant is in use. A WET-TOP HYDRANT (Figure 4-3) is a dry-barrel hydrant in which the threaded end of the

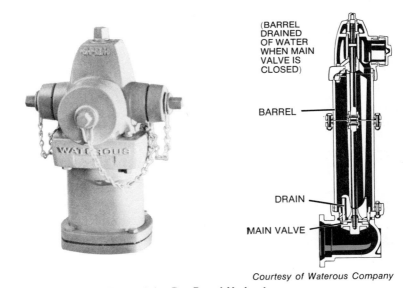

(BARREL DRAINED OF WATER WHEN MAIN VALVE IS CLOSED)

BARREL

DRAIN

MAIN VALVE

Courtesy of Waterous Company

Figure 4-1.   Dry-Barrel Hydrant

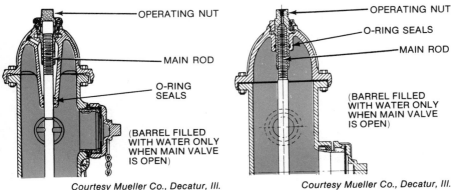

OPERATING NUT

MAIN ROD

O-RING SEALS

(BARREL FILLED WITH WATER ONLY WHEN MAIN VALVE IS OPEN)

Courtesy Mueller Co., Decatur, Ill.

Figure 4-2.   Dry-Top Hydrant

OPERATING NUT

O-RING SEALS

MAIN ROD

(BARREL FILLED WITH WATER ONLY WHEN MAIN VALVE IS OPEN)

Courtesy Mueller Co., Decatur, Ill.

Figure 4-3.   Wet-Top Hydrant

main rod and the operating nut are not sealed from water in the barrel when the main valve of the hydrant is open and the hydrant is in use.

Dry-barrel hydrants are sometimes classified according to the type of main valve. Several common types of main valves are shown in Figure 4-4.

- STANDARD COMPRESSION HYDRANT—The valve closes with the water pressure upwards for a positive seal.

- GATE HYDRANT—The valve is a simple gate valve, similar to one side of an ordinary rubber faced gate valve.

- COREY HYDRANT—The valve closes horizontally and the barrel extends well below the branch line.

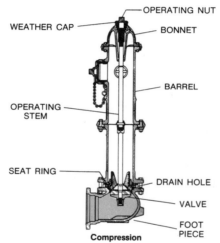

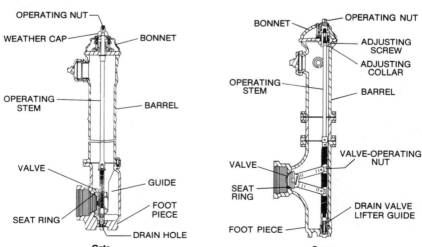

Figure 4-4. Common Hydrant Main Valve Types

Wet-barrel hydrants. Figure 4-5 shows a WET-BARREL HYDRANT. Under normal conditions, the barrel of the hydrant is filled with water supplied by the piping leading to the hydrant from the main. The gate valve on the piping is closed only to allow repair of the hydrant. The hydrant has no main valve. During fire emergencies, water is released from the hydrant through individual outlet nozzles, each of which is equipped with a valve. Because the barrel is always filled with water, the wet-barrel hydrant cannot be used in climates where temperatures fall below freezing—the water in the barrel would freeze and destroy the hydrant.

Warm-climate hydrants. A WARM-CLIMATE HYDRANT has a two-part barrel with the lower barrel below ground and the main valve located at the ground line. The main valve controls flow from all outlet nozzles; the lower barrel is always full and pressurized. There is no drain mechanism.

Post hydrants and flush hydrants. The upper barrel and head of a POST HYDRANT extend above the ground line, usually 24 in. (610 mm) or more. The hydrants shown in Figures 4-1 through 4-5 are post hydrants. A BREAKAWAY HYDRANT, or TRAFFIC HYDRANT, is a two-part, dry-barrel post hydrant with a stem coupling and a flange joining the upper and lower barrels at the ground line. The flange is designed to break when the unit is struck by a vehicle. The coupling joining the upper and lower main rods also breaks, releasing the upper section of the hydrant without damage to the lower section.

In contrast to the post hydrant, the entire standpipe and head of a FLUSH HYDRANT are below ground (Figure 4-6). The head of the flush hydrant, including the operating nut and outlet nozzles, is encased in a box that has a removable cover flush with the ground. Flush hydrants are used on aprons and taxiways at airports, at TANK FARMS, pedestrian malls, and other locations where post types are not considered to be suitable. Flush hydrants are usually a dry-barrel type, with a drain mechanism.

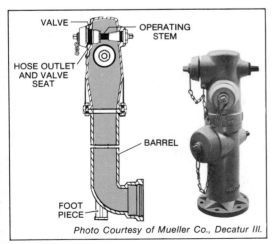

Photo Courtesy of Mueller Co., Decatur Ill.

Figure 4-5. Wet-Barrel Hydrant

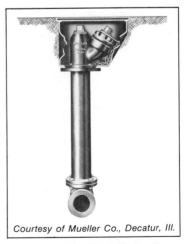

Courtesy of Mueller Co., Decatur, Ill.

Figure 4-6. Flush Hydrant

Hydrant Parts

The main parts of typical dry-barrel fire hydrants are shown in Figure 4-7.

Upper section: The UPPER SECTION, sometimes referred to as the UPPER BARREL, NOZZLE SECTION, or HEAD, consists of the operating nut, the bonnet, the upper barrel, the outlet nozzles, and the outlet-nozzle caps, described in the following paragraphs:

Operating nut: The operating nut, which is usually pentagonal, is rotated with a wrench to open and close the hydrant valve. The operating nut may be a single component or it may be in combination with a weather shield. Hydrants

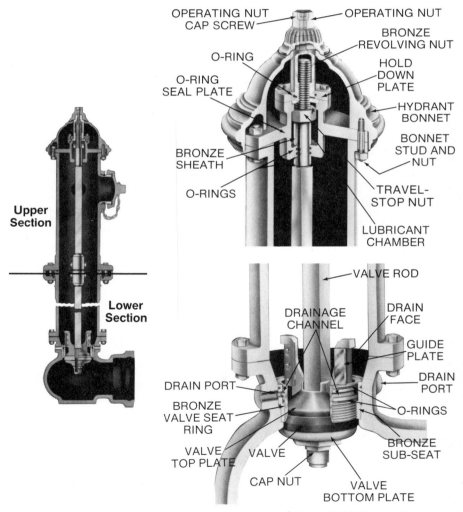

Courtesy of U.S. Pipe and Foundry Co.

Figure 4-7.   Main Parts of Typical Fire Hydrants

are usually opened by turning the operating nut counterclockwise. In any case, a visible arrow and the word *open* should be cast in relief on the top of the hydrant to designate the direction of opening.

Bonnet: The top cover or closure on the hydrant upper barrel is referred to as the BONNET. The bonnet may or may not be pressurized.

Upper barrel: The upper barrel is a cast gray or ductile iron section that carries water from the lower barrel (defined below) to the outlet nozzles.

Outlet nozzles: OUTLET NOZZLES are threaded bronze outlets on the upper barrel used to connect direct hose lines or suction hose from the hydrant to the fire department pumper (Figure 4-8). Outlet nozzles are usually either 2½-in. or 4½-in. nominal inside diameter. The smaller outlet nozzle is for fire hose connection. The larger outlet nozzle (called a PUMPER OUTLET or STEAMER OUTLET NOZZLE) is used to connect a fire department pumper. Outlet-nozzle threads are normally National American Standard threads; however, some utilities use special threads. Both utility and fire department personnel must be aware of the thread type used so that it will be compatible with the fire-fighting equipment.

Outlet-nozzle caps: OUTLET-NOZZLE CAPS are cast-iron covers that screw onto outlet nozzles, protecting them from damage and unauthorized access.

Courtesy of U.S. Pipe and Foundry Co.

Figure 4-8.   Outlet Nozzles

Lower section: The LOWER SECTION includes the lower barrel, the main valve assembly, and the base:

Lower barrel: The LOWER BARREL is a barrel section made as static-cast or centrifugally-cast gray or ductile iron. The section carries water flow between the elbow (see BASE, defined below) and the upper barrel and is usually buried in the ground with the connection to the upper barrel approximately 2 in. (50 mm) above ground line.

Main valve assembly: The MAIN VALVE ASSEMBLY is a subassembly including the LOWER MAIN ROD, UPPER VALVE PLATE, RESILIENT HYDRANT VALVE, LOWER VALVE PLATE, CAP NUT, and BRONZE SEAT RING. A TRAVEL-STOP NUT is found in some dry-barrel hydrants. The nut is screwed on the threaded section of the main rod. It stops at the base of the packing plate or revolving nut and ends downward travel (opening) of the hydrant valve.

Base: The BASE, also known as the SHOE, INLET, ELBOW, or FOOT PIECE, is the inlet structure of the hydrant with the inlet side at right angles to the outlet that discharges into the barrel when the valve is opened. The base is usually cast iron. When a hydrant mechanism does not include a travel-stop device to terminate downward travel of the main rod and valve, a stop must be built into the middle of the base against which the valve assembly will rest.

4-3.   Inspection and Installation

Hydrants should be inspected at the time of delivery in order to verify compliance with specifications and to check for possible damage during shipment. They should be reinspected before installation.

Specifications to be checked during the initial inspection include:

- Direction to open

- Size and shape of operating nut

- Depth of bury

- Size and type of inlet connection

- Main valve size

- Outlet-nozzle sizes and configuration

- Thread dimensions.

The hydrant should be cycled to full-open and full-closed positions to ensure that no internal damage or breakage has occurred during shipment and handling. All external bolts should be checked for proper tightness.

After inspection, the hydrant valves should be closed and the outlet-nozzle caps replaced to prevent foreign matter from getting into the hydrant. Stored hydrants should be protected from the weather whenever possible, with inlets facing downward.

Installation Procedures

Proper installation decreases operational problems and maintenance costs. Refer to Module 1, Pipe Installation, for a detailed description of proper methods of installation and testing.

Always install an auxiliary valve between the hydrant and its supply main to permit isolation of the hydrant for maintenance purposes (Figure 4-9). Remove foreign matter from the hydrant lateral before installing the auxiliary valve and hydrant. Connect fire hydrants only to water mains adequately sized to handle fire flows. Install hydrants away from the curb line at a sufficient distance to avoid damage to or from overhanging vehicles. A setback of 2 ft (0.6 m) from the curb line to the point on the hydrant nearest the curb is recommended. The pumper-outlet nozzle should face the street. Make sure that the outlet nozzles are high enough above the ground line for hose attachment and that there are no obstructions to prevent operation. In rural areas, protect hydrants from traffic

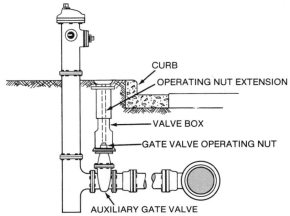

CURB
OPERATING NUT EXTENSION
VALVE BOX
GATE VALVE OPERATING NUT
AUXILIARY GATE VALVE

Figure 4-9. Hydrant With Auxiliary Valve

damage by large setbacks or other means, always being certain that the hydrant will be accessible to fire apparatus.

In setting a hydrant, use a firm footing, such as a stone slab or a concrete base on firm ground, in order to prevent settling and strain on the lateral line joints. Install hydrants as plumb as possible. Make provisions for anchoring the hydrant on the end of its lateral by strapping, blocking, or using a restraining type of joint. When pouring thrust blocks for dry-barrel hydrants with drains, exercise care not to plug or block the drain holes. Install traffic hydrants with extra care to ensure that there is adequate soil resistance to avoid transmitting shock movement to the lower barrel and inlet connection. In loose or poor load-bearing soil, a concrete collar, approximately 6 in. (150 mm) thick with a diameter of 2 ft (0.6 m), should be installed around the hydrant barrel at or near the ground line. In areas of substantial frost penetration, such collars are not recommended.

Make provisions for carrying off drainage from dry-barrel hydrants. One acceptable method is to excavate an area around the base of the hydrant large enough to permit the placement of approximately $1/3$ cu yd (0.25 m^3) of clean stone to a level several inches above drain openings. To prevent clogging of the drainage pit, the stone should be covered with 8-mil polyethylene or similar waterproof material before backfilling. This practice permits ready hydrant drainage after use. If it can be avoided, do not install hydrants where the water table is so high that the hydrant cannot drain. When water remains in a hydrant because of a high water table, it may be necessary to plug the hydrant drains. Mark hydrants that will not drain after use and must be pumped dry to maintain proper dry-barrel conditions. Do not connect hydrant drains to a sanitary or storm sewer.

Although the adoption of a capacity marking scheme is optional, Table 4-1 shows a suggested uniform color scheme for painting hydrant tops and/or caps to indicate the expected flow rate. An alternate scheme for color coding can be related to water main size.

Table 4-1.   Color Scheme to Indicate Flow Capacity

Flow gpm (L/min) at 20 psig* (140 kPa [gauge])	*Color*
Greater than 1000 (>3800)	Green
500–1000 (1900–3800)	Orange
Less than 500 (<1900)	Red

*Capacities should be rated by flow measurements of individual hydrants at a period of ordinary demand. When initial pressures are over 40 psig (280 kPa [gauge]) at the hydrant being tested, the rating should be based on 20 psig (140 kPa [gauge]) residual pressure, observed at the nearest hydrant connected to the same main and when no water is being drawn. When initial pressures are less than 40 psig (280 kPa [gauge]), residual pressures should be at least half of the initial pressure.

Hydrants must be highly visible and unobstructed at all times. Therefore, whether or not a color code is used, the hydrant barrel should be painted with colors that are easily visible both day and night. Hydrants likely to be covered with snow should have I.D. flags on 4–6 ft (1–2 m) long rods attached to the hydrants. This will help locate hydrants in emergencies and hopefully prevent them from being damaged by snow removal equipment. The rods should be installed so that they will not interfere with hydrant operation.

Testing

After installation and before backfilling (and after pressure testing a newly installed water main), test the hydrant as follows:

1. Open the hydrant fully and fill with water; close all outlets.

2. Vent air from the hydrant by leaving one of the caps slightly loose as the hydrant is being filled. After all air has escaped, tighten the cap before proceeding.

3. Apply a maximum pressure of 150 psig (1000 kPa). (City water pressure will suffice if it is impractical to apply higher pressure with a portable pump.)

4. Check for leakage at flanges, outlet nozzles, and operating stem.

5. If leakage is noted, repair or replace components or complete hydrant until the condition has been corrected.

A dry-barrel hydrant with unplugged drains should not be tested at the same time as the water main; the main should be tested with the hydrants closed. The water main cannot be tested satisfactorily if the hydrant is connected to it because of allowable leakage through the hydrant drains (see below). The gate valve in the hydrant lateral can be closed and the hydrant pressure-tested by introducing water under pressure through an outlet nozzle. Dry-barrel hydrants with unplugged drains may exhibit slight leakage (up to 5 fl oz/min [2.5 mL/s]) through the drains.

It is not uncommon for outlet nozzles and pressure bolting to become loosened due to rough handling in shipment, storage, and installation, causing leakage on a pressure test. Tightening of flange bolting, light caulking of leaded-in outlet nozzles, and tightening of threaded-in outlet nozzles will stop minor leakage and achieve a satisfactory test in most cases.

To test dry-barrel hydrants for drainage, proceed as follows: following the pressure test, close the hydrant; remove one outlet-nozzle cap; place the palm of one hand over the outlet-nozzle opening. Drainage rates should be sufficiently rapid to create a noticeable suction. Foreign material left in newly laid lines or hydrant laterals can damage valves and valve seats and affect pressure tests. After backfilling, operate the hydrant so that any foreign material is flushed out. Make sure dry-barrel hydrants have drained completely, then tighten outlet-nozzle caps and back them off slightly so that they will not be excessively tight; leave tight enough to prevent removal by hand. After installation, if necessary, apply a field coat of the paint preferred by the utility to the portion of the hydrant above ground.

4-4.   Operation and Maintenance

Hydrants are designed to be operated by one person using a 15-in. (380-mm) long wrench (Figure 4-10). The use of a longer wrench or an indefinite extender (cheater) operated by two or more persons is not good practice. If one person cannot open and close a fire hydrant with a 15-in. (380-mm) wrench, the hydrant is not in proper working order, and this condition should be corrected. Wrenches for fire hydrants should have no taper in the openings, so that they can be readily reversed. To prevent damage to the operating nut, wrenches not specifically designed for hydrant operation should not be used.

Courtesy Mueller Co., Decatur, Ill.
Figure 4-10.   Spanner Wrench

Hydrants should be opened and closed slowly to prevent pressure surges in the mains. These pressure surges, which reflect off closed valves and oscillate back and forth within the main, are called WATER HAMMER. If hydrants are opened or closed suddenly, the water hammer that results can easily damage mains and appurtenances.

A partially opened main valve in dry-barrel hydrants may allow substantial leakage through the drain valves, preventing hydrants from draining properly in soils that easily become saturated or undermining hydrants in soils that are easily washed away. The main valve of dry-barrel hydrants should always be completely opened to ensure that the drain valve has closed. Throttling the main valve may cause damage to the valve seat and rubber. Instructions to this effect should be given to all persons authorized to use the hydrants, including those authorized to use hydrants for purposes other than fire fighting, such as contractors, street cleaners, and summer playground supervisors.

Regularly scheduled inspections of hydrants are necessary to ensure satisfactory operating conditions. All hydrants should be inspected annually. In freezing climates, hydrants may require two inspections, one in the fall and another in the spring. If possible, maintenance should be performed at the time of inspection to avoid duplication of effort and reduce costs.

During freezing weather, dry-barrel hydrants should be inspected after each use. In fact, it is advisable to check all types of hydrants after any use. Dry-barrel hydrants installed with permanently plugged drains must be pumped out after each use.

Inspection Procedure

Check that there is nothing near the hydrant that would interfere with its operation, the removal of outlet-nozzle caps, or the attachment of hoses to the outlet nozzles. Visually check the hydrant to be sure it is not leaning. Remove one outlet-nozzle cap and check with plumb bob or other device for presence of water or ice standing in the barrel of dry-barrel hydrants. Check for seat leakage with a listening device on dry-barrel hydrants and visually at each valve on wet-barrel hydrants. Replace outlet-nozzle caps and open the hydrant to a full-open position, checking the ease of operation. To prevent damage, vent air from the hydrant. Wet-barrel hydrants require that each valve be so operated, using a special test outlet-nozzle cap. If stem action is tight, repeat the operation several times until the opening and closing action is smooth and free. Water conditions can cause hard water buildup on stem threads. A series of opening and closing operations is usually sufficient for removing this buildup.

While the hydrant is under pressure, check for leakage at joints, around outlet nozzles, at packing or seals, and past outlet-nozzle caps. If leakage is observed, tighten or recaulk outlet nozzles, lubricate and tighten compression packing or replace O-rings or similar seals, and replace gaskets. If leakage cannot be corrected with the tools at hand, record the nature of the leakage for prompt attention by the repair crew.

On dry-barrel hydrants, close the main valve to the position at which the drains open and allow flow through the drains under pressure for about 10 sec to flush the drain. Then close completely. Remove an outlet-nozzle cap and attach a section of hose or a flow diverter, if necessary, to direct the flow into the street. Open the hydrant and flush to remove foreign material from the interior and lateral piping.

Close the main valve and check dry-barrel hydrants for drainage from the barrel. Drainage should be sufficiently rapid to create suction when the hand is placed over an outlet nozzle during drainage. Check again for seat leakage with a listening device on dry-barrel hydrants and visually on wet-barrel hydrants. If dry-barrel hydrants do not drain, pump out any residual water in the barrel and check for leakage of the main valve with a listening device.

Remove all outlet-nozzle caps and inspect for thread damage from impact or cross threading. Clean and lubricate outlet-nozzle threads and use caps to check for easy operation of threads. Be sure outlet-nozzle cap gaskets are in good condition. Check outlet-nozzle cap chains for free action on each cap. If binding is observed, open the loop around the cap until the action is free, so that kinking during removal of the cap under emergency conditions is prevented. Replace caps, tighten with hydrant wrench, then back them off slightly so they will not be excessively tight; leave tight enough to prevent removal by hand.

Lubricate operating-nut threads in accordance with manufacturer's instructions. Some hydrants require oil in the upper operating-nut assembly. This lubrication is critical and must not be overlooked during routine maintenance.

On traffic hydrants, check the breakaway device and inspect couplings, cast lugs, special bolts, etc. for damage due to impact or corrosion.

Clean the exterior of the hydrant and repaint if necessary. Be sure the auxiliary valve is in the full-open position. If a hydrant is inoperable, identify it with a highly visible marking to prevent loss of time by fire-fighting crews if any emergency occurs before the hydrant is repaired (Figure 4-11). Report the condition to the fire department at once.

Figure 4-11. Inoperable Hydrant

Repairs

Any condition that cannot be corrected during the regular inspection should be recorded and reported for subsequent action by repair crews. Leakage, broken parts, hard operation, corrosion, and other major defects should be corrected by a crew as soon as possible after the defect is reported. Before any repair takes place, the fire department must be notified. If repairs are to be performed in the field, the repair crew should take a full complement of spare parts to the job site.

Close the auxiliary valve ahead of the hydrant or otherwise cut off flow and pressure to the hydrant. Disassemble the hydrant in accordance with the procedure supplied by the manufacturer of the hydrant. Replace damaged parts and parts that show indication of wear, corrosion, or signs of incipient failure. Always replace gaskets, packing, and seals. Reassemble the hydrant and open the auxiliary valve or otherwise apply pressure. Test the seats for leakage. With dry-barrel hydrants, apply pressure to the entire hydrant assembly, checking for leakage, ease of operation, and drainage. Be sure to vent air from the hydrant at the start of operation.

Record the fact that the hydrant has been repaired and is in operable condition. Remove any markings indicating that the hydrant is inoperable and notify the fire department that the hydrant has been repaired.

Flow Testing

As a municipality expands, the demand on the water distribution system for normal use and fire protection increases. At the same time, because of corrosion, scale, and sedimentation in the pipes, the carrying capacity of the system may slowly decrease. (See Module 1, Pipe Installation and Maintenance, for a description of procedures to help correct these problems.) It is therefore necessary to check the capacity of the system to determine the need for additional feeder or looping mains and the need to clean and line existing pipes.

Hydrants should be tested as necessary to obtain required data on distribution system flow and fire flow capabilities. Flow capability should be retested after major changes in the distribution system near the hydrant. The necessary test equipment includes:

- A 0–200 psi (0–1400 kPa) pressure gauge with 1-psi (10-kPa) gradations
- An outlet-nozzle cap tapped for a gauge
- A 0–60 psi (0–500 kPa) PITOT GAUGE with ½-psi (5-kPa) gradations (handheld or clamp-on type gauge) (Figure 4-12)
- A sill-cock gauge adapter (if RESIDUAL PRESSURE must be measured at a faucet)
- A special calibrated nozzle or tube used to reduce the turbulence caused by water passing around the discharge valve (for testing wet-barrel hydrants only).

All gauges must be accurately calibrated, and recalibration should be performed regularly. Instructions for constructing a pitot gauge are included in Appendix B. If a pitot gauge is not available, a 60-psi (500-kPa) pressure gauge mounted on a tapped outlet-nozzle cap can be placed on one of the nozzle outlets of the flowing hydrant (Figure 4-13); the reading on the gauge will be approximately equal to the reading that would be obtained with a pitot gauge at the flowing outlet.

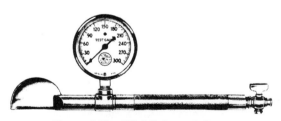

Figure 4-12. Pitot Tube and Gauge

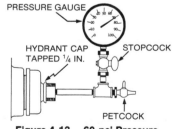

Figure 4-13. 60-psi Pressure Gauge

Test procedure.   Select a hydrant at or near the location for which flow-test data is required; this will be the flowing hydrant. Select another hydrant for the residual pressure readings, or locate a hose bibb or faucet at which the readings are taken. Residual pressure should be read at a location such that the flow runs from the source to the flowing hydrant and then to the hydrant, hose bibb, or faucet at which the readings are taken. If residual pressure is read at a hose bibb or faucet, be sure there is no use of water in the building while the test is being conducted.

At the start of the test, observe and record the normal system pressure. Then open the selected hydrant to allow flow. While the hydrant is flowing, determine the velocity pressure of the discharge by the use of a pitot gauge and record that pressure. Hold the pitot tube in the middle of the flowing stream about 1 in. (25 mm) from the end of the outlet nozzle. Observe and record residual pressure.

Test data reflecting the flow capabilities of the system under the pressure drop induced by the test can be converted to information regarding flow capabilities under a greater pressure drop (as would be influenced by fire pumping) by the use of tables contained in Appendix C, Flow Testing.

Records

In order to carry out a meaningful inspection and maintenance program, it is essential that records be kept on each hydrant. Location, make, type, size, and date of installation should all be recorded. Other information can be entered depending on the nature of the maintenance or inspection.

When a hydrant is inspected, an entry should be made in the record indicating the date of inspection and the condition of the hydrant. If repair work is necessary, then the nature of the work required should also be indicated. On completion of the repair work, the nature of repairs, date, and other relevant information should be recorded. Other information (such as testing, pumping, and opening and closing) is also important and should be carefully recorded. Samples of record and survey sheets are shown in Figures 4-14 through 4-17.

4-5. Safety

Specific areas where the need for safety precautions must be recognized and observed when dealing with hydrants include:

- Material handling (refer to Module 1)
- Barricades, warning signs, and traffic control (refer to Module 1)
- Hand tools (refer to Module 1)
- Portable power tools (refer to Module 1)
- Flushing.

Flushing

During hydrant flushing and flow testing, special precautions are necessary to prevent damage to private property and injury to utility personnel and pedestrians because of the force and volume of water used. If the temperature is below freezing, water from hydrants should not be allowed to flow into streets or pedestrian walkways since a dangerous icing condition could be created. If flow is diverted to a sewer, care must be taken not to create a cross connection.

HYDRANT INSPECTION REPORT
XYZ UTILITY

Hyd. No.	
Location	
Nozzle	
Pressures — Initial	
Pressures — Resid.	
Pressures — Pilot	
Flow gpm	
Time Flushed min	
Water Used gal	
Paint	
Chains	
Caps	
Stems	
Packing	
O Ring	
Top Nut	
Valve	
Valve Seat	
Cond. of Water	
Remarks	

By _____ Date _____

Figure 4-15. Inspection Report

FIRE HYDRANT MASTER RECORD
XYZ WATER UTILITY

Manufacturer _____ Date _____ Hydrant No. _____

Type _____ MVO _____ Inlet _____

Bury _____ Hose Nozzle Size _____ Thread Type _____

Pumper Nozzle Size _____ Thread Type _____

Installed by _____ Date _____ W/O No. _____ Cost _____

Operating Nut _____ Turns to Open _____

Location _____ Line Static Pressure _____

Date	Inspected	Tested	Repaired	Painted	Opened by	Cost	Remarks

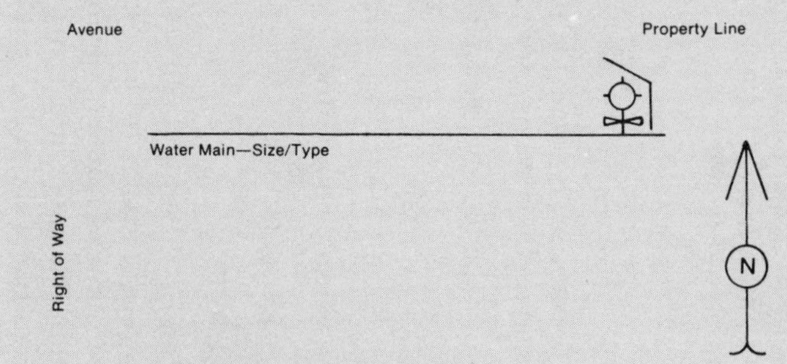

Avenue

Water Main—Size/Type

Property Line

Right of Way

N

Figure 4-14. Master Record

(3 in. × 5 in. Sheet)

HYDRANT MAINTENANCE REPORT

XYZ Water Utility _____ Hydrant No. _____

Location _____

Caps   Missing _____ Replaced _____ Greased _____

Chains Missing _____ Replaced _____ Freed _____

Paint O.K. _____ Repainted _____

Oper. Nut O.K. _____ Greased _____ Replaced _____

Nozzles O.K. _____ Caulked _____ Replaced _____

Valve & Seat O.K. _____ Replaced _____

Packing O.K. _____ Tightened _____ Replaced _____

Drainage O.K. _____ Corrected _____

Flushed _____ Minutes _____ Nozzle Open _____

Pressure Static _____ Residual _____ Flow _____ gpm

Branch Valve Condition _____

Any Other Defects _____

_____

_____

_____

Inspected _____ By _____

Defects Corrected _____ By _____

Figure 4-16. Maintenance Report

HYDRANT TEST
XYZ WATER UTILITY

Manufacturer _____   No. _____

Date, time	Nozzle Size	Pressure			Flow gpm	Flow 20 psi	Time min	Water Used gal
		Stat.	Res.	Pitot				

Figure 4-17. Test Report

Flow diverters or diffusers should be used where necessary to direct the flow into a gutter or drainage ditch in order to prevent flooding or erosion damage. When flushing or flow testing hydrants, a rigid diverter should never be used. A rigid diverter consists of a pipe screwed onto the hydrant outlet, which extends out to a desired length and bends by up to 90° to change the direction of the water flow, discharging the full flow into the atmosphere. The water discharged from the diverter generates a pushing force (thrust) that can be very dangerous. This force is magnified by the distance from the outlet to the bend in the pipe, generating what can be a very high torque on the hydrant due to the leverage. A rigid diverter several feet (metres) long can produce many hundreds of foot pounds (newton-metres) of torque on the fire hydrant, which can damage the hydrant and connections leading to the hydrant. The greatest danger exists when the rigid diverter is installed in such a manner that the line pressure would create sufficient torque to cause the hydrant head to unscrew from the spool or riser.

To prevent possible bodily injury, property damage, or damage to the fire hydrant and its supporting structures, only a diffuser or flexible hose should be used for flushing or flow testing. When a hose is used to divert flow, the discharge end of the hose should be securely anchored so the hose cannot jerk around under pressure, endangering operators as well as passersby.

Selected Supplementary Readings

Carl, K.J. Fire Fighters Rely On Usable Water Hydrants. *OpFlow*, 1:5:1 (May 1975).

Dry-Barrel Fire Hydrants. AWWA Standard C502-85. AWWA, Denver, Colo. (1985).

Fire Hydrant Vandals Cause Problems for Water Utilities. *OpFlow*, 7:7:1 (July 1981).

Fire Hydrants Improve With Age. *OpFlow*, 7:7:4 (July 1981).

Franklin, B.W. Maintaining Distribution System Valves and Hydrants. *Journal AWWA*, 74:11:576 (Nov. 1982).

Inspecting and Maintaining Fire Hydrants. *OpFlow*, 8:5:6 (May 1982).

Installation of Gray and Ductile Cast-Iron Water Mains and Appurtenances. AWWA Standard C600-82. AWWA, Denver, Colo. (1982).

Installation, Operation, and Maintenance of Fire Hydrants. AWWA Manual M17. AWWA, Denver, Colo. (2nd ed., 1980).

Lockwood, D.N. Testing Hydrant Flows in a Small Water Main Network. *OpFlow*, 2:2:1 (Feb. 1976).

Seevers, Dick. Keeping Hydrants in Fire-Fighting Condition. *OpFlow*, 8:5:1 (May 1982).

Seevers, F.R. (Dick). Hydrant Inspection—Part I. *OpFlow*, 3:8:1 (Aug. 1977).

Seevers, F.R. (Dick). Hydrant Inspection—Part II. *OpFlow*, 3:9:4 (Sept. 1977).

Wet-Barrel Fire Hydrants. AWWA Standard C503-82. AWWA, Denver, Colo. (1982).

Wilson, P.S. Hydrant Discharge Measurement Methods As Used By Operators. *OpFlow*, 1:11:1 (Nov. 1975).

Zimmerman, Robert. Techniques of Hydrant Care in Sandusky. *OpFlow*, 2:3:2 (Mar. 1976).

Glossary Terms Introduced in Module 4

(Terms are defined in the Glossary at the back of the book.)

Barrel	Nozzle section
Base	Operating nut
Bonnet	Outlet nozzle
Breakaway hydrant	Outlet-nozzle cap
Bronze seat ring	Pitot gauge
Cap nut	Post hydrant
Corey hydrant	Pumper outlet nozzle
Dry-barrel hydrant	Residual pressure
Dry-top hydrant	Resilient hydrant valve
Elbow	Shoe
Fire flow	Standard compression hydrant
Fire hydrant	Steamer outlet nozzle
Flush hydrant	Tank farm
Foot piece	Traffic hydrant
Gate hydrant	Travel-stop nut
Head	Upper barrel
Inlet	Upper section
Lower barrel	Upper valve plate
Lower main rod	Warm-climate hydrant
Lower section	Water hammer
Lower valve plate	Wet-barrel hydrant
Main rod	Wet-top hydrant
Main valve assembly	

Review Questions

(Answers to Review Questions are given at the back of the book.)

1. What is the principal purpose for hydrants in a distribution system?

2. Give examples of secondary purposes associated with the use of hydrants.

3. How can hydrant operation cause water quality problems?

4. On the figure below, indicate which hydrant is wet-barrel and which is dry-barrel, and label the component parts.

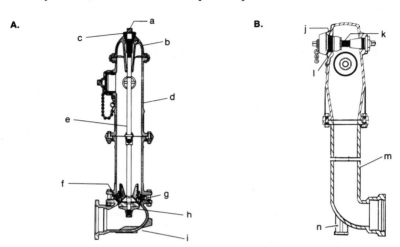

5. Explain the principal difference between a dry-barrel and wet-barrel hydrant.

6. What type of fire hydrant is recommended for use in cold climates?

7. List two precautions that should be taken during the installation of dry-barrel hydrants.

8. What is the purpose of color coding hydrant tops and/or caps?

9. In relation to the street, what direction should the pumper outlet nozzle be pointed?

10. List two ways hydrants can be protected from damage by traffic.

11. How should hydrant valves be operated, and why is the speed of valve operation important?

12. Explain the correct procedure for throttling the flow of water from a hydrant.

13. Explain the proper procedure for flushing a hydrant.

14. What items should be routinely checked during a hydrant inspection?

15. What preventive measures should be taken in cold climate areas to ensure hydrants will remain operable during the winter?

16. What procedure should be followed for flushing the drain of a dry-barrel hydrant?

17. What are the possible causes of water found standing in a dry-barrel hydrant, and how is the actual cause determined?

18. List six items of information that should be included on a hydrant record-keeping form.

19. List appropriate operating procedures for identifying out-of-service hydrants.

20. What safety precautions should be taken during hydrant flushing and testing to prevent:
    (a) Injury to water department personnel and pedestrians?
    (b) Damage to private property?

Study Problems and Exercises

1. Complete a hydrant master record card, which includes all necessary information pertaining to the receipt, inspection, and testing of a new hydrant. Also, make any recommendations that seem appropriate for modifying the record-keeping form or procedures based on local practices.

2. Under the supervision of the instructor or an experienced operator, conduct a hydrant flow test and determine the flow available from the test hydrant at 20 psi (140 kPa) residual pressure.

3. Under the supervision of the instructor or an experienced operator, inspect an installed hydrant and record the appropriate information.

Module 5

Services and Meters

Water SERVICE LINES are the pipes that lead from the water main to the customer's plumbing. Service lines vary in size depending on the pressure at the main, the distance from the main to the meter, the quantity of water required, and the residual pressure needed for the required quantity of water. Plastic and copper pipe are commonly used for residential service lines, whereas larger industrial service lines may be cast-iron, asbestos–cement, or other common piping material used in distribution systems.

In most cases, each customer is served through an individual service line. However, sometimes multiple service connections within a short section of main provide for large service demands, or one large service line is used to serve all occupants of a multifamily apartment complex.

Because most service lines are underground, water suppliers must maintain accurate records on the size, location, and components used, in order to quickly and efficiently repair, replace, or shut off a service.

WATER METERS are used to measure and record the volume of water flowing through a line. The primary function of metering is to aid a water utility in equitably charging customers for the water they use. Different types of water meters are available for various applications. This module discusses the function and upkeep of several types commonly used in distribution systems.

After completing this module you should be able to

- Explain what is meant by a service connection, and identify the components that make up a typical residential connection.

- Identify the principal factors that govern the size of water service lines, and explain what information is needed to properly determine the size of various services.

- Identify factors that should be evaluated in the selection of pipe material used for service connections and relate them to local conditions and practices.

- Discuss the importance of metering water, and describe the basic operating principles of meters commonly used in distribution systems.

- Describe what factors should be taken into consideration in selecting the correct meter size for a particular service and what conditions should be examined in determining meter locations.

- Recommend appropriate records that should be kept for service connections and meter installations, testing, and repair.

- Describe important safety procedures to be observed when installing and maintaining services and meters.

5-1. Services

A water service line is basically a small diameter pipe that runs from the distribution main to the customer's plumbing system. As shown in Figures 5-1 and 5-2, the complete service line installation consists of a connection (CORPORA-TION STOP) at the main, a shutoff valve (CURB STOP) or a meter with a shutoff valve near the customer's property line, and the connecting piping from the main to the customer's plumbing.

In colder climates, the meter is usually placed inside the house or building for protection against freezing. A shutoff valve (curb stop, Figure 5-3) is installed in a CURB BOX (Figure 5-4) near the customer's property line. Where freezing is not a problem, the curb stop can be eliminated and the shutoff valve and meter can be incorporated in a meter box near the customer's property line (Figure 5-5).

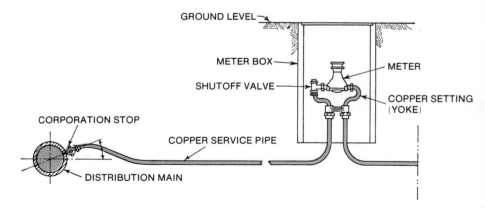

Figure 5-1.   Small Service Connection (Outdoor Meter)

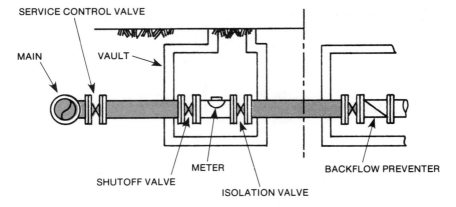

Figure 5-2. **Large Service Connection**

Courtesy of Mueller Company, Decatur, IL

Figure 5-3. Curb Stop

Courtesy of Mueller Company, Decatur, IL

Figure 5-4. Curb Boxes

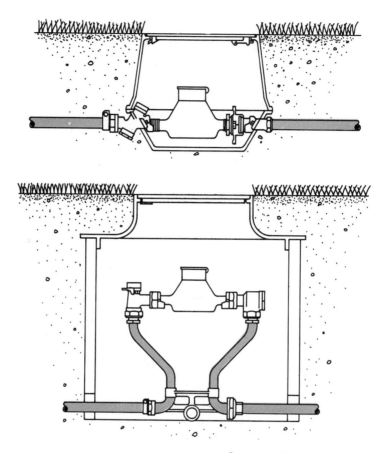

Courtesy of Ford Meter Box Co.

Figure 5-5. Meter Boxes Containing Meter and Valve

The pipe material and type of installation for a service varies depending on whether the customer is a residential, commercial, or industrial user. Residential service lines are usually constructed of ¾- to 1½-in. (19- to 38-mm) flexible pipe, usually copper or plastic. Commercial and industrial service lines can be as large as 4 in. (100 mm) in diameter and are usually a rigid pipe, such as ductile iron.

As shown in Figure 5-1, a bend in the flexible piping, known as a GOOSENECK, is used near the connection of a residential service with the main. The purpose of the gooseneck is to prevent the service line from breaking or pulling away from the main when the line's position shifts due to soil movement or expansion and contraction of the line with temperature changes.

For commercial and industrial services, a gooseneck is not required since the rigid, larger diameter pipe used for these services is quite strong. A valve vault is used to house the meter and shutoff valve for these larger services (Figure 5-6).

Courtesy of Badger Meter, Inc.

Figure 5-6.   Valve and Meter Vault for Large Services

Service connections are typically installed by the water utility, rather than the customer, since installation involves attachment to a water main, working in public streets or highways, and crossing other utility lines. It is logical that such responsibility should be assumed by one qualified party rather than by many individuals. The cost of installing the portion of the service from the main to the curb or property line may or may not be charged to the customer, depending on local practice or regulatory requirements. The portion of the service from the curb or property line to the building is usually installed by a contractor or plumber at the customer's expense.

Specifications for installation of the service are often prescribed either in plumbing codes or by the water supplier. These specifications (sometimes in the form of drawings) describe how the property portion of the service line is to be installed, the sizes and kinds of approved service-line materials, and any other information necessary to provide a complete understanding of installation requirements in order to ensure proper installation by the plumber or contractor.

Many water suppliers have tried to standardize the materials and components used for services. Standardization helps to simplify material ordering and inventory practices. It also enables water suppliers to take advantage of economic bulk ordering, and it saves time and money because the utility does not need to train field crews in a variety of installation techniques.

Pipe Materials

A variety of pipe materials are available for use as service lines, and no single material is the best selection for all installations. Economic and environmental conditions both influence the selection, and successful performance of one pipe material in a community does not necessarily guarantee similar results in another community. Factors a water supplier should consider before selecting a pipe material include:

- Weather
- Soil conditions
- Water quality characteristics
- Cost of installation
- Durability
- Standardization
- Ease of locating after installation
- Minimum AWWA standards.

Residential service connections.   Pipes used for residential services are typically copper or plastic. In the past, lead, brass, and galvanized wrought iron were used. Lead and brass are not used today primarily due to the high cost of labor required for jointing. Lead pipe can also cause public health concerns when used in communities with corrosive water. Black and galvanized wrought iron require threading, making them difficult to install properly. In addition, they can quickly develop leaks because of the corrosive action of soil or water and because of GALVANIC CORROSION resulting from contact with bronze corporation or curb stops.

Copper services.   Copper pipe is popular for service line installations because it is flexible, easy to install, corrosion-resistant in most soils, and can withstand high pressures. It is not sufficiently soluble in water to be a health hazard, and brass and bronze fittings can be used on copper pipe without causing appreciable galvanic corrosion. The ends of copper tubing can be readily flared to provide an easy method of attachment to fittings. Compression-type fittings can also be used.

Plastic services.   Of the many types of plastic pipe available, three are generally used. They are polyvinyl chloride (PVC), polyethylene (PE), and polybutylene (PB). All plastic pipes possess a smooth interior surface and, therefore, a high C value (low resistance to flow). Plastic pipe is lightweight and handles with ease. It is flexible in varying degrees, permitting deflection as the pipe is laid. Plastic pipe is resistant to corrosion and should provide a long service life.

Fittings used with both PE and PB pipe are normally conventional brass fittings with an adapter for connecting to the plastic. Tubing size PE and PB pipe can be cold flared and used with conventional copper flare fittings (Figure 5-7). (Heat may be applied to PE pipe to make the plastic easier to shape.) There are also a number of proprietary connections on the market for plastic pipe, including compression-type fittings (Figure 5-8) with clamps to prevent pull-out of the plastic. The various proprietary connections usually require a stainless-steel insert within the pipe to provide rigidity. Since PVC service pipe is rigid, a gooseneck is used for connection to the main, and adaptors are used for making connection to brass fittings. PVC-to-PVC connections are usually welded with special solvents.

Courtesy of Ford Meter Box Co.

Figure 5-7.   Flare Fitting

Courtesy of Ford Meter Box Co.

Figure 5-8.   Compression Fitting

Large service connections.   Large service connections are usually more complex than small connections, primarily because of the size and restrictive requirements imposed by rigid, larger diameter pipe. The size of some services may require special heavy equipment that can lift pipe, excavate sizeable trenches, or tap large mains. Large services can be expensive projects. Larger service connections (those sized 2 in. [50 mm] or larger, which normally serve industrial or commercial users) may be constructed of asbestos–cement pipe, ductile or cast-iron pipe, or PVC pipe. Refer to Module 1, Pipe Installation and Maintenance, for a discussion of the pipe materials used for large services.

5-2.   Sizing Services

To determine the proper size for a service line, the following data are required:

- Maximum demand of the customer
- Average daily demand
- Pressure at the meter outlet
- Main pressure at peak-demand periods
- Difference in elevation (in excess of 10 ft [3 m]) between the system's supply main and the customer's building
- Distance from the main to the proposed meter location.

Services must be sized with two major requirements in mind—flow rate and pressure, which are influenced by the previous factors. Where a small service line and meter are installed, pressure losses due to friction within the pipe may reduce the customer's water pressure to an unacceptably low level, even though the flow rate may be sufficient to supply all the water required.

To determine the effects of friction, pressure-loss tables have been developed. These tables allow fast and accurate selection of a pipe size that will supply a given pressure and flow rate. Where tables do not apply to a given situation, pressure losses can be determined by the Hazen–Williams formula.[1]

Pressure loss due to friction is generally the controlling factor in determining the size of the service and meter, but maximum velocity is also important. High water velocity through the service can cause water hammer and pipe noise, and it may also cause CAVITATION, which will wear away the inside of the pipe or meter and eventually lead to failure. Services generally should be sized so that the flow velocity in the service from the main to the meter can be maintained at less than 15 fps (4.6 m/s) even if the allowable pressure loss at the maximum customer demand could be maintained with a smaller line.

5-3. Installation of Services

Factors that must be considered when installing service lines include location, excavation procedures, tapping, and disinfection.

Location

Frost penetration and the location of sewer and other utility lines should be considered when determining the depth and location of water service lines. Service lines freeze more often than water mains because they contain a smaller volume of water to be cooled and are subject to longer periods of time during which there is little or no flow. The rate and extent of frost penetration is a function of the character of the soil, snow cover, and the duration and intensity of air temperatures below freezing. Figure 1-24 in Module 1 provides recommended depth of cover based on frost penetration depth throughout the United States. Weather records and local experience should also be consulted to determine any local variations.

In evaluating the location of sewers and other utility lines, water suppliers should consult local and/or state regulations for separation distances, and then ensure that service installation crews are familiar with such regulations.

Excavation

Procedures for excavation vary depending on whether the pipe is laid in an open cut, or washed or bored under a roadway or sidewalk. The excavation and installation of pipe for service lines is similar to the procedures for installation of larger lines covered in Module 1. Water suppliers should ensure that the excavation, installation, and area restoration are done in a manner that will not inconvenience the public.

Tapping

The method of making service connections at the main varies both with the size and material of the service and with the size and material of the main to which the service is attached. If the size and wall thickness of the main are sufficient to

---

[1] *Basic Science Concepts and Applications*, Hydraulics Section, Head Loss.

provide adequate full threads, ¾-in. to 2-in. (19-mm to 50-mm) services may be connected by direct drilling and tapping of the main.

Such connections to the main can be done either when the pipe is empty (a DRY TAP) or when the main is under pressure (a WET TAP). Other terms for a wet tap include hot tap, live tap, pressure tap, and pressure cut. For dry taps, depending on the pipe material, the pipe can be drilled and tapped to receive the corporation stop. For wet taps, a tapping machine designed for drilling, tapping, and inserting a corporation stop under pressure is used. The machine enables a worker to install a corporation stop without shutting down the main.

When the pipe wall is too thin or too soft, or the pressure is too great to permit directly tapping the pipe, a SERVICE CLAMP, or SADDLE, must be used. These fittings are attached around the pipe, and the corporation stop is threaded into the fitting rather than into the pipe. The main is then drilled through the corporation stop. Saddles can be used for both wet and dry taps into most types of pipe.

Tees, tapping sleeves, or saddle connections can be used for connecting larger service lines. Some systems use multiple taps to provide for large service demands. When multiple taps are used, they should be spaced at least 10 in. (250 mm) apart. For a more extensive discussion of tapping, refer to Module 2, Tapping.

Disinfection

Precautions are usually taken during the installation of service lines to prevent contamination. Flushing and disinfection should, however, be performed when the installation is complete, in order to ensure that safe, palatable water is delivered to the customer. Large service lines are disinfected in a manner similar to water mains, whereas smaller services are generally just flushed thoroughly with potable water. Refer to Module 1 for a complete discussion of disinfection practices.

5-4. Maintenance of Services

The major maintenance problems with service lines result from encrustation, corrosion, and freezing.

Encrustation. Encrustation is usually formed by deposits of calcium carbonate that precipitate out of unstable water. The deposits reduce the carrying capacity of the line and increase the resistance to flow. Damage to the water meter can also occur.

Although encrusted lines can be cleaned, the cleaning procedure is time-consuming and can damage the line. If encrustation is a continuing problem, water stabilization at the treatment plant is the best long-term solution. (See Volume 2, *Introduction to Water Treatment*, Module 9, Stabilization.)

Corrosion. Corrosion can considerably shorten the life of service lines. Even if the line does not fail completely, pinhole leaks can result in low pressure and less water to consumers, as well as endangering the quality of the water. Corrosion can be caused by soil conditions or by unstable water.

If corrosive soil conditions are known to exist, corrosion-resistant piping materials should be used for the service. Internal corrosion can best be prevented by stabilizing the water (Volume 2, Module 9).

Freezing. Freezing of service lines interrupts water service to customers. Freezing can also rupture lines and damage meters. The best way to prevent freezing is to ensure that service lines are installed a sufficient distance below the normal frost depth. A distance of 6–12 in. (150–300 mm), depending on the soil type, is usually sufficient.

If freezing of the service line occurs outside the meter box, there are two commonly used methods for thawing the line. If the service line is metallic, electric current can be run through the pipe, causing heat to be generated that will melt the ice. The current is usually supplied by a portable source of direct current, such as a welding unit.

Electrical thawing can be dangerous, and it can cause damage to the service line, the customer's plumbing, and the customer's electrical appliances. A major danger when electrically thawing a service is that there will be poor conductivity between sections of the service line or direct contact with other metal pipes or conductors, such as natural gas pipelines. When these conditions exist, the path of the applied electrical current may be diverted from the service line and the current may enter adjacent buildings. These stray currents can cause major damage to electrical appliances and may even be responsible for fires.

A second danger involved with electrical thawing is that the current may damage O-rings, gaskets, and soldered joints on the service, resulting in joint failure and leakage. The danger of stray current and the danger of damage to the service can both be reduced by using only low-voltage generators and by closely monitoring the voltage and amperage. Only experienced operators should attempt electrical thawing.

Before thawing, the property owner (and tenant, if any) must be informed of the risks involved and be required to sign a waiver form, approved by the utility's lawyer, absolving the utility of liability in case of accident. Thawing is a service performed for the customer on the customer's property, and the utility must have written confirmation from its insurer stating that adequate liability insurance is in effect to cover possible consequences of the thawing.

Because of the problems associated with electrical thawing, and because plastic pipe cannot be thawed electrically, techniques using hot water are becoming common. These techniques typically involve pumping hot water through a small flexible tube that is manually fed into the frozen service line. Since this introduces a cross connection into the potable water system, all equipment and water used must be kept sanitary at all times. Additional information on two techniques for hot water thawing can be found in the Supplementary Readings section of this module.

If the service line or meter is frozen in the meter box or vault, an effective method of thawing is to circulate warm air through the box. A safe method for generating the hot air is with a portable heat gun (a device resembling a hair dryer), which can be operated with a portable generator. A butane torch could also be used, but it is not recommended because of the possibility of explosive gases in the meter box, either from sewers or from natural gas lines. The intense, localized heat of a torch can also cause rapid expansion and rupture of meters or pipelines.

5-5. Meters

Public and private water suppliers meter the water they distribute for a number of reasons, the most common of which is to determine billing charges according to use. Meters installed on each service line enable utilities, municipalities, and sanitary districts to charge each customer for the exact amount of water used, which encourages customers to use water wisely.

Larger water meters are used to measure water flowing into different zones or areas to provide system control. Knowing how much water is supplied helps in the production of water; it informs administrators how to apportion water rights and costs among counties or districts; and it provides the engineering staff with information needed to check on the capacity of pipelines or detect loss of water by comparing service meter readings with production readings. Meters also aid in the blending of water from several sources so that the mix in reservoirs or storage tanks is satisfactory.

Types of Meters

A number of different types of water meters are available for specific applications. Meters commonly used at small water distribution systems include:

- Positive-displacement meters
- Current meters
- Compound meters.

Larger systems may also use:

- Proportional meters
- Venturi meters
- Orifice meters
- Pitometers
- Magnetic meters
- Sonic meters
- Detector-check meters.

Positive-displacement meters. The most common type of meter for measuring water use through a customer service is a POSITIVE-DISPLACEMENT METER, which uses a MEASURING CHAMBER of known size to determine the volume of water flowing through it. There are two types of positive-displacement meters—the piston type and the nutating-disc type. In the PISTON METER (Figure 5-9), water flows into a chamber housing a piston. As it flows through the chamber, the

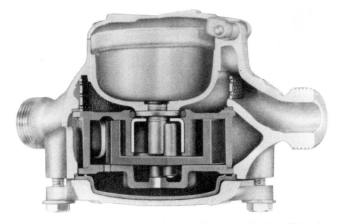

Courtesy of Badger Meter, Inc.

Figure 5-9.   Piston Meter

Courtesy of Badger Meter, Inc.

Figure 5-10.   Nutating-Disc Meter

piston is displaced. The motion of the piston is then transmitted to a REGISTER (via magnets in newer models or gears in older models) that records the volume of water flowing through the meter. The NUTATING-DISC (or WOBBLE) METER (Figure 5-10) uses a measuring chamber containing a hard rubber disc instead of a piston. When water flows into and through the chamber, the disc wobbles in proportion to the volume of water. This motion is then transmitted to a register that records the volume of water flowing through the meter.

Positive-displacement meters range in size from ⅝ in. to 6 in. The use of the smaller sizes (⅝ in. to 3 in.) is more widespread because of their excellent sensitivity at low rates of flow. In the larger sizes of positive-displacement meters,

Table 5-1.   Maximum Flow Rates—Displacement-Type Meters

Meter Size in.	Safe Maximum Operating Capacity gpm	Recommended Maximum Rate for Continuous Operations gpm
5/8	20	10
5/8 × 3/4	20	10
3/4	30	15
1	50	25
1½	100	50
2	160	80
3	300	150
4	500	250
6	1000	500

*Source: AWWA C700-77, Standard for Cold-Water Meters—Displacement Type.

sensitivity at low flows decreases to such an extent that it often becomes advisable to use another type of meter.

Positive-displacement meters, like most other types of water meters, will often underregister when excessively worn. UNDERREGISTRATION is a condition where the meter reading indicates less water than has actually passed through the meter. To avoid excessive wear and underregistration, positive-displacement meters should not be operated at a rate in excess of their stated maximum operating capacity. The safe maximum operating capacities listed in Table 5-1 are the maximum rates of flow at which water should be passed through the positive-displacement meter. These flow rates should be attained only for short periods of time at infrequent intervals. Continuous maximum flow would be destructive.

Positive-displacement meters are generally sized to one-half of the safe maximum operating capacity of the meter. Figure 5-11 shows a typical accuracy curve for a positive-displacement meter.

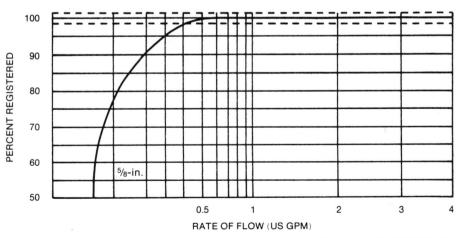

Courtesy of Badger Meter, Inc.

Figure 5-11.   Typical Accuracy Curve for Positive-Displacement Meter

Figure 5-12.   Turbine Meter

Current meters.   CURRENT METERS are generally used for measuring flows through 3-in. (76-mm) and larger lines. Current meters, also called VELOCITY METERS, include turbine, multijet, and propeller meters.

TURBINE METERS (Figure 5-12) have a rotor (bladed wheel) housed in a measuring chamber or passage; the wheel is turned by the flow of water. Over the optimum range of the meter, the volume of water recorded on the register of the meter is almost in direct proportion to the number of revolutions made by the rotor. Turbine meters have little friction loss and, as with other meters, must be sized to the flow. With turbine meters, water must be moving at a sufficient speed before the rotor will start to rotate. Turbine meters can overregister when the blades of the wheel become partially clogged or coated with sediment or foreign material. This causes a decrease in the cross-sectional area of the openings between the blades; as a result, for any given rate of flow, the water must pass through these openings at a higher velocity, causing overregistration. This condition can easily be kept under control by periodic inspection and testing. Present-day horizontal turbine meters have improved low-flow accuracy and higher maximum flows than previous instruments.

In MULTIJET METERS, the moving element is a multiblade rotor mounted on a vertical spindle within a cylindrical measuring chamber. Water enters the measuring chamber through several tangential orifices around the circumference and leaves the measuring chamber through another set of tangential orifices placed at a different level in the measuring chamber.

PROPELLER METERS (Figure 5-13) are like turbine meters in that the propeller is turned by the flow of the water and this movement is then transmitted to a register. Especially in larger sizes, the propeller may be small in diameter in relation to the internal diameter of the pipe. Propeller meters are primarily used for main-line measurement where flow rates do not change abruptly, since the propeller has a slight lag in starting and stopping. Propeller meters can be built

Courtesy of Badger Meter, Inc.

Figure 5-13.   Propeller Meter

within a section of pipe or they can be saddle mounted. Friction loss with a propeller meter is generally less than with a turbine meter.

Compound meters.   COMPOUND METERS (Figure 5-14) are used for customers whose water demand varies considerably, such as schools, hospitals, hotels, or office buildings. A compound meter consists of three parts—a turbine meter, a positive-displacement meter, and an automatic valve arrangement. The automatic valve enables high flows to be metered through the turbine side of the meter and low flows through the displacement side, thereby combining the favorable characteristics of turbine and displacement meters into one unit. Compound meters may have one or two registers, depending on the design. A compound meter is usually accurate over a wide range, with only a slight loss in accuracy at the changeover point between the two meters within the unit. The range of flows over which the inaccuracy occurs in the better meters is narrow and of no serious consequence, especially if the majority of the flow does not occur at the changeover point.

Proportional meters.   A PROPORTIONAL METER (Figure 5-15) uses a slightly reduced section in the line to divert a portion of water into a loop holding a turbine or displacement meter. Because the diverted flow in the loop is proportional to the flow in the main line, the turbine or displacement meter measuring the diverted flow can be adjusted by a multiplying factor to record the flow in the pipeline. Proportional meters are relatively accurate but are difficult to maintain. However, they have little friction loss and offer little obstruction to the flow of water.

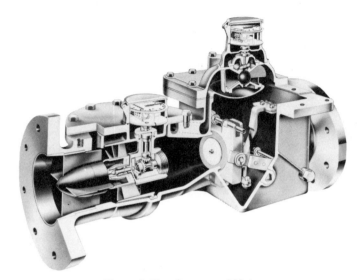

Figure 5-14.   Compound Meter

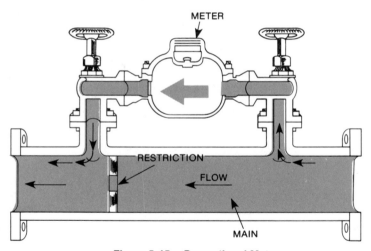

Figure 5-15.   Proportional Meter

Venturi meters.   A VENTURI METER (Figure 5-16) looks like an hourglass. The meter consists of an upstream reducer, a short throat piece, and a downstream expansion section for increasing the diameter from that of the throat to that of the downstream pipe. The amount of water passing through the meter is determined by comparing the pressure at the throat and at a point upstream from the throat. Electronic or mechanical instruments are used to record the flow rate and to totalize the flow. Venturi meters are accurate for a range of flows and have little friction loss.

A. Ready for Installation

B. Schematic

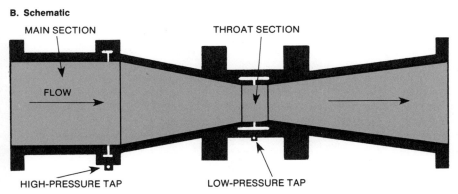

MAIN SECTION

THROAT SECTION

FLOW

HIGH-PRESSURE TAP

LOW-PRESSURE TAP

Courtesy of Badger Meter, Inc.

Figure 5-16.   Venturi Meter

Orifice meters.   The ORIFICE METER (Figure 5-17) is often called a THIN-PLATE ORIFICE METER because a thin plate with a circular hole in it is installed in the line between a set of flanges. As is the case with the venturi meter, the flow is determined by comparing the upstream line pressure with the reduced pressure at the restriction of the orifice. Pressure losses through the meter are more severe than for the venturi device, but an orifice occupies considerably less space.

Pitometers.   A PITOMETER indicates flow by comparing the height of water in a tube having an orifice facing upstream with the height of water in a second tube having an orifice pointing downstream. The water in the upstream tube will be higher because of velocity. Pitometers may be designed for easy field use, and they are often used by water suppliers to locate leaks or to measure the flow from fire hydrants.

Magnetic meters.   In a MAGNETIC METER, commonly called a "mag meter," a magnetic field is generated around an insulated section of pipe. Water passing through the magnetic field induces a small electric-current flow, proportional to

HIGH PRESSURE TAP   LOW PRESSURE TAP

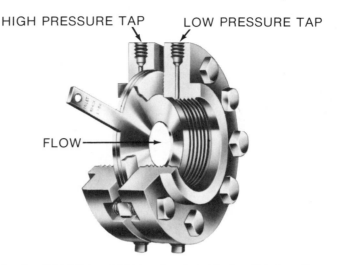

FLOW

Figure 5-17.   Orifice Meter

the water flow, between electrical contacts set into the pipe section. The electric current is measured and converted to a measure of water flow.

Sonic meters.   SONIC METERS contain sound generating and receiving sensors (transducers) attached to the sides of the pipe. Sound pulses are alternately sent in opposite diagonal directions across the pipe. Because of a phenomenon called the DOPPLER EFFECT, the frequency of the sound changes with the velocity of the water through which the sound waves are propagated. The difference between the frequency of the sound signal traveling with the flow of water and the signal traveling against the flow of water is an accurate indication of the water flow.

Detector-check meters.   DETECTOR-CHECK METERS are designed for services where daily use is to be measured but where emergency main-line flow will bypass the meter. The "detector check" consists of a weight-loaded check valve in the main line that remains closed under normal usage and a bypass around the valve containing a displacement-type meter to measure daily use. When the outlet of the service is opened to allow full flow, the weight-loaded valve detects the increased flow and opens; resistance in the main line from the open valve is negligible.

Meter Characteristics

Whatever type of meter is employed for a given situation, water suppliers generally look for certain characteristics in a meter. Desirable meter characteristics include the following:

- Accuracy within the range of anticipated flows
- Minimum head loss
- Durability

- Ease of repair
- Availability of spare parts
- Freedom from noise
- Reasonable cost.

Specific requirements in terms of accuracy, capacity, and head loss are found in AWWA standards for meters.

Meter Selection

There is no hard and fast rule for selecting the correct size of meter for a particular service. Proper meter selection involves both size and type considerations. The size of the meter is usually dependent on several factors, including:

- Expected customer demand for water
- Pressure considerations
- Friction losses in service line, meter, and customer plumbing
- The range of flow rates
- The allowable pressure losses.

Too often size is simply chosen to match pipe size. This can cause problems when the service is installed oversize to allow for possible future increases in water use or to reduce pressure loss in a long length of pipe.

Metering small services. The standard meter for 1-in. (25-mm) and smaller services (the usual residential sizes) is the positive-displacement meter—either disc or piston. Most residential services are metered with a ⅝-in. size meter having ¾-in. connections. Sometimes local conditions may dictate the need for a ¾-in. or 1-in. size meter. However, today's smaller (⅝-in.) magnetic drive meters have a very low pressure loss, and their life is much longer than that of older models since they withstand increased water usage better.

For services that can be metered with 1½-in. meters, the positive-displacement meter is also standard. The low-flow accuracy of modern 1½-in. displacement meters is excellent. Compound meters are not usually manufactured in sizes less than 2 in., and turbine meters are typically used only for larger flows.

Metering large services. For large (2-in. to 4-in. [50-mm to 100-mm]) services, the choice among displacement, compound, and turbine meters will depend on what average flow rates are to be measured. If the meter will usually operate at between 5 and 35 percent of maximum rated capacity and if accuracy at extremely low rates is not too important, then the displacement type should be used. If close accuracy at very low flows is important but large capacity is also needed, then the compound type would be the best choice, because it usually has lower pressure loss at high flow rates. If large capacity is of primary importance and if flows are usually above 10 or 15 percent of maximum rating with low flow accuracy secondary in importance, then the turbine type should be selected, especially one of the newer models with very low pressure loss.

5-6.   Meter Installation

Meters can be installed either indoors or outdoors. Indoor locations for small meters are often found in basements, utility rooms, garages, or undesirable locations such as crawl spaces. Outdoor installations are usually located underground in a meter pit or box at the curb end of the service line.

Large meters may be installed in vaults (precast or poured-in-place concrete), on pads, or in basements. Large meter installations are expensive and require considerable planning.

Determining Size and Type

Indoor settings are used in northern states where harsh winter weather can cause frost damage to the meter. Outdoor settings are used in warmer, more temperate, climates. Whether the meter is installed indoors or out, there are several general requirements for acceptable installation.

- The meter should not be subject to flooding with nonpotable water.
- The installation should provide an upstream and a downstream shut-off valve of high quality to isolate the meter for repairs.
- The installation should position the meter in a horizontal plane for optimum performance.
- The meter should be reasonably accessible for service and inspection.
- The location should provide for easy reading either directly or via a remote-reading device.
- The meter should be reasonably well protected against frost and mechanical damage.
- The meter should not be an obstacle or hazard to customer or public safety.
- The meter should be sealed to prevent tampering.
- There should be sufficient support for large meters to avoid stress on the pipe.
- There should be a bypass or multiple meters on large installations.

Good operating practice dictates that meters be tested for accuracy before installation to ensure that there is no shipping damage. Such testing not only ensures the accuracy of the meter but also provides a complete meter history for use in eventual maintenance and repair.

Some service conditions, such as a hospital or a phased-construction situation, may require that two or three meters be manifolded, or installed as a battery (Figure 5-18). For 3-in. (76-mm) lines, two 2-in. meters are used; for 4-in. (100-mm) lines, three or four 2-in. meters may be used. Meters used in batteries are most often the displacement type. For large lines, such as 6- to 10-in. (150- to 250-mm) lines, the battery may be composed of compound meters as needed. Details of such installations should be worked out with a representative of the

Figure 5-18.   Battery of Three Meters

meter manufacturer. A manifolded battery of 2-in. displacement meters is often used in 3-in. or 4-in. (76-mm or 100-mm) services for the following reasons:

- One meter can be removed for servicing simply by closing its valves, leaving the water supply in metered operation.

- Meters can be added and the system expanded if required in the future; this growth can best be provided for at the time of initial installation.

- There is no need to buy or stock parts for several different size meters; all meters and valves are the same size.

- The battery can be mounted on a sidewall to conserve floor space.

In a manifold unit with two or three meters, all but one of the meters will have a lightly loaded back-pressure valve on their outlets. When flow is small, only one meter will operate, ensuring the best accuracy. As flow increases, the back-pressure valves open to permit flow through the other meters.

Few displacement meters are made larger than 3 in., so when determining the size of meter for large services, the choice usually lies between a compound and a turbine meter. In the past, general commercial and industrial services were metered with compound meters because those units registered well over the widest flow ranges and had a relatively low pressure loss at fast flows. Today,

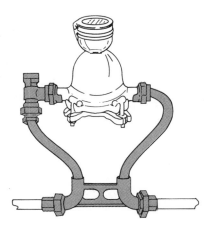

Figure 5-19.   Meter Yoke

modern horizontal-turbine meters are quite sensitive to low flows and often have a lower initial cost; therefore, many water suppliers are now installing turbine meters in situations where medium to high flow rates predominate and low flows are only occasionally encountered. In situations where low flow rates are common, the compound meter is advisable because of its accuracy. When using a large turbine or compound meter, water suppliers generally size the meter to have a maximum capacity of twice the expected flow. Tables are available showing the capacity of large meters and should be consulted when size determinations are made.

For main-line or pump-station measurement, propeller, venturi, proportional, or turbine meters are generally used, because the breakdown or failure of these meters does not interfere with the free flow of water. For temporary measurements in main lines, pitometers can also be used.

Yokes.   Meter connections vary with the size of the meter. Meters smaller than 1½ in. usually have screw type connections while meters larger than 1½ in. usually have flanged connections. Many water suppliers use a special device to simplify meter installation. This device is called a meter YOKE or HORN (Figure 5-19). Its purpose is to hold the stub ends of the pipe in proper alignment and spacing. Yokes cushion the meter against stress and strain in the pipe and provide electrical continuity when metal pipe is used. Electrical bonding around the meter reduces the chance of an electrical accident during meter changes. The electrical continuity around the meter must be broken when thawing frozen services using electricity.

Indoor installations.   The installation of a small meter indoors is a relatively simple job. When indoor meters are to be read directly, a drawing should be made showing basic installation requirements (Figure 5-20). This drawing should include minimum and maximum elevations above the floor, the type of recommended connections to be used (screwed versus flanged), the required valving (preferably before and after meter), and the minimum access space

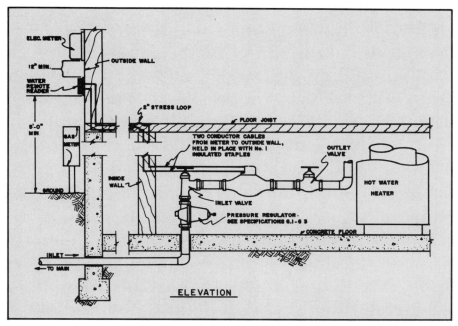

Courtesy of Colorado Springs Water Division, Colorado Springs, Colo.

Figure 5-20.   Sample Installation Drawing (Indoor Meter)

required for reading and servicing. The drawing should specify that the meter be located in the supply line as near as practical to the point where it enters the basement or building. If a yoke is not used, the drawing should require an electrical ground connection across the meter. Electrical grounding is a requirement specified by the National Electrical Code. An AWWA policy statement discourages the use of water systems as an electrical ground. Where the system is used as a ground, the current can flow through the service line creating an electrical hazard to employees removing the meter or repairing the service line and increasing the possibility of corrosion to the service line and connections.

The drawing of the meter location ensures proper installation for the mutual benefit of the water supplier and the customer. A copy of the drawing should be furnished to applicants for service, and compliance with the drawing should be made one of the conditions for acceptance to serve the premises.

Outdoor installations.   The installation of a small meter outdoors requires the use of a METER PIT (METER BOX, see Figure 5-21) to protect the meter. Many factors influence the design, materials, installation details, and overall performance of an outdoor setting. Considerations include soil conditions, groundwater level, maximum frost penetration, and accessibility for ease of reading and servicing. Regardless of pit depth, the meter itself should be located at a depth from the surface that makes reading and accessibility convenient. It is also desirable to provide 2–4 in. (50–100 mm) of clearance between the service line and the bottom of the meter pit to avoid any damage or strain that might occur if

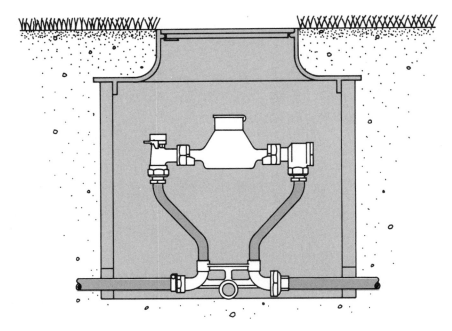

Figure 5-21.   Outdoor Meter Box

the meter box settles after installation. Consideration should also be given to the location of the curb stop, or service control valve. This valve may be made an integral part of the meter setting, or it may be located elsewhere and housed in a separate curb box.

The wide variations in ground-frost penetration throughout the country make it impossible to detail a universally practical outdoor setting. In areas where frost penetration is more than a few inches, serious attention to frost protection is required. Knowledge of local conditions is necessary to select a pit of sufficient size and depth to provide frost protection. Differences in rate of frost penetration depending on soil conditions complicate the problem. Experience with outdoor installations under a given set of conditions is the best guide for avoiding problems with freezing. Suggestions for meter pit design, including recommended size and depth, can be obtained from meter-box manufacturers.

When sufficient experience is available to establish standards, a suitable drawing should be prepared as a guide for further meter settings. As with the indoor setting, the drawing should be furnished to applicants for service, and compliance should be made one of the conditions for acceptance to serve the premises. In preparing the drawing, the following guidelines should be observed:

- The meter pit should be located as close as possible to the property line. The pit is the point at which the customer's responsibility starts.

- The lid of the pit or box should be placed flush with the ground surface.

- No portion of the riser piping or meter should be less than 1 to 2 in. (25 to 50 mm) from any portion of the meter box (more if required for frost protection).

- The distance below ground surface at which the meter spuds or couplings are to be located must be specified.

- The dimensions of the meter box to be used for each size meter must be specified.

- The location of the curb stop or service control valve must be specified.

There is no uniform standard for large meter installations. Some water suppliers prefer single meter installations and have had good success, whereas others use battery installations for measuring large flows. Whether single or battery, large meters are heavy and removal for maintenance or testing can be costly and time consuming. Large meter settings can have a variety of designs, depending on specific requirements, code specifications, and individual preferences. These settings represent sizeable investments that will provide long and satisfactory service if adequate planning is done in advance.

When large meters are installed in a vault, provisions should be made for at least 20 in. (510 mm) of clearance to the vertical vault walls and at least 24 in. (610 mm) of head space from the highest point on the meter to the vault cover. Test valves should be installed to permit volumetric tests, and provisions should be made for discharging test water. The meter and valves should be supported, and thrust blocking should be provided when necessary. All valves within the vault should rotate in the same direction for on–off operation. Valuable aid and installation recommendations can be obtained by contacting the meter manufacturer for recommendations as to material compatibility, compression couplings, check valves, and general vault installations. Such aid should be sought in the preliminary planning stages of any unusual installation.

5-7. Meter Reading

Meters record the flow of water in gallons, cubic feet, or cubic meters through the use of a register. Water meter registers are usually of two types—circular (Figure 5-22) or straight.

Direct Reading

The most common method of meter reading is the direct readout, which involves an individual going from one meter to another and reading the register directly. To read a circular register, the reader starts with the scale labeled with the highest number and successively reads the scales with lower numbers around the register until the last scale is read. When the hand on a scale is between two numbers, the lower number is read. When a hand is near a number, the next scale is checked—if the hand for that scale is not on or past zero, then the lower number on the dial in question should be read, since the hand is actually still between the two numbers. One complete revolution of each dial will equal the

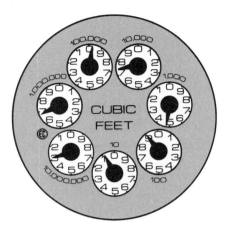

Figure 5-22.   Circular Meter Registers

number designated by that dial. On a 10,000,000 dial, one revolution equals 10,000,000; so numbers on the dial represent 1,000,000 through 9,000,000. On a 10 dial, the numbers correspond to one through nine. It is not necessary to read the dial below 10 (sometimes called the unit dial) if it exists, because the individual numbers are less than one. A system may choose to round off readings to the 100 or 1000. The number of digits in the total meter reading will be equal to the number of zeros on the largest numbered dial label.

Straight registers are much easier to read. On these units, the reader simply reads the number indicated on the counting wheels, including any fixed zeros to the right of the counting wheel window. Reading the circle around the circumference of the register is not necessary.

Occasionally a meter may have a MULTIPLIER on it. If so, it will be noted on the meter or the register face by a number such as "10X" or "1000X." This marking indicates that the reader must multiply the reading by the multiplier in order to obtain the correct reading.

Remote Reading

Direct reading is an inefficient system for many water suppliers, especially in northern climates where domestic meters are installed indoors and access problems cause skipped readings, call backs, and estimated billing. Furthermore, customers are often concerned with strangers entering their home, dirty shoes tracking up floors, and the general inconvenience of direct reading.

To eliminate the majority of the problems with direct reading and establish a more efficient means of meter reading, devices for reading meters using a register located at a distance from the meter were developed. A typical REMOTE-METER READING SYSTEM is shown in Figure 5-23.

The most common type of remote setup is an electric one, although pneumatic and flexible cables are also available. The electric-type remote reader has a pulse generator, or transmitting head, installed directly on the meter. When water is

Courtesy of Badger Meter, Inc.

Figure 5-23.   Remote-Reading Meter

flowing through the meter, the motion of the measuring element operates a gear train that stores energy in a spring-release mechanism. This energy is accumulated very slowly over a period of either 10 cu ft (0.3 m³) or 100 gal (380 L). When this point is reached, the energy is released by a mechanical mechanism that creates rapid relative motion between a permanent magnet and a copper coil. This sudden motion passes copper conductors through a permanent magnetic field, generating an electric impulse that is transmitted through a two-conductor electric cable to the remote register. The remote register contains a counter that advances one digit for each pulse received.

Remote reading devices are quite popular in northern states, and the use of remote registration has been increasing in areas where pit settings have been traditional, both in residential and in commercial and industrial installations. For example, in areas where large meters are installed under the street, it is often necessary to send a truck and a crew to direct traffic, manipulate pit entries, and then read the meters. Conveniently located remote registers make it possible for one person to obtain these meter readings without disruption of traffic or exposure to personal injury.

When considering the installation of remote-registration systems, the water supplier should make a thorough study of all the related factors. Much of this information is available on request from meter manufacturers, including:

- Comparisons of new meter setting costs with and without remote reading

- Desirability and economics of retrofitting existing installations

- Interchangeability within meter models and brands used by the water supplier

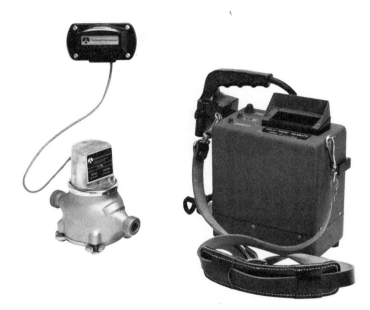

Figure 5-24.   Remote-Reading Meter With Plug-in Receptacle

- Compatibility of the reading data obtained if both direct and remote systems are intermixed in the same route
- Original equipment purchase price
- Installation costs
- Probable maintenance costs (if any)
- Ultimate result on the cost of obtaining and processing meter-reading data.

Several manufacturers offer remote-meter-reading systems with a plug-in receptacle (Figure 5-24). In this type of system, the meter reading is converted into an encoded electrical message within the meter head, which is connected by a multiconductor cable to a receptacle located on an outside wall of the house. A meter reader can then plug a battery-operated reading device into the remote receptacle. On connection, electrical circuits are completed and the battery-operated reading device obtains a record of the meter registration. This reading may be recorded on magnetic tape, in punched card form, or translated directly into a visual reading. (Receptacle designs differ with manufacturers, and reading devices made by one manufacturer may not fit plug-in receptacles made by another.)

Automatic Readings

The ultimate system for meter reading is one where readings are obtained directly from a central reading station. Several manufacturers have demonstrated that meter readings can be obtained through the telephone system. Other approaches to centralized meter reading include the use of electric-power distribution networks, sound transmission through water lines, and radio-equipped trucks that would pick up meter readings while driving down the street. The potential for such systems is enormous; however, their use is still in the developmental stage.

5-8. Meter Testing, Maintenance, and Repair

Water meters should be tested before use, on removal from service, on customer request or complaint, and after any repair or maintenance.

Testing New Meters

New water meters are susceptible to damage from rough roads, mishandling, or general negligence from the time they leave the factory to the time of arrival at the water utility. Testing new meters enables water suppliers to identify damaged meters and limit registration error. Because of the time involved in testing new meters, some state regulatory commissions have adopted regulations that allow a certain number of random meters in a shipment to be tested to determine the accuracy of the entire shipment. For example, if meters are shipped in boxes of 10, a water supplier might test only one random meter in that box. If that meter tests out accurately, then the whole box could be assumed accurate. If that meter does not test out accurately, then the water supplier would have to test all the other meters in the box or return the box of meters to the meter supplier for replacement. The state public utility regulatory commission should be consulted concerning such a regulation; if no such regulation is in force, all new meters should be tested before installation.

Frequency Requirements for Testing

Water meters within the system are subject to wear and deterioration from corrosion, deposition, or the chemical character of the water. Over a period of time, meter wear and deterioration lead to efficiency loss, and meters must, therefore, be field tested or brought in periodically for testing and repair. The normal result of meter deterioration is underregistration. This, in turn, leads to loss of revenue for the water supplier. Very rarely will a meter overregister. To help control meter deterioration and limit registration problems, many state regulatory commissions have adopted requirements for the frequency of meter testing. Because of the variability of water characteristics, the cost of testing, and water costs, a nationwide test frequency cannot be established. Most states with meter test-frequency requirements have based the test intervals on the size of the meter—the larger the meter the more often it should be tested. AWWA recommends that $\frac{5}{8}$-in. meters be tested every 5 to 10 years. The state public utility regulatory commission should be consulted for testing requirements in a given area.

Whether a system tests its own meters or has someone else do it depends on the staff size, time available, and facilities. Some water distribution systems have facilities to test all their meters, but the majority of systems generally test only smaller meters (⅝-in. to 2-in.). Larger meters may be field tested or sent to the manufacturer for testing. Some small water distribution systems do not have the facilities or time to test meters. These systems send meters to a nearby system or to the meter factory for testing.

Testing Procedures

There are three basic elements to a meter test:

- Running a number of different rates of flow over the operating range of the meter to determine overall meter efficiency
- Passing known quantities of water through the meter at various test rates to provide a reasonable determination of meter registration
- Meeting accuracy limits on different rates for acceptable use.

Rates of flow generally used in testing positive-displacement meters are maximum, intermediate, and minimum. For current and compound meters, four or five flow rates are usually run. State regulations should be consulted for suitable flow rates.

The quantity of water used to determine the registration accuracy depends on the accuracy of the test equipment and the accuracy with which the meter can be read. For recommended accuracy limits, consult state public-utility regulations or AWWA Manual M6, Water Meters—Selection, Installation, Testing, and Maintenance.

Equipment for meter testing does not have to be elaborate. A setup like the one shown in Figure 5-25 is suitable. For a complete discussion of meter testing, refer to AWWA Manual M6 or the literature accompanying a meter test bench.

It is generally more economical to field-test large meters, especially current and compound types. Field-testing of meters is essentially the same as shop testing, except that, instead of using a tank to measure the test water, a comparison is made between the meter to be tested and one that has been previously calibrated. The two meters are connected in series and the test water is discharged to waste. Since the calibrated meter is not 100 percent accurate at all flows, it is necessary to use a special calibration curve to adjust for different rates of flow in order to compute the accuracy of the meter being tested.

Some water suppliers do not field-test large meters over their entire operating range, but rather only check the lower portion (5–10 percent) of the rated or operating range. Such situations arise because discharging large quantities of water used in high test-flows presents a problem. The assumption is that the test curve will flatten out after reaching a peak registration that is approximately 10 percent of rated capacity and then stay within required limits of registration.

For field-testing to be accurate, both meters must be full of water and under positive pressure. The control valve for regulating flow should always be on the discharge side of the calibrated meter. A control valve on the inlet side of the meter configuration or one located between the two meters should not be used.

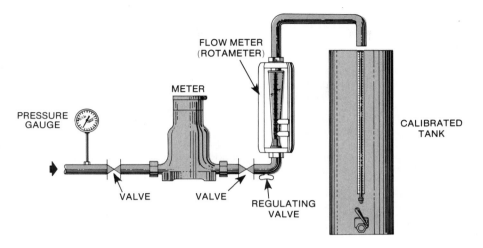

Figure 5-25.   Meter Testing Equipment

Maintenance and Repair

Maintenance and repair of positive-displacement meters involves dismantling the meter; thoroughly cleaning it; inspecting all parts for wear, pitting, and distortion; replacing or repairing parts as necessary; reassembling the meter; and then testing it. Maintenance and repair of other meters generally follows the same procedure taking into account variations in design.

Before undertaking a maintenance and repair program, water suppliers should evaluate the cost effectiveness of a meter replacement program versus a meter repair program. Some systems find it more economical to simply replace old parts with new parts or old meters with new meters rather than checking tolerances, shimming discs, or repairing registers.

For those systems that find it cost effective to repair old meters, the steps outlined in Table 5-2 are general guidelines for maintenance and repair of positive displacement meters. Meter manufacturers can supply instructions specific to a given unit.

5-9.   Record Keeping for Services and Meters

It is important to maintain complete records of all services installed so that the services can be located in the future. There are several methods of maintaining service records. A service application book can be used, in which the applications are numbered and the location of the premises and the applicant's name are given. Usually, however, an index card file is maintained. Pertinent information concerning each service installation is recorded on a card as it is received. Such information would include the permanent service number, the applicant's name and address, the dates of application and installation of service, sizes of corporation and curb stops used, size and kind of pipe used, depth at which the

Table 5-2.   Meter Repair Procedures

1. Remove the meter from service and perform an initial test. Record test results.

2. Clean the exterior of the meter. Bronze meter exteriors can be cleaned by shot-blasting, buffing, brushing, or hard scrubbing. If buffing is used, make sure the work area is well-ventilated. Plastic meter exteriors can be cleaned by hand scrubbing. Meters that will be serviced immediately should be filled with water to keep the inside deposits soft.

3. After cleaning the exterior, dismantle the meter into its basic components—casing, bottom, chamber, gear train, and register. Pneumatic tools are preferred over electric because they involve no hazard of electric shock. Keep the meter components together.

4. Examine the casing and meter bottom for distortion, cracks, or signs of fatigue. Replace as necessary.

5. Examine the chamber. Check the inside surfaces and wall socket for pit holes or wear.

6. On nutating-disc meters, check the disc for warpage, deposition, and any friction or slippage during movement. Check the ball on the disc for wear and ease of movement. (New meters can be used for tolerance and ease-of-movement comparisons.) On oscillating-piston meters, check for piston friction or slippage and deposition. Check the disc or piston spindle for wear.

7. Clean, replace, or repair any parts as necessary.

8. Examine mechanical and magnetic gear trains for ease of operation. Older mechanical gear trains may be changed to magnetic gear drive if desirable. This changeover can be done either by water supply personnel or the meter manufacturer.

9. Examine the registers. Straight registers should be checked for loose test hands, worn worm gears, worn wheels, broken tumblers, and bent main and tumbler shafts. On circular registers, cracked or faded register faces should be replaced and loose hands should be resoldered.

10. Reassemble the meter.

11. Test the meter again.

---

service was laid, and detailed measurements of the service location. This file can be supported by a large-scale map on which the service is plotted in varying degrees of detail as necessary. A sample service card can be found in Module 10, Maps, Drawings, and Records.

Some water suppliers use a computer system for record keeping. In such a system, information generally kept on a service card can be fed into a computer, using a control number to establish a permanent record. Information concerning the service can be added or changed via the control number.

Meter records are an essential part of any water distribution system that meters its supply. Meter records should provide information on the installation, repair, and testing of each meter. Such records enable field crews to easily locate meters, help repair personnel with meter testing, and aid managers in assessing the value of the system.

One method for maintaining meter records is the use of a meter history card. Meter history cards provide information on the size, make, type, date of purchase, location, and any testing or repair on the meter. Basic data is inserted at the top of the card and the remaining portion of the card is divided up to record, in chronological order, the location of the meter and any test and repair work (Figure 5-26). Each line of the test and repair section is usually divided into two segments, the upper being used to record test results when the meter was removed and the lower being used to record test results before the meter is reinstalled after maintenance or repair.

In a small to medium-size shop, test and repair information should be entered on the meter history card by personnel in the meter repair shop. After recording all information, meter history cards should be stored in a safe, permanent file in the shop area. Meter history cards should be filed in sequence, either by the manufacturer's serial number or by the water system's number (if the system has its own numbering arrangement) to enable quick access to the history card.

Many water suppliers are now using computers in their operation. Information commonly kept on a service or meter history card can just as easily be entered into a computer using a control number to establish a permanent record. Any future information concerning work on customer service lines or meter testing and repairs can also be entered using the appropriate control number.

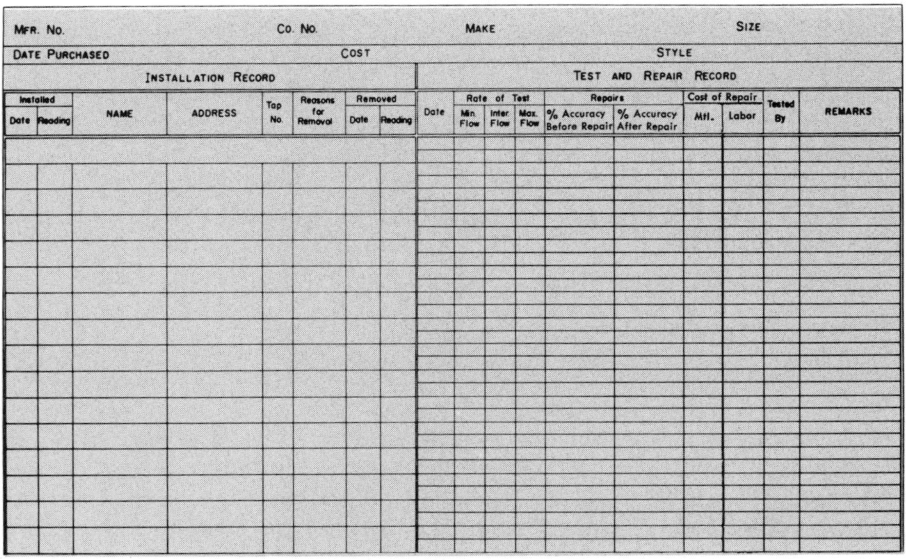

MFR. NO. CO. NO. MAKE SIZE

DATE PURCHASED COST STYLE

INSTALLATION RECORD									TEST AND REPAIR RECORD									
Installed		NAME	ADDRESS	Tap No.	Reasons for Removal	Removed		Date	Rate of Test			Repairs		Cost of Repair		Tested By	REMARKS
Date	Reading					Date	Reading		Min. Flow	Inter. Flow	Max. Flow	% Accuracy Before Repair	% Accuracy After Repair	Mtl.	Labor		

Figure 5-26. Meter History Card

5-10.   Safety

Specific areas where the need for safety precautions must be recognized and observed when dealing with services and meters include:

- Barricades, warning signs, and traffic control (refer to Module 1)
- Personal-protection equipment (refer to Module 1)
- Hand tools (refer to Module 1)
- Portable power tools (refer to Module 1)
- Confined spaces (refer to Module 3)
- Manholes and vaults (refer to Module 3)
- Grounding
- Thawing frozen services.

Grounding

It is common for home electrical circuits to be grounded to the home plumbing system. If any electrical problems exist in wiring or appliances being used in the home, then a current flow can exist through the meter (Figure 5-27). When the meter is removed, this current flow is broken ,and a dangerous potential voltage exists between the inlet and outlet connections where the meter was removed. An operator who bridges the connections by touching the inlet and outlet at the same time can receive a fatal shock.

To prevent this type of injury, many meters are installed using a yoke (see Figure 5-19), which maintains the electrical connection even after the meter is removed. Where a yoke is not installed, the operator should use a cable to jumper between the inlet and outlet pipes before removing the meter. The jumper will maintain the electrical connection and reduce the chances of injury.

As a further safety precaution, even where a yoke or jumper has been installed, the operator should take care to not bridge the connection by touching the inlet pipe with one hand while touching the outlet pipe with the other hand.

AWWA, by an official policy statement, has opposed the grounding of electrical systems to water piping. Unfortunately, the problem still exists in many buildings, both new and old.

Thawing Frozen Services

Electrical thawing of frozen service lines is a very dangerous operation. The path of the current flow must always be verified, because poor electrical conductivity of the service pipe can cause the current to find an alternate path. These stray currents can be very dangerous. They can cause nearby electrical system wire to overheat, creating fires; they can overheat sewer laterals, causing leaks; they can follow gas piping, creating a potential gas-explosion hazard; they can damage electrical wiring and appliances; and they can cause personal injury.

When electrical thawing must be used, it is extremely important that the following precautions are observed:

- A trained supervisor must be on hand at all times during the thawing operation.

- Occupants of the building must be warned not to use electrical appliances or plumbing during the thawing operation, since disconnection of the external ground may cause hazards.

- Operators should use a temporary jumper cable when removing the water meter, and they should take care to avoid personal contacts bridging the open section of water pipe, since stray currents could cause electrocution. Owners and tenants should be warned to stay away from the open section as well.

- All ground connections from the house to the service must be removed before thawing so the frozen service will be the only element in the circuit. If ammeter readings indicate that not all the current generated by the welder is passing through the service, the thawing operation must be stopped until the service can be fully isolated.

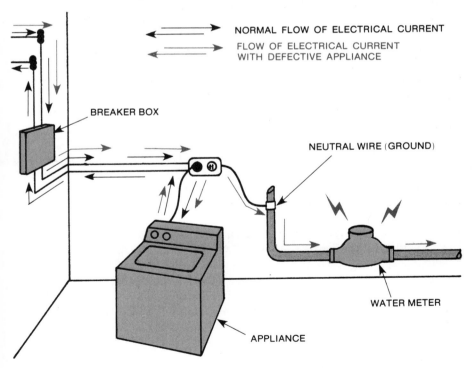

NORMAL FLOW OF ELECTRICAL CURRENT

FLOW OF ELECTRICAL CURRENT WITH DEFECTIVE APPLIANCE

BREAKER BOX

NEUTRAL WIRE (GROUND)

WATER METER

APPLIANCE

Figure 5-27. Possible Electrical-Current Flow Through Water Meter

Selected Supplementary Readings

Accurate Flow Measurement A Necessity. *OpFlow*, 11:2:4 (Feb. 1985).

All You Ever Wanted to Know About Meter Installation. *OpFlow*, 7:8:4 (Aug. 1981).

Anderson, G.D. Large Meters: Their Selection, Installation, and Maintenance for Maximum Service. Proc. AWWA DSS, Birmingham, Ala. (Sept. 1983).

Benson, E.M. Meter Size and Selection for Large Commercial and Industrial Services. Proc. AWWA DSS, Los Angeles, Calif. (Feb. 1980).

Christenson, E.A. Select Meters to Fit Needs Based on Expected Demand. *OpFlow*, 5:7:1 (July 1979).

Cold-Water Meters—Compound Type. AWWA Standard C702-78. AWWA, Denver, Colo. (1978).

Cold-Water Meters—Displacement Type. AWWA Standard C700-77. AWWA, Denver, Colo. (1977).

Cold-Water Meters—Fire Service Type. AWWA Standard C703-79. AWWA, Denver, Colo. (1979).

Cold-Water Meters—Multi-Jet Type for Customer Service. AWWA Standard C708-82. AWWA, Denver, Colo. (1982).

Cold-Water Meters—Propeller Type for Main Line Applications. AWWA Standard C704-70(R75). AWWA, Denver, Colo. (reaffirmed, 1975).

Cold-Water Meters—Turbine Type for Customer Service. AWWA Standard C701-78. AWWA, Denver, Colo. (1978).

Direct-Reading Remote Registration Systems for Cold-Water Meters. AWWA Standard C706-80. AWWA, Denver, Colo. (1980).

Edgar, J.T. Maintaining Your Commercial/Industrial Meters Doesn't Cost—It Pays. *OpFlow*, 4:1:3 (Jan. 1978).

Electrical Thawing of Frozen Pipe Systems. *OpFlow*, 3:11:5 (Nov. 1977).

Encoder-Type Remote-Registration Systems for Cold-Water Meters. AWWA Standard C707-82. AWWA, Denver, Colo. (1982).

Evaluating Your Water Utility. *OpFlow*, 3:8:3 (Aug. 1977).

Holloway, L.J. Copper Service Line Installation. Proc. AWWA DSS, Los Angeles, Calif. (Feb. 1980).

Hot H_2O—Winter Potion for Thawing Out Pipes. *OpFlow*, 8:12:1 (Dec. 1982).

Hudson, W.D. Field Testing of Large Meters. *Jour. AWWA*, 58:7:867 (July 1966).

Lacina, W.V. and Coel, J.B. Plastic Water Meters. *Jour. AWWA*, 68:5:246 (May 1976).

Lee, R.L. Plastic Service Line Installations. Proc. AWWA DSS, Los Angeles, Calif. (Feb. 1980).

Maz, A.C. Reducing Meter Leaks Cuts Claims for Water Damage. *OpFlow*, 2:5:1 (May 1976).

Moodhe, N.S. Correct Meter Sizing. *Jour. AWWA*, 59:1:43 (Jan. 1967).

Moore, Floyd. Meter Test Bench Ensures Accuracy. *OpFlow*, 11:5:3 (May 1985).

Nelson, L.M. Arc Welder Used to Thaw Frozen Pipes. *OpFlow*, 10:1:6 (Jan. 1984).

Newman, G.J. and Noss, R.R. Domestic ⅝ Inch Meter Accuracy and Testing, Repair, Replacement Programs. Proc. AWWA Ann. Conf., Miami Beach, Fla. (May 1982).

Power, J.A. Written Procedures and Records Needed for Routine Connections. *OpFlow*, 6:11:3 (Dec. 1980).

Shierry, R.S. Apartment and Condominium Metering. *Jour. AWWA*, 69:2:73 (Feb. 1977).

Sizing Water Service Lines and Meters. AWWA Manual M22. AWWA, Denver, Colo. (1975).

Underground Service Line Valves and Fittings. AWWA Standard C800-84. AWWA, Denver, Colo. (1984).

Walker, M.J. Water Service Connections. Proc. AWWA DSS, Los Angeles, Calif. (Feb. 1980).

Water Meters—Selection, Installation, Testing, and Maintenance. AWWA Manual M6. AWWA, Denver, Colo. (1973).

Young, M.B. Water Metering Accuracy Vital to Water Industry. *OpFlow*, 9:8:1 (Aug. 1983).

Glossary Terms Introduced in Module 5

(Terms are defined in the Glossary at the back of the book.)

Cavitation
Compound meter
Corporation stop (cock)
Curb box
Curb stop (cock)
Current meter
Detector-check meter
Doppler effect
Dry tap
Galvanic corrosion
Gooseneck
Horn
Magnetic meter
Measuring chamber
Meter box
Meter pit
Multijet meter
Multiplier
Nutating-disc meter
Orifice meter

Piston meter
Pitometer
Positive-displacement meter
Propeller meter
Proportional meter
Register
Remote-meter reading system
Saddle
Service clamp
Service line
Sonic meter
Thin-plate orifice meter
Turbine meter
Underregistration
Velocity meter
Venturi meter
Water meter
Wet tap
Wobble meter
Yoke

Review Questions

(Answers to Review Questions are given at the back of the book.)

1. What is the purpose of a water service line?

2. What is the function of a gooseneck in a service line?

3. What is the function of a curb stop?

4. What is the purpose of standardizing water service lines?

5. What factors should a water supplier consider when selecting service line materials?

6. Identify the two most popular materials used for residential water services.

7. Explain why lead and wrought iron are no longer used for residential service lines.

8. What materials are commonly used for large water services?

9. What is a possible problem when iron services are installed with bronze curb stops?

10. Explain the difference between a wet and a dry tap.

11. What are two controlling factors in determining service line sizes?

12. What two factors must water suppliers consider when determining the depth and location of a service line?

13. What are three reasons for metering water customers?

14. Identify three meters commonly used in the water distribution system.

15. Name and describe the operation of two major types of positive-displacement meters.

16. What is the most common application for a small positive-displacement meter?

17. Compound meters are generally used under what conditions?

18. What types of meters might be used for main line or pump station measurement?

19. What factors must be considered in selecting the correct size and type of meter?

20. What are the requirements for acceptable meter installations?

21. What is a meter yoke?

22. Explain the need for maintaining electrical continuity around the meter during removal.

23. List four reasons for a manifold, or battery, meter installation.

24. What are the major advantages of remote-reading meters?

25. When should water meters be tested?

26. List three basic elements in a meter test.

27. What items should be recorded on a service connection record card?

28. What items should be recorded on a meter history card?

29. What hazards are associated with electrically thawing a frozen service line?

Study Problems and Exercises

1. Summarize any state and/or local regulations governing separation of water service lines and other utility lines (such as sewer, gas, and electric). Discuss why these separations are necessary.

2. Compare the advantages and disadvantages of using plastic and copper tubing for residential service lines.

3. From historical information, determine the frost penetration in your locale or region. On the basis of the information collected, recommend the depth for installing service lines in your system's service area.

4. Design and sketch a typical residential service line for your system showing all the necessary features.

5. Locate your state meter-testing-frequency requirements and determine how many meters your system should test each year to meet the requirements. As a matter of good operating practice, how many meters would you recommend be tested annually and why?

Module 6

Cross-Connection Control

As potable water is transported from the treatment facility to the user, opportunities exist for unwanted substances to contaminate it. One common means for such contamination is by backflow of nonpotable fluids through cross connections into the potable water system. It is the utility operator's responsibility to ensure that water is protected from contamination from the time it leaves the treatment plant until it reaches the user. Operators of all utilities need to be concerned with this problem, because cross connections and backflow can and do occur in systems ranging in size from those serving only a few customers to large municipal water systems. The degree of an operator's involvement may vary from little or no responsibility to sole responsibility for planning and operating a program designed to prevent the backflow of contaminating material into the distribution system. To meet this responsibility, the operator needs information on cross-connections and how they occur, how backflow can occur, and what preventive measures can be taken to ensure a safe, potable water supply.

After completing this module you should be able to

- Identify public health hazards that are created by cross connections.

- Explain the principles of pressure and state what other physical factors are involved that allow or cause backflow, back pressure, or backsiphonage to occur.

- List various installations or facilities that are likely to have cross connections and recognize cross connections according to their potential hazard.

- Recommend appropriate backflow-prevention devices according to the degree of hazard involved.

- Perform a routine cross-connection inspection and complete a cross-connection survey with recommendations.

- Outline steps to establish and administer a cross-connection control program.

6-1. Cross Connections

The following definitions are important since the conditions they describe aid in the understanding and elimination of potential hazards posed by cross connections. State or local statutes may define these conditions more specifically.

Backflow. BACKFLOW is the flow of any water, foreign liquids, gases, or other substances back into a potable water system. Two conditions that can cause backflow are:

- BACK PRESSURE—a condition in which a pump, elevated tank, boiler, or other means results in a pressure greater than the supply pressure.

- BACKSIPHONAGE—a condition in which the pressure in the distribution system is less than atmospheric pressure.

Cross connection. A CROSS CONNECTION is any (actual or potential) connection or structural arrangement between a potable water system and any other water source or system through which backflow can occur.

The existence of a cross connection does not always result in backflow; but where a cross connection exists, the potential for backflow is always present, and given either of the other conditions (back pressure or backsiphonage), backflow may occur.

Where and How Cross Connections Occur

Cross connections can be found in almost any type of facility where water is used, including houses, factories, restaurants, hospitals, laboratories, and water and wastewater treatment plants. Many pieces of equipment within these facilities withdraw water from the potable water distribution system for use in cooling, lubricating, washing, or as a SOLVENT. Because water is used, these pieces of equipment are susceptible to cross connections. An operator should be able to identify these actual or potential problems during inspections and surveys. Table 6-1 provides a partial list of the types of fixtures that can have cross connections and could pose a hazard to the potable water supply.

Cross connections are frequently created by individuals who are not familiar with the hazards involved, even though they may be otherwise well-trained and experienced in plumbing, steam-fitting, or water distribution work. Many such connections are made simply as a matter of convenience without regard for the problems that may result. Often, there is also a tendency to rely on simple protection such as a single valve whose reliability may be suspect or short lived.

Table 6-1.   Some Cross Connections and Potential Hazards

Description	Hazard	Description	Hazard
Sewage pumps	High	Sprinkler systems	High
Boilers	High	Water systems	Low to high
Cooling towers	High	Swimming pools	Moderate
Flush valve toilets	High	Plating vats	High
Garden hose (sill cocks)	Low to high	Laboratory	High
Auxiliary water supply	Low to high	glassware	
Aspirators	High	washing equipment	
Dishwashers	Moderate	Pump primers	Moderate
Car wash	Moderate to high		to high
Photographic developers	Moderate to high	Baptismal founts	Moderate
Commercial food	Low to moderate	Manhole flush	High
processors		Agricultural	High
Sinks	High	pesticide mixing	
Chlorinators	High	tanks	
Solar energy systems	Low to high	Irrigation systems	Low to high
Sterilizers	High	Watering troughs	Moderate

Description of a Cross Connection

A cross connection will appear to be simply a pipe, hose connection, or any water outlet with no specific or outstanding features. The real distinction is where the connection leads. It is a cross connection if it leads from a potable line to anything other than potable service. Figures 6-1 through 6-9 show cross connections that can be found in many different places, such as homes, farms, laboratories, and factories. Each of these connections leads to some vessel containing materials that should not be allowed to enter the potable water system. Recommendations on how the cross connections shown in these figures can be eliminated or protected against are discussed in Section 6-4, Backflow Control Methods and Devices.

Cross connections can also be described as ACTUAL (where the connection is in existence at all times) or POTENTIAL (where something must be done to complete the cross connection). Examples of an actual cross connection would be solid piping to an AUXILIARY SUPPLY (Figure 6-1) or into a boiler (Figure 6-2). The

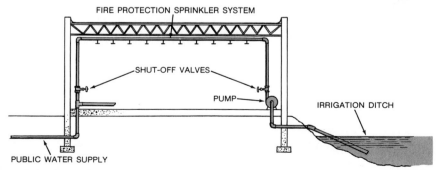

Figure 6-1.   Cross Connection Between Potable Water Supply and Auxiliary Water for Fire Protection Sprinkler System

Figure 6-2. Boiler Feed-Water Cross Connection

Figure 6-3. Slop-Sink Hose Cross Connection

auxiliary supply in Figure 6-1 does not have to be for fire control; it could be a nonpotable auxiliary used for emergencies or for reducing the cost of cooling, washing, etc. The diagram could also apply to older homes that have retained their private water supply but have also been connected to a municipal system.

Figure 6-2 shows a FEED WATER connection to a boiler. This connection is through a single CHECK VALVE, which is not considered adequate protection. Therefore, it must be described as a potentially hazardous cross connection. In Figures 6-3 and 6-4, the slop sink and water tank are examples of potential cross connections. In both cases water has to be added to the vessel before the connection is completed. Although the cross connection shown in Figure 6-4 appears to pose a very low hazard, the tank could be used for the mixing of a pesticide, weed killer, or fertilizer, which would greatly increase the danger. The results of two such connections are described in Section 6-2, Public Health Significance.

Figure 6-5 depicts a very common cross connection, formed by adding the chemical dispenser to a garden hose and attaching the hose to a sill cock. Figure 6-6 shows an unprotected metered hose from the dock to the ship. Ships in port are a possible source of hazard to the potable water supply because their on-board high pressure fire systems use sea water. Cooking vessels in hospitals, restaurants, and canneries and chemical reaction tanks such as those shown in Figure 6-7 use potable water. If these inlets are below the OVERFLOW RIM, they can become a cross connection.

In hard-water areas, many buildings, particularly homes, have individual water softeners. The softeners form a cross connection that is not normally hazardous. However, when the drain is connected directly to the sanitary sewer, as shown in Figure 6-8, it becomes potentially very hazardous to both the residents and the community water supply. At least 36 persons were struck by a HEPATITIS epidemic in California as a result of an arrangement similar to that shown in the photograph.

Figure 6-4. Water Tank Cross Connection

Figure 6-5. Garden Hose Cross Connection

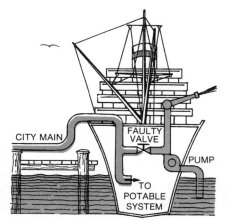

Figure 6-6. Unprotected Potable Water Supply From Dock to Ship in Port

Figure 6-7. Cooking Vessel With Water Inlet Below Overflow Rim

Figure 6-9 illustrates a cross connection where the water supply has been protected. An atmospheric vacuum breaker is attached between the soap dispenser and the water faucet. (Atmospheric vacuum breakers will be discussed in detail in the protection-equipment section.)

6-2. Public Health Significance

A water utility goes to great lengths to ensure that the water leaving a treatment plant is free from disease-causing microorganisms, odors, colors, and other unwanted materials. A municipality constantly repairs the distribution system for the same reasons. In spite of these efforts, disease outbreaks, poisonings, and water quality complaints still occur, many of which are attributable to backflow.

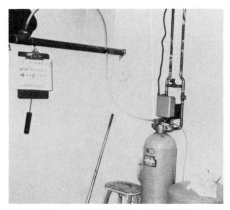

Figure 6-8.   Water Softener Cross Connection

Figure 6-9.   Atmospheric Vacuum Breaker Installed Between Soap Dispenser and Water Faucet

These outbreaks still occur as a result of

- Improper design and construction of water distribution and plumbing systems
- Modifications to existing systems and structures
- Efforts to economize by using less expensive (and less reliable) plumbing and backflow-prevention devices
- Failure to test and repair installed backflow-prevention devices.

Diseases Attributable to Cross Connections and Backflow

A number of diseases are known to be carried or spread by water. Of these, several have been traced to water contamination through cross connection and backflow. Table 6-2 lists the diseases and their incidence. Only those outbreaks where the number of cases and their outcomes were documented are listed. The general consensus is that many more incidents have occurred than are reported or definitely traced to backflow.

Several case histories are described in the following paragraphs to illustrate the various types of backflow and the seriousness of their consequences. In each case, a public water system was involved in spreading the disease. It is useful to note that all sizes of systems are vulnerable to these problems and that such incidents continue to occur.

Dysentery.   One of the most publicized outbreaks of DYSENTERY ever recorded in the United States was attributed to cross-connected water distribution systems in two large Chicago hotels during the 1933 World's Fair. This incident resulted in over 1400 persons being infected with amebic dysentery, and 98 of those stricken are known to have died directly from the illness. A special committee conducted an investigation of this outbreak and determined that cross connections were the probable cause. Numerous cross connections were found as a result of improper connections in toilet bowls, submerged bathtub

Table 6-2.   Backflow Related Disease Outbreaks (1920-1980)

Disease	Cases	Deaths	Notes
Typhoid fever	3934	292	
Dysentery and gastroenteritis	55 738	98	
Salmonellosis	750	0	
Polio	26	2	18 Paralyzed
Hepatitis	126	0	Incapacitated a football team
Brucellosis (undulant fever)	801	0	

inlets (some of which were found to be clogged), and connections between potable and nonpotable water lines.

Gastroenteritis.   A homeowner in Shreveport, La., installed and then modified a lawn sprinkling system over a period of years. When the system was finally completed, it was only partially in conformance with the local plumbing code. The approved portion of the system was supplied with potable water from the municipal system, while other portions were supplied from a nearby bayou. In 1954, the entire system was interconnected using only a manually operated valve for separation. The sections supplied from the bayou operated at a higher pressure than the public system. In 1956, a broken sewer line near the home drained raw sewage into the bayou. Within four days an outbreak of GASTROENTERITIS occurred in the neighborhood, starting with the owner of the interconnected sprinkling system. Investigation found the manual valve to be open and city water in the area to be heavily contaminated. A total of 72 cases (none fatal) of gastroenteritis resulted from the cross connection and the resulting backflow. In this case the cross connection had existed for about two years before a combination of circumstances resulted in a serious disease outbreak.

Hepatitis.   A hepatitis outbreak resulting from a cross connection affected most of the members of a college football team in September 1969. The disease CARRIERS evidently urinated in irrigation pits at the practice field. The same water line that supplied water for irrigation also supplied drinking water to a nearby equipment building. Approximately 25 days before the first case appeared, a fire occurred nearby and large amounts of water were drawn for firefighting. The fire demand created a backsiphonage condition at the practice field with presumed contamination of the team's drinking water. The 25-day time delay coincided with the INCUBATION PERIOD of infectious hepatitis. Ninety cases developed from this outbreak, and the college had to cancel their remaining football games for that season.

Salmonellosis.   In 1976, 750 cases of SALMONELLOSIS were reported at a large catering facility in Suffolk County, N.Y. As reconstructed by investigators, this incident was caused by booster pumps backsiphoning water from a clogged slop sink into the drinking water system.

Food contamination.   Events that occurred at a large meat packing plant in Iowa, in April 1979, demonstrate that contamination of water supplies does not always result in a disease outbreak, but it can still be very costly. In this case, contaminated water was recycled through a cross connection from the kill floor and rendering operation into the carcass and meat wash operation. Product losses amounted to 1.1 million pounds of fresh pork at a total cost of over $3 million to the company. Another 1.3 million pounds of processed pork was suspect because it contained some of the contaminated meat. In addition to prompt action by the processor, it is believed that an outbreak was avoided because the pork was subsequently cooked to a "well done" condition that would destroy bacteria.

Chemical Contaminations Attributable to Cross Connections and Backflow

Although the number of reported disease outbreaks has dropped over the years, chemical contamination appears to be on the rise. Technology is constantly developing new chemical products and processes to meet market demand. One result of this effort has been the increased use of water for industrial processes such as cooling, mixing, washing, and diluting. This increase in water use in turn increases the risk of materials entering the water distribution system. The home use of chemicals, such as pesticides, also adds to the risk. The following examples illustrate the types of hazards that have been identified and that continue to occur.

Fertilizer poisoning.   Fertilizers, pesticides, and herbicides can be expected to contaminate public water supplies because they are often mixed with water for spraying operations. The water may be obtained from fire hydrants or public water outlets. Generally, no provisions are made for backflow prevention at such facilities.

In 1971, a case was reported where a contractor filled a tank rig designed to mix fertilizer, seed, and wood pulp in preparation for spray-seeding a large roadway area. The wood pulp plugged a circulating line and forced the mixture to backflow into the city water system while the contractor was drawing water and mixing. Fortunately, only a few people became ill. The water utility, however, was still obliged to flush, test, and disinfect all of the water lines in a nearby subdivision.

Endrin contamination.   In 1975, endrin (a pesticide) was backsiphoned from an applicator's truck into a small public supply serving 21 homes in the state of Washington. The incident occurred when three tank trucks were being filled from different hydrants at the same time. One of the trucks was situated nearly 200 ft above the well, while the other two were at lower elevations. The flows were so large that line pressure was lowered at the high end, resulting in the pesticide being SIPHONED into the water lines. Since this incident occurred in a very small community that did not have a full-time operator monitoring the system, the contamination was not noticed for two days. During this time, many families used the contaminated water for drinking and bathing. Fortunately, no illness was reported. This was attributed to the extreme dilution of the poison. However, costly efforts were required to decontaminate the system.

Boiler-chemical contamination.   In 1974, the manager of a major fast-food chain restaurant in North Carolina began receiving complaints of bitter-tasting soft drinks. His investigation showed that the water being supplied to the restaurant was at fault. Further investigation by the water utility revealed a direct connection to a boiler in a fertilizer plant. In this case, the potable water source was thought to have been protected by a single check valve; however, back pressure from the boiler caused the backflow of water treatment chemicals. There were complaints of illness among children who had consumed the soft drinks, but numbers and severity were not recorded.

Only a few examples have been described in this section to illustrate some of the more serious exposures resulting from cross connections. In addition to the public health aspects associated with cross connections, the water supplier has a legal responsibility to protect the water supply.

Utility Responsibility

The passage of the Safe Drinking Water Act has made each water utility responsible for the quality of water at the consumer's tap. Legal proceedings have also established that the utility is responsible for cross-connection control in some jurisdictions. In addition, many states assign the cross-connection control responsibility directly to the WATER PURVEYOR; others assign it to subordinate municipalities or health authorities; and some maintain it at the state level. Where it has not been specifically assigned, the responsibility for providing potable water clearly rests with the water utility. Likewise, the size of the utility has no bearing on the degree of risk posed by cross connections, which can occur in all sizes of systems. Therefore, all water utilities, large and small, public or private, should maintain an active cross-connection control program.

Legal consequences. Violations of statutory water regulations are generally classed as misdemeanors. California laws are a typical representation—a person who violates the California Pure Water Laws can be found guilty of a misdemeanor punishable by a fine not exceeding $1000 or by imprisonment in the county jail for a period not to exceed one year, or both.

Beyond any specific statutes and ordinances, the courts have long held that a water purveyor is liable when it furnishes unfit water. The courts have ruled that a water purveyor is not an insurer and, therefore, is not liable for incidents over which it has no knowledge or control; however, they have further ruled that since a purveyor has sole possession and control over the water system, it should have complete knowledge of the system's condition and operation; and, therefore, any damage resulting would not occur other than through the purveyor's negligence.

Cross-connection control program. One means through which a water utility can fulfill its responsibilities is by developing and operating a cross-connection control program or jointly participating in such a program with other municipal agencies. The specific nature of any program depends on state and local laws and regulations, municipal agencies, and size of the community.

An effective program deals with the two major sources of problems. These are:

- Plumbing within the customer's premises
- Auxiliary water sources.

Plumbing within the customer's premises is usually under the supervision of state and local health departments or local building–engineering departments. Auxiliary water sources can be under the supervision of a health department or environmental protection agency or its equivalent. When different agencies have an interest, the program will have to be administered as a cooperative effort. The situation where the utility is a single program manager is more likely to occur in a smaller community. Many small communities have a public water distribution system but depend on health and sanitary support from a higher authority, such as a county or district.

A control program will give the water utility (or responsible agency), the legal authority to take actions for protection of the water supply, to provide a systematic procedure for locating, removing, or protecting all cross connections in the distribution system, and to establish the procedures for obtaining the cooperation of the customers and the public. Anyone attempting to establish such a program should consult the state water program agency or other authorities for legal direction. A step-by-step description of how to proceed when establishing and administering a cross-connection control program is discussed in Section 6-5, Cross-Connection Control Program.

6-3.   Principles of Backflow

There are two conditions under which backflow can occur through a cross connection:

- Back pressure
- Backsiphonage.

Both conditions occur when the absolute pressure in the nonpotable system is higher than the absolute pressure in the potable system to which it is cross connected.[1]

Backflow Due to Back Pressure

If a pressurized nonpotable system is cross connected with a lower-pressure potable water system, then the pressure in the nonpotable system can force nonpotable material into the potable supply. This situation is referred to as backflow due to BACK PRESSURE.

The common cross connections where high pressures in a nonpotable system create a risk of backflow due to back pressure are illustrated in Figures 6-10 through 6-12. The first cross connection occurs where potable water enters a boiler system (Figure 6-10). When pressures inside the boiler exceed those of the water supply, backflow can result. A similar situation occurs in Figure 6-11, where a chemical storage tank, pressurized by an air compressor, is connected to a potable water supply line. The third type of cross connection involves the recirculation of potable or nonpotable water on a premises for the purpose of meeting fire demand or processing requirements (Figure 6-12). When the auxiliary system pressures exceed the pressures in the potable supply, then any feeder connections become potential sources of backflow.

---

[1] *Basic Science Concepts and Applications*, Hydraulics Section, Pressure and Force.

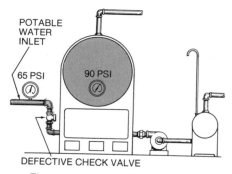

Figure 6-10. Backflow From High Pressure Boiler Due to Back Pressure

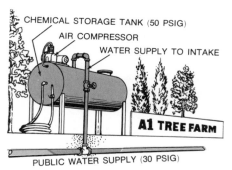

Figure 6-11. Cross Connection Between Pressurized Nonpotable System and Lower Pressure Potable System

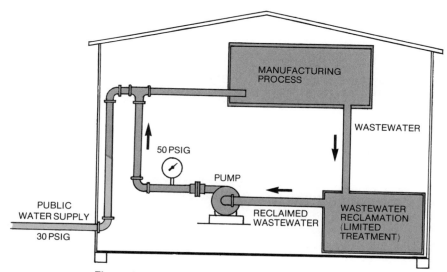

Figure 6-12. Backflow From Recirculated System

Backflow Due to Backsiphonage

If a potable water distribution system is cross connected to a nonpotable source that is open to the atmosphere, and if the pressure in the potable system falls below atmospheric pressure, then the pressure of the atmosphere can force the nonpotable material into the potable supply (Figure 6-13). Another way of describing the situation is that the partial vacuum in the potable system sucks nonpotable materials into it. This situation is called backflow due to BACKSIPHONAGE.

Because backflow due to backsiphonage is actually caused by atmospheric pressure, the lift of the siphon (at sea level) is limited to 33.9 ft (10.3 m), which is the point at which the downward pressure caused by the weight of a column of water equals the pressure of the atmosphere forcing it upward (Figure 6-14).

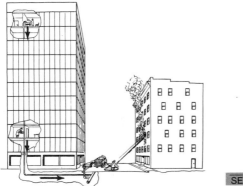

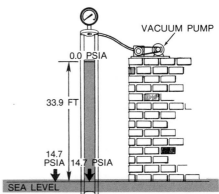

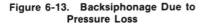

Figure 6-13. Backsiphonage Due to	Figure 6-14. Effect of Evacuating Air
Pressure Loss	From a Column

Figures 6-15 through 6-17 illustrate examples of backsiphonage. In Figure 6-15, a hose to the bottom of the sink siphons into the supply pipe. If the supply to the slop sink also serves a booster pump downstream, as shown in Figure 6-16, then starting the pump may drop pressure significantly. Given a slop sink 1 ft (0.3 m) in depth, a negative pressure of only 0.433 psig [2.99 kPa (gauge)] could draw liquid from the sink into the water distribution system. Such a situation evidently caused the salmonellosis incident described previously.

A common cause of the less-than-atmospheric pressure (negative gauge pressure) needed to create a backsiphonage condition is overpumping by fire or booster pumps. Undersized distribution piping can also create negative pressures. Undersized piping creates a high velocity of water, which in turn causes severe pressure drops. An outlet in an undersized area may have negative gauge pressure when water is flowing through the main supplying that outlet.

A broken main can have the same effect as a booster pump, except that both upstream and downstream connections may backflow. Breaks in the water distribution system, particularly in low elevations, can contribute to the same pressure conditions as the opening of the lower faucet in Figure 6-15. In the situation illustrated, only the water being drawn from the lower faucet could be contaminated. However, when a break occurs in a main, the entire distribution system could become contaminated between the break and the cross connections (Figure 6-17). Subsequent reestablishment of pressure could in turn contaminate all of the system downstream of the break.

6-4. Backflow Control Methods and Devices

When cross connections are found in an area, one of two actions must be taken. Either the cross connection must be removed or some means must be devised to protect the potable water supply from possible contamination. Where removal is impractical, a protective device should be installed.

The preventive measure chosen depends on the degree of hazard involved, the

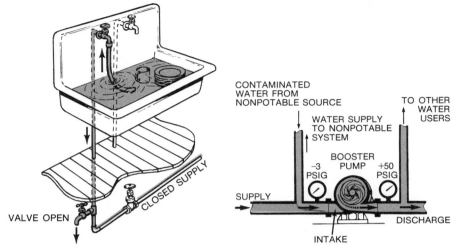

Figure 6-15. Backsiphonage—Hose Forms Cross Connection

Figure 6-16. Backsiphonage From a Booster Pump

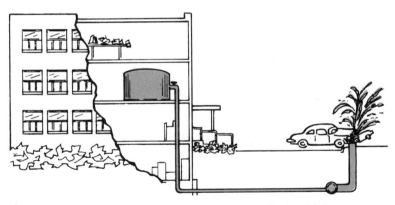

Figure 6-17. Backsiphonage Due to a Broken Main

accessibility to the premises, and the type of internal water distribution system.

Protective methods and backflow-prevention devices are discussed in the following paragraphs in the order of decreasing effectiveness:

- Air gap
- Reduced-pressure-zone backflow preventer (RPZ)
- Double check-valve assembly
- Vacuum breakers (atmospheric and pressure)
- Barometric loop.

Air Gap

When correctly installed and maintained, an AIR GAP, shown in Figures 6-18 and 6-19, is the most positive method available for protecting against backflow. It is acceptable in all cross-connection situations and for all degrees of risk. Another advantage is that there are no moving parts to break or wear out; only surveillance is needed to ensure that no BYPASSES are added. The only requirement for installation is that the gap between the supply outlet and the OVERFLOW LEVEL of the downstream receptacle measures at least two times the internal diameter of the outlet's tip, but not less than 1 in. (25 mm). Fixtures or services with simple gravity drains, such as the water truck shown in Figure 6-4, can be modified to add a simple air gap, as shown in Figure 6-18. In situations where the air gap isolates the water system, a SURGE TANK may be required. Pumps may be added at the tank's outlet (Figure 6-19) to provide necessary pressure to distribute the water throughout the premises. Further protective devices may be needed beyond the air gap to protect personnel using the isolated system. Also, more frequent visual inspection of the gap is needed when pumps are required to boost pressure downstream of the air gap because of the increased risk that a bypass may be installed during power outages or pump repair.

Reduced-Pressure-Zone Backflow Preventer

The second device that can be used in every cross-connection situation and with every degree of risk is the REDUCED-PRESSURE-ZONE BACKFLOW PREVENTER (RPZ

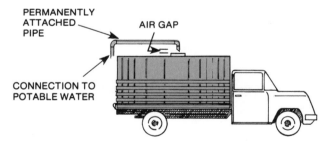

Figure 6-18.   Water Truck Cross Connection Prevented by Air Gap

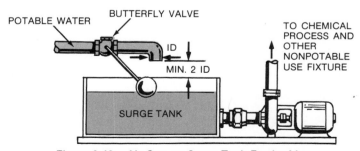

Figure 6-19. Air Gap on Surge-Tank Feeder Line

Courtesy of Cla-Val Co., Backflow Preventer Division

Figure 6-20.   Reduced-Pressure-Zone Backflow Preventer

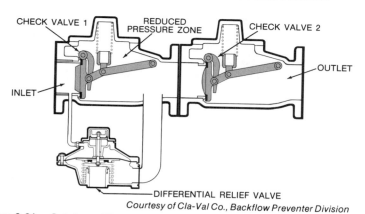

Courtesy of Cla-Val Co., Backflow Preventer Division

Figure 6-21.   Cut-Away View of a Reduced-Pressure-Zone Backflow Preventer

or RPBP). This device consists of two spring-loaded check valves with a pressure-regulated relief valve located between them, as shown in Figures 6-20 and 6-21. In the figures, flow from the supply at left enters the central chamber (reduced-pressure zone) against the pressure exerted by the loaded check valve. The amount of pressure loss through this check varies with the size, the flow rate, and the manufacturer. The second check valve is loaded considerably less [causing about 1 psi (7 kPa) further pressure drop] to keep the total pressure loss within reason. Two check valves, even though well designed and constructed, are not considered sufficient protection, because all valves can leak from wear or obstruction. For this reason, a relief valve is positioned between the two checks. When the unit is operating correctly and supply pressure exceeds the down-stream pressure, the supply pressure opposes the spring tension and keeps the valve closed. If either check valve fails or if downstream pressure approaches supply pressure, then the pressure in the reduced pressure zone will approach the supply pressure. When the two pressures reach a difference of 2 psi (14 kPa) or lower, the spring opens the relief valve and drains the reduced pressure zone. A definite sign of malfunction is continuous drainage from the relief port. Because of the multiple protective systems, the probability of backflow occurring across an RPZ backflow preventer is very small. Figures 6-22A through 6-22D show valve positions, pressures, and water flow under normal flow, backsiphonage, back pressure, and back pressure with leakage conditions.

Most of the RPZ units on the market today are of the general shape shown in Figure 6-20, with the flow in a direct line. A typical installation of an RPZ device is illustrated in Figure 6-23.

Even though the RPZ backflow preventer is highly dependable, it is a mechanical device that must be maintained. Valve faces, springs, and diaphragms deteriorate with age, and solids can lodge in or damage the check valves, causing

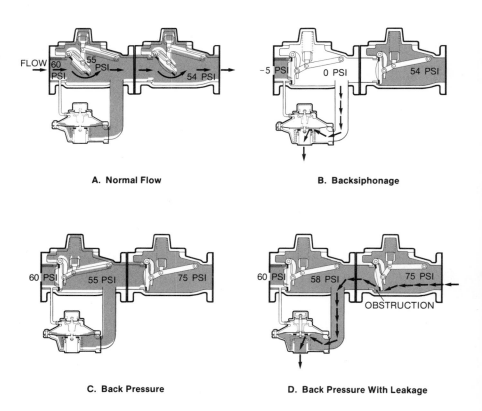

A. Normal Flow

B. Backsiphonage

C. Back Pressure

D. Back Pressure With Leakage

Courtesy of Cla-Val Co., Backflow Preventer Division

Figure 6-22.   Valve Position and Flow Direction in an RPZ

Figure 6-23.   Typical Installation of Reduced-Pressure-Zone Backflow Preventer

leakage. For these reasons, each installed unit must have periodic inspection, testing, and maintenance. RPZ devices should be tested in accordance with the manufacturer's specifications.

The backflow preventer's performance can be affected by how and where it is installed. To achieve continued satisfactory performance, certain installation procedures should be followed:

- Install the unit where it is accessible. Accessibility will provide for ease of testing and maintenance. Installation at least 12 in. (300 mm) above the floor is one criterion for accessibility. If the device is installed in a noticeable position, it will increase the probability that any malfunction (leakage) will be noticed and repaired.

- Install the unit so that the relief-valve port cannot be submerged. Submergence of the port creates another cross connection and may also prevent the unit from operating properly.

- Protect from freezing, which will damage the unit.

- Protect from vandals. Accessible units and particularly the associated gate valves are a temptation to vandals.

- If a drain is needed, an air gap must be provided below the relief-valve port. Most manufacturers have designed attachments to fit the unit.

- Install a screen upstream of the preventer to eliminate the problems of debris becoming lodged in the unit, which could render it inoperative. Screens are available commercially.

- Most models must be installed horizontally. Check with the manufacturer to determine the allowable positions for each model.

Double Check-Valve Assembly

The double check-valve backflow preventer is designed basically the same as the RPZ but without the relief valve (Figure 6-24). The absence of the relief valve

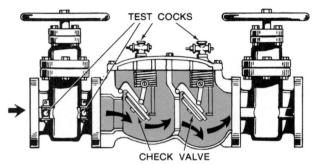

TEST COCKS

CHECK VALVE

Figure 6-24. Double Check-Valve Assembly

significantly reduces the level of protection provided. The unit will not give any indication that it is malfunctioning (whereas a malfunctioning RPZ will leak or discharge water), and drainage of the reduced-pressure zone will not occur to bypass any possible backflow.

A double check-valve assembly is not recommended as protection in situations where a health hazard may result from failure. Before installing such a unit in a potable water line, contact the agency having statutory jurisdiction to determine whether or not its use is permitted.

Like the RPZ, the installation and maintenance of a double check-valve assembly should follow certain procedures:

- The unit must be an approved model; two single check valves in series will not be satisfactory.
- The installation must be protected from freezing and vandals.
- The unit must be tested and maintained on a periodic basis. Frequency of testing and maintenance is governed by state and local codes.
- The unit should be installed in a position that will allow testing and maintenance to be performed.
- The unit should be installed in an area that does not flood. The test cocks can form a cross connection when submerged.

Vacuum Breakers

As described previously, backsiphonage occurs when a partial vacuum pulls nonpotable liquids back into the supply lines. If air enters the line between a cross connection and the source of the vacuum, then the vacuum will be destroyed (broken) and backsiphonage will be prevented. This is the function of a VACUUM BREAKER. Figure 6-25 shows an ATMOSPHERIC VACUUM BREAKER, consisting of a check valve operated by water flow and a vent to the atmosphere. When flow is forward, the valve lifts and shuts off the air vent; when flow stops or reverses, the valve drops to close the water-supply entry and open an air vent. A version of the unit designed for a HOSE BIBB is shown in Figure 6-26.

Atmospheric vacuum breakers are not designed to protect against back pressure, nor are they reliable under continuous use or pressure, since the gravity-operated valve may stick. For these reasons, any shut-off valve must be placed upstream from the breaker, and the flow from the protected fixture must

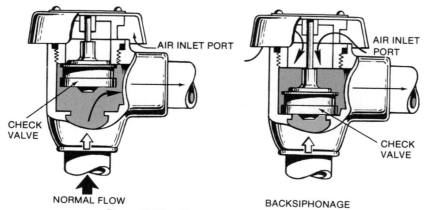

Figure 6-25. Atmospheric Vacuum Breaker

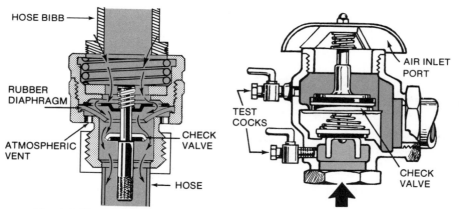

Figure 6-26. Hose-Bibb Type Atmospheric Vacuum Breaker

Figure 6-27. Pressure Vacuum Breaker

be to the atmosphere. Installation of these breakers must be at least 6 in. (150 mm) above the highest point of the downstream outlet.

Figure 6-27 shows a PRESSURE VACUUM BREAKER. The unit illustrated has a spring-loaded check valve that opens during forward flow and is closed by the spring when flow stops. When pressure drops to a low value, a second valve opens and allows air to enter this breaker. With this arrangement, the breaker can remain under supply pressure for long periods without sticking and can be installed upstream from the last shut-off valve. The placement and use of the pressure vacuum breaker is restricted to situations where no back pressure will occur and where it can be installed 12 in. (300 mm) above the highest point of the downstream outlet.

Barometric Loop

The barometric loop is a simple installation that prevents only backsiphonage. To create a barometric loop, an inverted U is inserted into the supply pipe upstream of the cross connection as shown in Figure 6-28. Figure 6-14 illustrated how the height of a column of water, open to the atmosphere at the bottom is limited to about 33.9 ft (10.3 m). If the barometric loop is taller than the 33.9-ft (10.3-m) limitation, siphonage or backsiphonage will not occur. Although the barometric loop is effective against backsiphonage and requires no maintenance or surveillance other than to ensure it remains leak-free, the space requirement [about 35 ft (11 m) above the highest liquid level] is a very serious disadvantage. Also, the barometric loop is completely ineffective against backflow due to back pressure. For these reasons, barometric loops are no longer being installed.

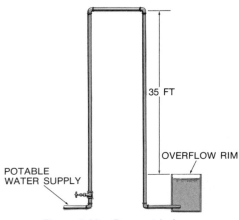

35 FT

OVERFLOW RIM

POTABLE WATER SUPPLY

Figure 6-28.   Barometric Loop

Other Methods and Devices Used With Auxiliary Water Systems

One of the first methods for cross-connection control was the complete separation of potable and nonpotable piping systems. This is a most effective approach when separate water sources are available, adequate color codes can be maintained, and adequate surveillance is provided to prevent connections between systems. Separate systems have been interconnected by such devices as spool pieces, flexible-temporary connections, and swing connections (Figure 6-29). None of these interconnections is recommended for use regardless of the degree of risk involved, and they should be removed from any system.

Level of Protection

The type of backflow-prevention device that is to be used for each installation depends on the degree of hazard present and whether or not backflow could result from back pressure or backsiphonage. Table 6-3 provides a listing of the minimum protection suggested for various applications. Local codes should also be consulted before deciding on a method for protecting an installation. Table 6-4 lists the types of cross connections discussed earlier and suggests the level of protection appropriate for each.

AUXILIARY SUPPLY

POTABLE SUPPLY

A. Spool Piece

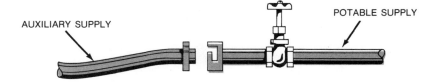

AUXILIARY SUPPLY

POTABLE SUPPLY

B. Temporary Connection

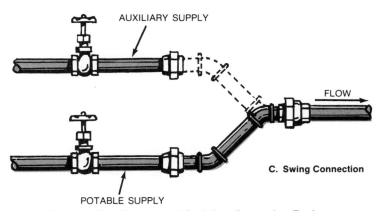

AUXILIARY SUPPLY

FLOW

C. Swing Connection

POTABLE SUPPLY

Figure 6-29. Unapproved Backflow Prevention Devices

6-5.   Cross-Connection Control Program

The degree to which the operator can become involved in a cross-connection control program will vary from almost no involvement, to being an inspector, to providing major assistance in establishing and operating the program. Knowledge of program content and procedures are valuable at every level of responsibility.

Program Content

An effective cross-connection control program has the following elements:

- An adequate plumbing and cross-connection control ordinance

Table 6-3.   Recommended Minimum Protection Requirements

Type of Hazard on Premises	Minimum Protection* at Meter					Minimum Options to Isolate Area of Plant Affected†					Comments
	AG	RPBD	DCVA	AVB or PVB	None	AG	RPBD	DCVA	AVB or PVB	None	
1. Sewage treatment plant	X	X				See comments					RPBD at meter with air gap in plant also
2. Sewage lift pumps	X	X								X	
3. Domestic water booster pumps			X					X			
4. Equipment or containers manufactured for industrial use without proper backflow protection											
A. Dishwashing			X			X			X		
B. Clothes washing									X		Normally machine has built-in AG
C. Food processing	†	†				†	†				If no health hazard exists, DCVA is acceptable
D. Pressure vessels		X					X				
E. Tank or vat containing a nonpotable or objectionable solution	X	X				X	X				A vacuum breaker or DCVA may be used if no health hazard exists
F. Sinks with hose threads on inlet									X		
G. Any dispenser connected to a potable water supply									X		
H. Aspirator equipment									X		
I. Portable spray equipment											
5. Reservoirs, cooling towers, circulating systems		X				X	X	X			Systems where no health hazard exists or potential for a health hazard exists, a DCVA may be used
6. Commercial laundry	X	X								X	
7. Steam generating facilities and lines			X				X	X			
8. Equipment under hydraulic test or hydraulically	X	X				X	X				

Type of Hazard on Premises	Minimum Protection* at Meter					Minimum Options to Isolate Area of Plant Affected†					Comments
	AG	RPBD	DCVA	AVB or PVB	None	AG	RPBD	DCVA	AVB or PVB	None	
9. Laboratory equipment											
A. Health hazard	X	X								X	
B. Nonhealth hazard			X						X		
10. Plating facilities	X	X								X	
11. Irrigation systems	X	X									Chemicals are often used in such systems
12. Fire fighting systems	X	X									
13. Dockside facilities	X	X				X	X				DCVA may be used on dockside if outlet is protected
14. Tall buildings			X								Two devices should be installed in parallel; if a health hazard exists within the building, two RPBD's in parallel may be required
15. Unapproved auxiliary supply	X	X				X	X				
16. Premises where inspection is restricted	X	X									
17. Hospitals, mortuaries, clinics	X	X				X	X				
18. Laboratories	X	X				X	X				DCVA, if no health hazard exists
19. Chemical plants using a water process	X	X				X	X				
20. Petroleum processing or storage plants	X	X				X	X				
21. Radioactive material processing plants or nuclear reactors	X	X				X	X				
22. Swimming pools	X	X				X					

NOTE: The list is not all inclusive, and does not necessarily conform to state and local codes. These codes must be checked before installation of any protective device.

* The following abbreviations are used:
air gap—AG; reduced-pressure-backflow device—RPBD; double check-valve assembly—DCVA; pressure vacuum breaker—PVB; atmospheric vacuum breaker—AVB.

† In areas where no health hazard or potential for a health hazard exists, a vacuum breaker properly installed in the line to the problem area may be adequate protection.

Table 6-4. Suggested Protective Means

Application	Reference Figure	Minimum Protection		Notes
		At Meter	At Cross Connection	
Auxiliary water system	6-1	RPZ or air gap	RPZ	
Boiler feed water	6-2	RPZ or air gap*	RPZ	Chemicals may be used in boiler
Sink	6-3	RPZ or air gap*	Atmospheric or pressure vacuum breaker	
Water truck	6-4	RPZ or air gap*	Air gap, RPZ	See Figure 6-18
Garden hose	6-5	RPZ or air gap*	Atmospheric or pressure vacuum breaker	A double check-valve assembly may be acceptable if no health hazard exists The hose-bib type of breaker is used
Ship	6-6	RPZ or air gap	RPZ	
Cooking vessel	6-7	RPZ or air gap*	RPZ	If no health hazards, double-check valve assembly may be used
Water softener	6-8	RPZ or air gap*	RPZ	

*Protection at the meter may not be necessary if protection at the cross connection is ensured.

- An organization or agency with overall responsibility and authority for administering the program with adequate staff
- Systematic inspection of new and existing installations
- Follow-up procedures to ensure compliance
- Backflow-prevention device standards, and standards for inspection and maintenance
- Cross-connection control training
- A public awareness and information program.

Procedures

The procedure for initiating a program begins with planning. This is normally a function of management. Assistance from the health, building, or plumbing inspection departments should be obtained at the beginning. Planning should include identifying a tentative organization, the appropriate authority, and internal procedures for executing the program.

Planning should be followed by informing the municipal government and the public of the nature of cross connections and backflow and the steps needed to protect the public. This can be done through newspaper announcements, interviews, and public presentations. The information provided will give the municipal government the background needed to understand and enact the authorizing control ordinance, which is the legal basis of a local program. The public should continue to be informed even after the program has been implemented.

The authorizing control ordinance provides authority for establishing the program and operating it. How much detail is included depends on state laws and codes. In areas where the state has an extensive and detailed code, this can be very simple. In other locations, details to completely describe the program may need to be developed. These details should include at least the following:

- Authority for establishment
- Responsibilities and organization for operating the program
- Authority for inspections and surveys
- Prohibitions and protective requirements (if the state code is not specific, then this can become very detailed)
- Penalty provisions for violations.

After establishing the program, the water utility or other agency designated by the ordinance begins inspections of premises. Those premises that, by nature of activity, present the greatest risk to public health should be inspected first. The inspector, who must be trained to identify cross connections and actions needed to protect the potable water supply, should be accompanied by the owner or the owner's representative so that there is an understanding of the procedures and results. The complete plumbing system should be inspected. Particular attention should be given to pipelines leading to process areas, laboratories, liquid storage, and similar facilities. A check list to help the inspector is shown in Figure 6-30. This list of installations is not complete; however, it should serve as a guide to indicate the types of equipment to be inspected.

When cross connections are found during the inspection, they should be pointed out to the owner and their significance discussed. Areas to be covered include: What is the material that could backflow? How hazardous is the cross connection? Can it be eliminated or must it be protected? How can it be protected?

The water utility is responsible for the water distribution supply to the owner's

---

Cross-Connection Survey Form

Place: _____ Date: _____

Location: _____ Investigator(s): _____

Building Representative(s) and Title(s):
_____
_____
_____

Water Source(s): _____
_____
_____

Piping System(s): _____
_____
_____

Points of Interconnection: _____
_____
_____
_____

Special Equipment Supplied with Water and Source:
_____
_____
_____
_____

Remarks or Recommendations: _____
_____
_____
_____
_____

NOTE: Attach sketches of cross connections found where necessary for clarity of description. Attach additional sheets for room-by-room survey under headings:

Room Number	Description of Cross Connection(s)

Figure 6-30. Sample Survey Form

premises. Therefore, for the purpose of a cross-connection program, the inspector does not need to require that each cross-connection be protected, only that protection be provided at the point of entry to the premises. However, the inspector should recommend that some action be taken on each connection, both for the owner's protection and to fulfill the requirements of the Occupational Safety and Health Administration (OSHA). Also, local plumbing or health codes may require that each connection be protected. Only in those cases where the codes are specific should the inspector specify the item needed for a certain application; otherwise a minimum level of protection should be recommended in accordance with the code and Table 6-3. On completion of the

---

State Department of Public Health
Water Department—Cross-Connection Survey

_____ , _____   Account No. _____
(City)   (State)

OWNER: _____
ADDRESS: _____ Tel. No. _____
OCCUPANT: _____
ADDRESS: _____ Tel. No. _____
Party Interviewed: _____ Type of Establishment: _____

The purpose of this plumbing survey, as requested by the State Health Department, is for the location of CROSS CONNECTIONS with the potable water supply. Your health and that of the community may be affected, therefore it is everyone's civic duty to eliminate CROSS CONNECTIONS. Drinking Water Standards prohibit any cross connection.

	Number				Number		
	Disapproved	Approved	Total		Disapproved	Approved	Total
BATHROOMS AND RESTROOMS:				LAWN SPRINKLING SYSTEM:			
Bath Tub				Syphon Breaker			
Bidet				PRESSING ESTABLISHMENT:			
Floor Drain or Shower				Dye Vat			
Lavatory				Solvent Machine			
Urinal				Still			
Water Closet				Steamer			
COMPRESSORS:				SHAMPOO BOWL			
Air Cooled				SINK:			
Water Cooled				Bar			
Cooling Tower				Dipper			
Well Water				Floor			
DENTAL CUSPIDOR				Glass,			
DEVELOPING TANK				Janitor			
DRINKING FOUNTAIN				Kitchen			
DRINKING FOUNTAIN DRAIN				Rinse			
EMBALMING TABLE				Scullery			
FOOD HANDLING EQUIPMENT:				Vegetable			
Carbonator				STERILIZER			
Coffee Urn				SUMP			
Deep Freeze				SWIMMING POOL			
Dishwashing Machine				SYPHON BREAKER			
Disposal (Garbage)				WATER FILTER			
Egg Boiler				WATER SOFTENER			
Extractor				OTHER:			
Glass Filler							
Ice Cube Machine							
Ice Cube Machine Drain							
Ice Box							
Ice Box Drain							
Potato Peeler							
Soda Fountain							
Steam Table							
REFRIGERATING COILS							
WATER COOLED COILS							
HEATING PLANT:							
Steam							
Water							
Syphon Breaker							
LAUNDRY EQUIPMENT:							
Commercial				SEWAGE TO:			
Domestic				City system			
Tub				Cesspool			
				Septic Tank			

Cross Connection is any arrangement that provides a means whereby impure water, gas, or liquid may gain entrance to and contaminate or pollute a safe water supply system. Owner will correct any changes noted as a disapproval of any item checked within ____ days.

NOTE: Attach sketches of cross connections found where necessary to clarify a description. Use reverse side of this form or additional sheets for a room-by-room survey.

Date _____

_____
Inspector or Sanitarian

Figure 6-30. Sample Survey Form (continued)

survey, a report should be prepared and the owner should be notified of the corrective actions needed, the approvals required, and the time limits.

In the case of new installations, municipalities normally require submission and approval of building and plumbing plans. Review of plans presents an opportunity to identify potential problem areas. The review should identify cross connections with a view toward their elimination or protection. Only after the plans are changed to ensure adequate protection should they be approved. Most municipalities inspect new buildings during and after completion. A survey for cross connections should be included in these inspections.

Repeat inspections are needed both to ensure that all corrective actions required from previous inspections have been completed and to look for new cross connections. These inspections are conducted in the same manner as the initial inspection. Timing of repeat inspections will vary according to their nature and whether compliance checks are routine or special.

Routine inspection intervals will vary according to the degree of hazard involved and manpower resources available; however, they should be conducted as often as feasible. The state code may recommend appropriate inspection intervals, which could be every 3 to 6 months for high-risk installations and annually for others. Special compliance checks should be made immediately following notification of a plumbing change or a change of activity at an installation (upgrading of protection may be required if risk has increased) and on notification of a violation that could endanger public health.

When a customer is not convinced of the seriousness of the potential hazard or refuses to cooperate, then the water purveyor must have the legal authority to shut off the customer's water. Before the water is shut off, the purveyor needs to make a final effort to gain the customer's cooperation and warn of the consequences of noncompliance. Discontinuance of water service should be a last-resort measure, and it is recommended that legal counsel be consulted first. It is vital that all warnings and notifications be made exactly as specified by ordinance.

One corrective action resulting from inspections is the actual elimination of cross connections. These actions will only require follow-up inspections to ensure that they are not reconnected. A second result is the protection of the water supply by installing backflow-prevention devices. If records of testing and maintenance are not maintained by the agency performing the field surveys, repeated inspections should include a check of the owner's records to ensure that the backflow equipment is being tested and maintained.

Backflow-prevention devices must operate properly for long periods of time. In order to accomplish this, two conditions must be met:

• Quality backflow-prevention units must be installed.

• Inspection and maintenance must be performed periodically.

The first condition is met by setting minimum standards for design, construction, and performance, followed by testing to evaluate conformance. This testing and acceptance is done by special laboratories and in some cases by the state. A list of approved devices should be consulted before the customer installs any backflow preventer.

The second condition—periodic inspection and maintenance—is met by systematic inspection and testing of all units that have been installed. Inspection consists of visually checking a unit weekly or biweekly for leaks and external damage. The owner should be required to make the inspections and keep a record of the findings. If the inspection shows defects, qualified repair personnel should be called to perform further examination and repair. Each unit should be tested for correct operation annually by an individual specifically qualified for such testing and repair. Any devices that fail must be repaired immediately and a report of the test and repair made when completed. (In some states, vacuum breakers are exempt from annual testing.)

Education and training for cross-connection control should be performed at several levels. The first is a continuous program of public education. This program should describe applicable codes and cross connections and the hazards they pose, in order to obtain the consumer's cooperation and reduce the overall risk. At the next level, water utility personnel, plumbing inspectors, and health personnel need special training to develop and operate the control program. Finally, specialized training in maintenance and testing is needed for those individuals assigned to maintain backflow preventers. The names of schools with programs for training in backflow-preventer repair can be obtained from the manufacturers of the devices.

6-6. Records and Reports

Records and reports play an essential role in administering a cross-connection control program. Complete records are a utility's first line of defense against potential legal liability in case of public health problems resulting from a cross connection.

This section discusses the importance of maintaining accurate and up-to-date records on installing, inspecting, testing, and repairing backflow-prevention devices as a shared responsibility of the water customer, the water utility or agency operating the control program, and personnel performing tests and repairs.

Water customer.   In order to fulfill the responsibility for maintaining safety within the premises, the water customer needs to be informed about the water system, any hazardous conditions that are creating potential cross connections, and the condition of the backflow-prevention devices installed. The information about the water system is available from building plans. Information about cross connections and their protection comes from field inspection surveys, periodic visual observations, and test reports.

Water utility or agency operating the control program.   Effective control dictates a time-based formal record system—that is, records that are organized so that each inspection, survey, test, and corrective action is performed as scheduled. Records maintained should consist of inspection reports, test/repair reports, reports of corrective actions, authorized tester/repair-personnel lists, approved backflow-preventer lists, and backflow-preventer installation locations.

Testing and repair personnel.   To ensure continued satisfactory operation of each backflow preventer, only qualified authorized individuals should make the periodic test or perform needed maintenance. A report of work performed should be recorded on a form similar to that shown in Figure 6-31. The distribution of the report by the control agency is specified by local regulations and practices.

An additional means for ensuring that each preventer is adequately maintained is to make sure that each unit installed is assigned to a single, qualified individual

Backflow Device Test Report

Return No Later Than _____

Name of Premises _____

Service Address _____

Location of Device _____

Device _____   _____   _____   _____
      Manufacturer   Model   Size   Serial No.

Line Pressure at Time of Test _____ lb

Pressure Drop Across First Check Valve _____ lb

	Check Valve No. 1	Check Valve No. 2	Differential Pressure-Relief Valve
INITIAL TEST	1. Leaked ....□ 2. Closed Tight .... □	1. Leaked ....□ 2. Closed Tight .... □	1. Opened at _____ lb Reduced Pressure 2. Did Not Open ....□

R E P A I R S	Cleaned ....□
	
	Replaced:
	  Disc ....□
	  Spring ....□
	  Guide ....□
	  Pin Retainer ....□
	  Hinge Pin .....□
	  Seat ....□
	  Diaphragm .....□
	  Other, Describe . □
		Cleaned ....□
		
		Replaced:
		  Disc ....□
		  Spring ....□
		  Guide ....□
		  Pin Retainer ....□
		  Hinge Pin .....□
		  Seat ....□
		  Diaphragm .....□
		  Other, Describe . □
			Cleaned...................□
			
			Replaced:
			  Disc, Upper ....□
			  Disc, Lower ....□
			  Spring ....□
			  Diaphragm, Large
			    Upper ....□
			    Lower.................□
			  Diaphragm, Small
			    Upper ....□
			    Lower.................□
			  Spacer, Lower............□
			  Other, Describe ....□

FINAL TEST	Closed Tight ......□	Closed Tight ......□	Opened at _____ lb Reduced Pressure

Remarks: _____

_____

_____

_____

_____

The Above Report is Certified To Be True

Initial Test Performed by _____ of _____ Date _____

Repaired by _____ Date _____

Final Test Performed by _____ of _____ Date _____

Figure 6-31.   Backflow Device Test Report Form

or firm for testing and repair. The assignment can take the form of a contract between the owner and an independent repair service or it may be a letter of instruction where the qualified individual is an employee of the owner. A record of the assignment should be kept by the owner and made available to the field inspector. It is advisable that the cross-connection control operating agency be informed of the name of the individual to whom each backflow preventer is assigned. To ensure continued satisfactory performance, repair personnel should have servicing instructions available for all backflow preventers they must maintain. Records of previous servicing and performance tests are also important.

Selected Supplementary Readings

Accepted Procedure and Practice in Cross Connection Control Manual. Cross Connection Control Committee, Pacific Northwest Section, AWWA, and Cross Connection Control Committee, British Columbia Section, AWWA. (1980).

Anderson, D.C. *Cross-Connections—Their Importance and Control.* AWWA Annual Conf. Proc., Paper No. 18-5. AWWA, Denver, Colo. (1981).

Anderson, G.D. The Basics of Cross-Connection Control. *OpFlow*, 4:11:3 (Nov. 1978).

Angele, G.J., Sr. *Cross Connections and Backflow Prevention.* AWWA, Denver, Colo. (2nd ed., 1974).

Asay, S.F. Cross Connection Monitoring and Control Program. *OpFlow*, 10:2:1 (Feb. 1984).

Backflow Prevention and Cross-Connection Control. AWWA Manual M14. AWWA, Denver, Colo. (1966).

Backsiphonage: A Hazard to Public Health. *OpFlow*, 2:5:3 (May 1976).

Craun, G.F. Outbreaks of Waterborne Disease in the United States: 1971-1978. *Jour. AWWA*, 73:7:360 (July 1981).

Cross-Connection-Control Guide for Operators—I. Cross-Connection Terminology. *OpFlow*, 9:1:4 (Jan. 1983).

Cross-Connection-Control Guide for Operators—II. Devices That Protect Against Cross Connections. *OpFlow*, 9:2:3 (Feb. 1983).

Cross-Connection-Control Guide for Operators—III. Cross-Connection Control Programs. *OpFlow*, 9:3:5 (Mar. 1983).

Cross-Connection Control In Your Own Back Yard. *OpFlow*, 6:6:6 (June 1980).

Cross-Connection Control Manual. Division of Sanitary Engineering, Tennessee Department of Public Health, Nashville, Tenn. (1975).

Cross-Connection Control Manual. US Environmental Protection Agency, Washington, D.C. EPA 430/9-73-002. (1975).

Cross Connection Rules Manual. Michigan Department of Public Health, Water Supply Services Division, Lansing, Mich. (2nd ed., 1982).

DDT in Water Supply. *OpFlow*, 5:3:1 (Mar. 1979).

List of Approved Backflow-Prevention Devices. University of Southern California, Foundation for Cross Connection Control and Hydraulic Research, Los Angeles, Calif. (updated periodically).

Manual of Cross-Connection Control. Foundation for Cross Connection Control and Hydraulic Research, Los Angeles, Calif. (6th ed., 1979).

Markwood, I.M. Cross-Connections—Legal or Lethal? *OpFlow*, 1:9:3 (Sept. 1975).

Miller, R.S. Cross-Connection Control Down on the Farm. *OpFlow*, 5:11:1 (Nov. 1979).

Roller, J.A. Cross-Connection Control Practices in Washington State. *Jour. AWWA*, 68:8:407 (Aug. 1976).

Springer, E.K. *A Sip Could Be Fatal.* Proc. AWWA Distribution System Symposium, Paper No. 3-1. AWWA, Denver, Colo. (Feb. 1980).

Springer, E.K. Cross-Connection Control. *Jour. AWWA*, 68:8:405 (Aug. 1976).

Springer, E.K. Viewpoint—Wanted: Comprehensive Cross Connection Control. *Jour. AWWA*, 72:8:17 (Aug. 1980).

The Manual of Cross Connection Prevention in Public Water Supplies. Missouri Section, AWWA (no date).

Glossary Terms Introduced In Module 6

(Terms are defined in the Glossary at the back of the book.)

Actual cross connection
Air gap
Atmospheric vacuum breaker
Auxiliary supply
Backflow
Back pressure
Backsiphonage
Bypass
Carrier
Check valve
Cross connection
Dysentery
Feed water

Gastroenteritis
Hepatitis
Hose bibb
Incubation period
Overflow level
Overflow rim
Potential cross connection
Pressure vacuum breaker
Reduce-pressure-zone backflow
    preventer
RPBP
RPZ
Salmonellosis

Siphon
Solvent
Surge tank

Vacuum breaker
Water purveyor

Review Questions

(Answers to Review Questions are given at the back of the book.)

1. What is a cross connection, and explain what is meant by backflow.

2. Where are cross connections found?

3. Explain what is meant by the terms backsiphonage and back pressure.

4. List five waterborne diseases that have occurred as a result of cross connections.

5. Explain what is meant by a negative gauge pressure and indicate another common term used to refer to this condition.

6. Is a single check valve positive protection against backflow? Why or why not?

7. What are the conditions needed to cause backsiphonage?

8. Identify three causes of negative gauge pressure in a potable water supply line.

9. What conditions are needed to cause backflow due to back pressure through a cross connection?

10. Describe two situations where backflow due to back pressure could occur.

11. What is the most reliable backflow prevention method?

12. List four mechanical backflow-prevention devices in the order of their use, based on the degree of hazard.

13. Of the above devices, which one cannot be used to protect against backflow due to back pressure?

14. Describe how a RPZ device prevents backflow.

15. Explain why the double check-valve assembly should be used or not used where a potential health hazard exists.

16. In what position, relative to a shut-off valve, should an atmospheric vacuum breaker be installed? Why?

17. How does the vacuum breaker prevent backsiphonage?

18. What are the differences between an atmospheric vacuum breaker and a pressure vacuum breaker?

19. Identify the elements that are essential to successfully implement and operate a cross-connection program.

20. How does a water utility or cross-connection control agency ensure that its corrective orders are followed and the potable water system remains free of cross-connection hazards?

21. How often should a reduced-pressure-zone backflow preventer be tested?

Study Problems and Exercises

1. Your city has just annexed a subdivision that has had its own wells and septic tanks for years. After the hookups are completed the weather turns very dry and water demand increases. You notice that water samples from the water system near the subdivision are showing high coliform counts. The local health officer informs you that she is receiving numerous complaints of stomach disorders and illness. Describe what you would do to investigate this situation and include an explanation as to the most likely cause of the problem and how it can be remedied.

2. A new hospital is being built in your city and the director of public works has asked you to recommend the cross-connection control devices for both the connection to the hospital and the inside water services. Prepare a report outlining your recommendations.

3. Describe the cross-connection control program in your community. (If your community does not have a cross-connection control program, describe features of a program that you would propose.)

Module 7

Pumps and Prime Movers

Pumping facilities are required wherever gravity cannot be used to supply water to the distribution system under sufficient pressure to meet all service demands. Where pumping is necessary, it accounts for most of the energy consumed in water supply operations.

Pumping equipment represents a major part of a utility's investment in equipment and machinery. If properly operated and maintained, quality pumps and PRIME MOVERS—the electric motors or engines that drive them—can give decades of efficient and reliable service. This module covers the basics of pump and prime-mover design, operation, and maintenance. Appendices to this volume discuss pump and prime-mover selection and installation. Typical uses of pumps in a water system are listed in Table 7-1 and diagrammed in Figure 7-1.

After completing this module you should be able to

- Describe the operating principles of common types of velocity and positive-displacement pumps.

- Explain the general operating procedures for centrifugal pumps and understand the reasons for those procedures.

- Describe the regular preventive-maintenance procedures required by centrifugal pumps.

- Compare the advantages of various types of electric motors and combustion engines as prime movers.

- Describe operating and maintenance procedures for electric motors, with particular attention to energy- and cost-saving factors.

- Describe general start-up, operating, and preventive-maintenance procedures for diesel and gasoline engines.

Table 7-1. Pump Applications in Water Systems

Application	Key to Figure 7-1	Function	Pump Type
Low service	1,3,7	To lift water from the source to treatment processes, or from storage to filter backwashing system.	Centrifugal
High service	8	To discharge water under pressure to distribution system.	Centrifugal
Booster	9	To increase pressure in the distribution system or to supply elevated storage tanks.	Centrifugal
Well	2	To lift water from shallow or deep wells and discharge it to the treatment plant, storage facility, or distribution system.	Centrifugal or Ejector (Jet)
Chemical feed	4	To add chemical solutions at desired dosages for treatment processes.	Positive displacement
Sampling	6	To pump water from sampling points to the laboratory or automatic analyzers.	Positive displacement or Centrifugal
Sludge	5	To pump sludge from sedimentation facilities to further treatment or disposal.	Positive displacement

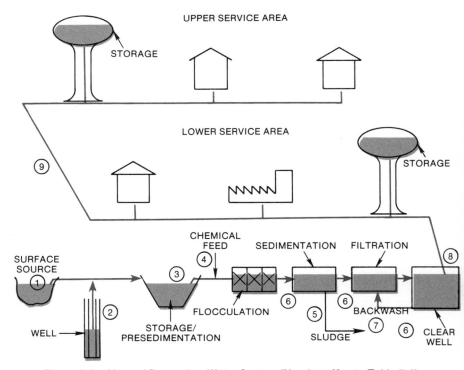

Figure 7-1. Uses of Pumps in a Water System (Numbers Key to Table 7-1)

Additional information related to this module is contained in Appendix D, Selection of Pumps and Prime Movers; Appendix E, Pump and Prime-Mover Installation; and Appendix F, Pump Trouble-Shooting Chart.

7-1. Types of Pumps

There are two basic categories of pumps used in water supply operations: VELOCITY PUMPS (also called KINETIC PUMPS) and POSITIVE-DISPLACEMENT PUMPS. Velocity pumps, which include centrifugal and vertical turbine pumps, are used for most distribution system applications. Positive-displacement pumps, which are commonly used in water treatment plants for chemical metering, are used only occasionally in distribution system operation.

Velocity Pumps

VELOCITY PUMPS use a spinning IMPELLER, or propeller, to accelerate water to high velocity within the pump body, or CASING (Figure 7-2). The high-velocity, low-pressure water leaving the impeller can then be converted to high-pressure, low-velocity water by shaping the casing so the water moves through an area of increasing cross section. This increasing cross-sectional area may be achieved with the VOLUTE (expanding spiral, Figure 7-3A) shape used in the common CENTRIFUGAL PUMP, or it may be the result of specially shaped DIFFUSER VANES or channels (Figure 7-3B), such as those built into the bowls of vertical turbine pumps.

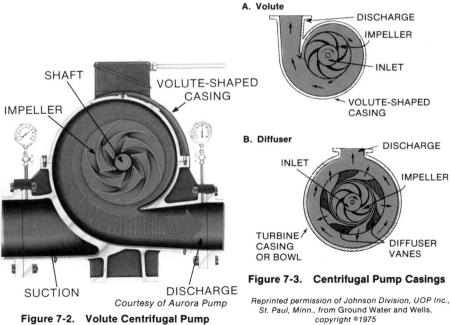

Figure 7-2. Volute Centrifugal Pump

Courtesy of Aurora Pump

Figure 7-3. Centrifugal Pump Casings

Reprinted permission of Johnson Division, UOP Inc., St. Paul, Minn., from Ground Water and Wells, copyright ©1975

A feature distinguishing velocity pumps from positive-displacement pumps is that velocity pumps will continue to operate undamaged, at least for a short period, when the discharge is blocked. When this happens, a shut-off HEAD (pressure) builds up that is typically greater than the pressure generated when pumping water, and water recirculates within the pump impeller and casing. This flow condition is referred to as SLIP.

Velocity Pump Designs and Characteristics

Depending on the casing shape, impeller design, and direction of flow within the pump, velocity pumps can be manufactured with a variety of operating characteristics.

Radial-flow pumps.   In the RADIAL-FLOW PUMP (Figures 7-2 and 7-3), water is thrown outward from the center of the impeller into the volute or diffusers that convert the kinetic energy (velocity) to pressure. The centrifugal pump commonly used in water supply practice is a radial-flow, volute-case type of pump. Centrifugal pumps of this type generally develop very high heads and have correspondingly low flow capacities.

Axial-flow pumps.   An AXIAL-FLOW PUMP (Figure 7-4) is often referred to as a propeller pump, and has neither a volute nor diffuser vanes. A propeller-shaped impeller develops head by the lifting action of the vanes on the water. As a result, the water moves parallel to the axis of the pump rather than being thrown outward as with a radial-flow pump. Axial-flow pumps develop very high flow (volume) capacity with limited heads. The low head capability creates a correspondingly low suction-lift capacity. Pumps of this design must have the impeller submerged at all times because the pump is not self-priming.

Mixed-flow pumps.   The MIXED-FLOW PUMP is a compromise in features between radial and axial flow pumps. The impeller is shaped so that centrifugal force will impart some radial component to the flow, as shown in Figure 7-5. This type of pump is useful for moving water with solids, as in raw-water intakes. The vertical turbine pumps commonly used for well and booster service are specially designed mixed-flow pumps using diffusers and multiple stages.

Multistage centrifugal pumps.   In general, any centrifugal pump can be designed with a multistage configuration. Each stage requires an additional impeller and casing chamber in order to develop increased pressure, which is additive to the pressure developed in the preceding stage. Although the pressure increases, the flow capacity of the pump does not increase beyond that of the first stage. There is no theoretical limit to the number of stages that are possible; however, mechanical considerations such as casing strength, packing leakage, and input power requirements do impose practical limitations.

Velocity Pumps Used in Water Distribution

Three designs of velocity pumps are widely used in water systems:

- Centrifugal pumps (volute pumps)
- Turbine pumps
- Jet pumps.

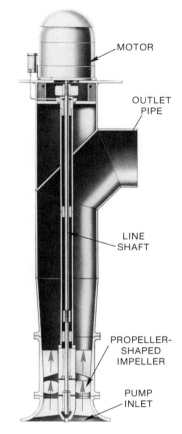

MOTOR

OUTLET
PIPE

LINE
SHAFT

PROPELLER-
SHAPED
IMPELLER

PUMP
INLET

*Courtesy of Worthington Pump Division,
Dresser Industries, Inc.*

Figure 7-4.   Axial-Flow Pump

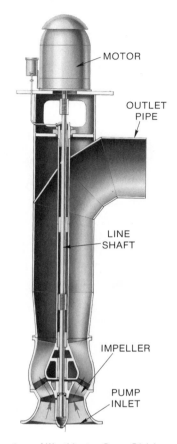

MOTOR

OUTLET
PIPE

LINE
SHAFT

IMPELLER

PUMP
INLET

*Courtesy of Worthington Pump Division,
Dresser Industries, Inc.*

Figure 7-5.   Mixed-Flow Pump

Centrifugal pumps.   The volute-casing type of centrifugal pump (Figure 7-6, also Figures 7-2 and 7-3A) is used in the majority of water utility installations. A wide variety of service requirements can be satisfied by variations in the width, shape, and size of the impeller, and in the clearance between the impeller and casing. Similarly, a wide range from low to high pump flow capacities is available depending on pump size. Head capabilities range up to 250 ft (76 m) per stage, with efficiencies ranging from approximately 50 percent for small pumps to 75 or 85 percent for larger units. As the clearance between the impeller and casing is increased or as the impeller surfaces are enlarged, INTERNAL BACKFLOW will increase while efficiency will decrease.

Initial cost is relatively low for a given pump size and maintenance is not an important factor. However, periodic checks are advised to monitor impeller wear and packing condition. The operating and maintenance sections of this module focus on horizontal centrifugal pumps. Advantages and disadvantages vary with the type of centrifugal pump used. A list of some of the more general advantages and disadvantages follows:

Advantages:

- Wide selection of capacity, head-fractional gpm to 50,000 gpm (190,000 L/min) and heads of 5–700 ft (1.5–210 m)
- Uniform flow at constant speed and head
- Simple construction; small amounts of suspended matter in water will not jam pump
- Low to moderate initial cost for a given size
- Adaptable to several drive types—motor, engine, turbine
- Moderate to high efficiency at optimum operation
- No internal lubrication required
- Little space required for a given capacity
- Noise level relatively low
- Can operate against a closed discharge valve for short periods without damage.

Disadvantages:

- Best efficiency limited to a narrow range of discharge flow and head
- Flow capacity greatly dependent on discharge pressure
- Generally not self-priming
- May run backward if stopped with the discharge valve open
- Impeller may be damaged by abrasive matter in water or clogged by particulate matter.

Vertical turbine pumps. VERTICAL TURBINE PUMPS have an impeller rotating in a channel of constant cross-sectional area, imparting mixed or radial flow to the water (Figure 7-7). As liquid leaves the impeller (Figure 7-8), velocity head is converted to pressure head by diffuser guide vanes. The guide vanes form channels that direct the flow either into the discharge or through DIFFUSER BOWLS into succeeding stage inlets.

Turbine pumps are manufactured in a wide range of sizes and designs, combining efficiency with high speeds to create the highest heads obtainable from centrifugal pumps. The clearance between diffuser and impeller is usually very small, limiting or preventing internal backflow and improving efficiency. Efficiencies in the range of 90–95 percent are possible in large units. However, the closely fitting impeller prohibits pumping of any solid sediments, such as sand, fine grit, or silt. Turbine pumps have a higher initial cost and are more expensive to maintain than centrifugal volute pumps of the same capacity.

For deep-well service, a shaft-type vertical turbine pump requires a lengthy pipe column housing, a drive unit, drive SHAFT, and multiple stages. In this type

Courtesy of Aurora Pump

Figure 7-6. Volute-Casing Centrifugal Pump

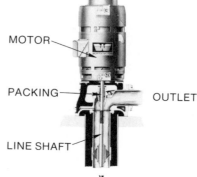

MOTOR

PACKING   OUTLET

LINE SHAFT

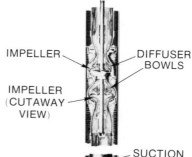

IMPELLER   DIFFUSER BOWLS

IMPELLER (CUTAWAY VIEW)

SUCTION PIPE

STRAINER

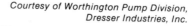

Courtesy of Worthington Pump Division, Dresser Industries, Inc.

Figure 7-8. Turbine Impeller

Courtesy of Worthington Pump Division, Dresser Industries, Inc.

Figure 7-7. Vertical Turbine Pump

of pump, a drive unit is located at the surface, with the lower shaft, impeller, and diffuser bowls submerged (Figure 7-7). This type of pump requires careful installation to ensure proper alignment of all shafting and impeller stages throughout its length. Deep-well turbines have been installed in wells with lifts of 2000 ft (610 m) and higher.

Multistage mixed-flow centrifugal pumps or turbine pumps with an integral or close-connected motor may be designed for operation while completely submerged, in which case they are termed SUBMERSIBLE PUMPS. These units are basically well pumps and may be installed directly into a well and supported by discharge piping, as shown in Figure 7-9.

Vertical turbine pumps are often used for inline booster service to increase pressure in a distribution system. Installing a CLOSE-COUPLED turbine pump with an integral sump or pot as shown in Figure 7-10 has become another increasingly popular configuration. The sump receives fluid and maintains an adequate level

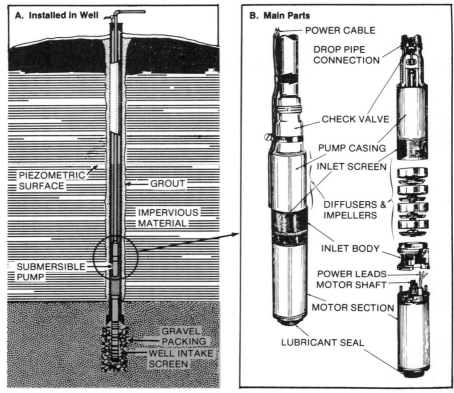

Figure 7-9.   Submersible Pump

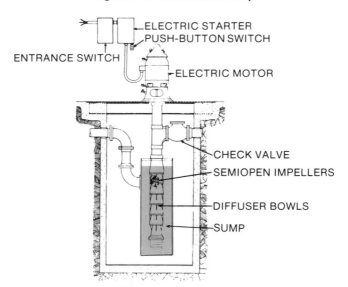

Figure 7-10.   Close-Coupled Turbine Pump With Integral Sump

above the turbine pump suction. Such a unit requires little space and can be installed under a street or in other locations without extensive supporting facilities.

Advantages and disadvantages of turbine pumps include the following:

Advantages:

- Uniform flow at constant speed and head

- Simple construction

- Individual stages capable of being connected in series, thereby increasing the head capacity of the pump

- Adaptability to several drive types—motor, engine, turbine

- Moderate to high efficiency under the proper head conditions

- Little space occupied for a given capacity

- Low noise levels.

Disadvantages:

- High initial cost

- Expensive to repair

- Support bearings located within casing require lubrication

- Cannot pump water containing any suspended matter (sand, sediment, etc.)

- Best efficiency limited to a very narrow range of discharge flow and head conditions.

Jet pumps.   The JET PUMP is classified as a velocity pump, but it operates in a somewhat different manner than the velocity pumps covered in the preceding paragraphs. The jet pump (Figure 7-11) uses a high-velocity flow of water, usually generated by a centrifugal pump. This stream of water is passed through an ejector, or VENTURI mechanism, creating a partial vacuum.

The Venturi cannot draw water to any great height by pure suction; such lifts can never practically exceed 25 ft (8 m) at sea level, with a decrease of 1 ft of lift for every 1000 ft in elevation (1 m for every 1000 m).[1] Instead, the suction brings water to a point where it is lifted to the surface by the "drive" water supplied by the centrifugal pump.

The discharge of the pump is split, with part of the water going to the distribution system and part of it returning to continue the operation of the Venturi. Because of this recycling, the jet pump does not have the efficiency of the centrifugal pump alone.

The jet pump finds few applications in most water supply pumping situations; however, it is used as a well pump for small and private water supplies. The jet may be housed within a well casing, as with the deep-well jet pump (which can raise water up to 2000 ft [610 m]), or it may be placed within the pump casing, as in shallow-well pumps.

---

[1] *Basic Science Concepts and Applications,* Hydraulics Section, Pump Head.

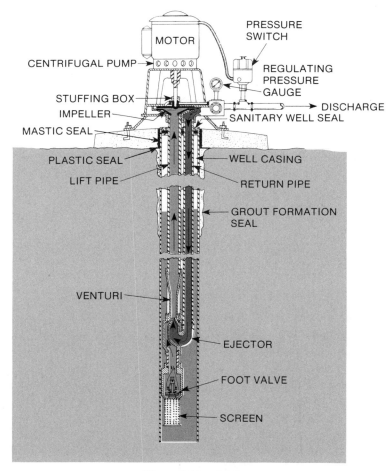

Figure 7-11.   Jet Pump

The advantages and disadvantages of a jet pump include:
Advantages:

- Low initial cost

- Low maintenance—no moving parts to wear out

- Moderate capacity with high potential suction lift.

Disadvantages:

- Discharge pressure limited to pressure of the water pressurizing the jet

- Requires high volume of supply flow available at low cost

- Low efficiency based on volume of water pumped.

Positive-Displacement Pumps

All POSITIVE-DISPLACEMENT PUMPS displace fixed volumes of fluid during each cycle or revolution of the pump. Many designs are available, with the reciprocating and rotary types being the most common for pumps pumping more than 1–2 gal/hr (4–8 L/hr). Positive-displacement pumps are no longer used for distribution system pumping in most water systems, but portable units may be used for dewatering excavations during construction. Smaller positive-displacement pumps, used for chemical feed, are discussed in Volume 2, *Introduction to Water Treatment,* Module 8, Fluoridation.

Reciprocating pumps.   RECIPROCATING PUMPS have a piston or plunger that moves back and forth in a closed cylinder or chamber fitted with necessary inlet and outlet valves. The piston alternately draws in and discharges fluid from the cylinder (Figure 7-12). Reciprocating pumps have single or multiple cylinders and are either single- or double-acting. Double-acting pumps displace liquid during both directions of piston movement.

Flow from reciprocating pumps pulsates; however, improved stability of flow can be obtained from multicylindered double-acting units. Reciprocating pumps are suited for applications where very high pressures are required. When fitted with hardened valves and valve seats, they work well for pumping abrasive slurries, such as sludge or muddy water from an excavation.

Rotary pumps.   ROTARY PUMPS have closely meshed gears, vanes, or lobes rotating within a close-fitting chamber. The two most common types are shown in Figure 7-13. As the gears or lobes rotate, fluid is moved from the pump inlet

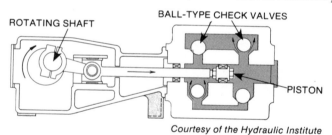

Courtesy of the Hydraulic Institute

Figure 7-12.   Double-Acting Reciprocating Pump

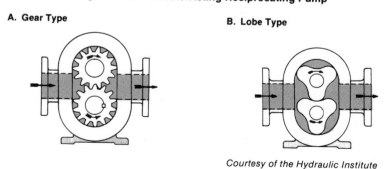

A. Gear Type

B. Lobe Type

Courtesy of the Hydraulic Institute

Figure 7-13.   Rotary Pumps

through the casing and pushed through the discharge pipe. Because of the close-fitting parts, high pressures are possible at the outlet and a partial vacuum is created at the inlet. Due to the partial vacuum, this type of pump can be self-priming under moderate lift conditions (up to 25 ft [8 m]).

7-2. Operation of Centrifugal Pumps

The procedures for centrifugal pump operation will vary somewhat from one brand of pump to another. Manufacturer's specific recommendations should be consulted before operating any unit. The procedures described in this module are typical and will serve as a guide if manufacturer's instructions are not available.

Pump Starting and Stopping

General procedures for pump starting and stopping are listed in Table 7-2. (Manufacturer's recommendations should be consulted and followed for each individual installation.) A major consideration in starting and stopping large pumps is the prevention of excessive surges and WATER HAMMER in the distribution system. Large pump-and-motor units have precisely controlled automatic operating sequences to ensure that the flow of water starts and stops smoothly. Battery-powered controls are often installed in large pump stations to ensure smooth shutdown of the pump in case of power failure.

Pump starting.   Note the importance of PRIMING—filling the pump casing and suction piping with water before starting the pump. Centrifugal pumps do not generate any suction when dry, so the impeller must be submerged in water for the pump to start operating. A FOOT VALVE is often provided on the suction

---

Table 7-2.   General Procedures for Centrifugal Pump Starting and Stopping

Pump Starting

Check pump lubrication.

Prime the pump—

    Where head exists on the suction side of the pump, open the valve on the suction line and allow any air to escape from the pump casing through the air cocks.

    Where no head exists on the suction side, if a foot valve is provided on the suction line, then fill the case and suction pipe with water from any source, usually the discharge line.

    Where no head exists on the suction side and no foot valve is provided, the pump must be primed by a vacuum pump or ejector operated with steam, air, or water.

After priming, start the pump with the discharge valve closed.

When the motor reaches full speed, open the discharge valve slowly to obtain the required flow. To avoid water hammer, do not open the valve suddenly.

Avoid throttling the discharge valve; this wastes energy.

Pump Stopping

Before shutting down the pump, close the discharge valve slowly to prevent water hammer in the system.

piping to hold the pump's prime. The foot valve is a one-way valve (a type of check valve) that prevents water from draining out of the pump when it is shut down.

Prevention of water hammer is important when starting a pump. A shut-off or stop valve on the discharge piping is used to create smooth changes of flow into the system. To accomplish this, the pump should be started with the valve fully closed. (This is true only for centrifugal and turbine pumps—positive-displacement pumps should never be started against a closed valve.)

Pump stopping.   A check valve should be installed in the discharge piping between the pump and shut-off valve to protect the pump from reverse flow and excessive backpressure. As soon as the pump stops discharging water, the check valve will swing shut and prevent reverse flow through the pump. On small pumps, where the swing-check valve closes rapidly enough to cause water hammer, relief valves or surge chambers may be used. On large pumps, the method of ensuring smooth shutdown is more complex. Under normal stopping conditions—that is, while main power is still applied to the drive motor—it is best to close the discharge valve slowly to approximately 95 percent of fully closed, then complete closure and stop the pump motor. In this manner, the pumping unit is eased off of the system. Some form of powered valve actuator (hydraulic, pneumatic, or electric) is used to obtain slow valve closure.

Whenever power failure occurs, the motor will stop while the discharge valve is still in the fully open position. In this case, it is desirable to close the discharge valve at a very rapid rate, 10–20 times faster than for normal closure, to approximately 95-percent closed, then close it the rest of the way at its normal slow rate of closure. Battery powered valves are often used to ensure smooth shutdown in case of power failure.

Whenever a pump must be shut down for more than a short period in freezing weather, the pump and associated suction and discharge piping should be drained of water to prevent freezing. If the pump will be out of service for some time, the pump and motor bearings should be flushed and regreased, and packing should be removed from the stuffing box. The units should be covered to prevent moisture damage to the motor electrical systems and to bearings.

Flow control.   Pumps are often operated at constant speed, with flow rate controlled by starting and stopping the unit. Flow rate can also be controlled by THROTTLING the discharge line, changing the configuration of the pump, or using variable-speed motors or pump drives.

The main advantages of flow control by starting and stopping a constant-speed pump include higher efficiencies (since constant-speed motors can be used) and standardization of motors and motor controls. Other advantages are lower initial cost per pumping unit, lower maintenance and repair costs for simple equipment, and fewer electrical circuits.

There are a number of disadvantages to starting and stopping the pump as a means of controlling output. To avoid excessive motor wear, the number of starts per hour may be generally limited to not more than one every 15 min for medium-size motors and less for larger motors. To comply with these motor-starting limitations, it is necessary to include storage facilities in the hydraulic system, or it may be necessary to allow for a greater range in level when pumping to an elevated tank. If a wide range in flow is required in the system, a station with only a few pumps will cause large changes in flow rate whenever a unit is started or stopped. A station designed to allow small incremental changes of flow rate, however, requires an excessive number of pumps.

Throttling is the partial closing of the pump discharge valve. It should be used only when elevated storage is not available or other, smaller sized pumping units are out of operation. It causes the entire flow to be lifted against a higher head than that required to meet the static head plus normal friction loss. In general, throttling should be avoided since it is an energy wasting practice.

Throttling the pump-discharge valve is usually the least efficient way to vary the rate of flow. However, under certain conditions over a limited range of flow, it can be shown that the efficiency of this means of control can be equal to or better than other means.

Changing a pump's configuration (such as adjusting the blade pitch of the impeller or the clearance between the impeller and suction cylinder) seems to offer some advantages since high-efficiency, constant-speed motors can be used. This type of equipment is, unfortunately, limited to large-capacity, low head, or special applications.

Adjustable-speed drives, also called variable-speed drives, provide the most common means for varying the pump-discharge flow rate. There are numerous variable-speed package drives on the market. These include continuously variable and stepped-speed motors, as well as constant-speed motors driving variable-speed electrical, hydraulic, and mechanical mechanisms coupled to the pump.

Monitoring Operational Variables

At any pumping station it is necessary to measure the amount of water pumped to provide a record of water delivered. Beyond this basic requirement, it is important to monitor pressure and tank-level gauges to ensure adequate and safe operation, and to provide a reasonable basis for system maintenance and control. Continual monitoring in combination with good records will provide a basis for a periodic maintenance schedule.

Module 9, Instrumentation and Control, covers the basic design and operation of the instruments used to monitor operations at a pump station. The following paragraphs discuss some of the specific types of data monitored and briefly note the types of equipment used to monitor them.

Suction-discharge heads.   Pressure gauges should be connected to both the suction and discharge lines of a pump, using the pressure taps supplied on the pump. The gauges should be mounted where the operator can monitor them conveniently, since they are necessary to accurately check pump performance.

Pump discharge- and suction gauge-readings must be referred to a reference point, normally the impeller eye or shaft centerline. The pressure gauges can also be used to determine static head by simply turning the pump off before taking the gauge readings.

Bearing and motor temperature. Special thermometers or bimetal temperature indicators are the usual means for measuring temperature on site. These instruments need to be checked to determine if overheating is occurring when only a local indication is required. Electronic temperature sensors are used where the temperature must be displayed at a distance from the pump station.

Vibration. Vibration detectors are sometimes used on large pump and motor installations to sense equipment malfunction, such as misalignment, bearing failures, single-phasing, or any condition that would result in excessive vibration. Vibration monitoring is discussed further under the Pump Maintenance section.

Speed. Monitoring pump speed is important, particularly with variable-speed pumps that may experience CAVITATION at low speeds. Centrifugal-speed switches can be installed on the pumping unit, or contacts can be provided on a mechanical speed-indicating instrument, to sound alarms or shut off the system in case of over- or underspeed. Other systems use a tachometer generator that generates a voltage in proportion to speed; this voltage is used to drive a standard indicator near the pump or at a remote location. Underspeed and overspeed alarms can be activated by the speed-sensing device.

General observations. Other items an operator should monitor include surge-tank air levels, recording meters, and intake-pipe screens. Packing should be checked to be sure it never runs dry. Idle pumps should be started and run weekly. All operations and maintenance should be recorded in log books.

Finally, although a variety of regular inspections should be performed as part of a preventive-maintenance schedule (as discussed in the next section), it is also important that the operator remain attentive to the general condition of the pump on a day-to-day basis. Unusual noises, vibrations, excessive seal leakage, hot bearings or packing, or overloaded electric motors are all readily apparent to the alert operator who is familiar with the normal sound, smell, sight, and feel of the pump station. Reporting and acting on such problems immediately can prevent major damage that might occur if the problem were allowed to remain until the next scheduled maintenance check.

7-3. Centrifugal Pump Maintenance

A regular inspection-and-maintenance program is important in maintaining the condition and reliability of centrifugal pumps. Bearings, seals, and other moving parts all require regular adjustment or replacement because of normal wear. General housekeeping is also important in prolonging equipment life. This section details the mechanical construction of the common centrifugal pump and gives general recommendations for maintenance schedules and procedures. Manufacturer's recommendations should be followed for best results with a specific installation.

Mechanical Details of Centrifugal Pumps

Size and construction may vary greatly from one volute-type centrifugal pump to another, depending on the operating head and discharge conditions for which they are designed. Basically, however, the principle of operation is the same.

Water enters the impeller eye from the pump suction inlet. There it is picked up by curved vanes, which efficiently change the flow direction from axial to radial. Both pressure and velocity increase as the water is impelled outward and discharged into the pump casing. The major components of typical volute-type centrifugal pumps are described in the following paragraphs.

Casing.   As the water leaves the impeller, it is traveling at high velocity in both radial and circular directions. To minimize energy losses due to turbulence and friction, the casing is designed to convert the velocity energy to additional pressure energy as smoothly as possible. This is accomplished in most water-utility pumps by casting the case in the form of a smooth volute, or spiral, around the impeller.

Casings are available in a variety of forms—single or double suction; side, bottom, or end suction; side or bottom discharge; and multistage. Cast iron is the most common construction material, but ductile iron, bronze, and steel are usually available on special order.

Single-suction pumps.   SINGLE-SUCTION PUMPS are designed with the water inlet opening at one end of the pump and the discharge opening placed at right angles on one side of the casing (Figure 7-14). Single-suction pumps, also called end-suction pumps, are used in smaller water systems that do not have a high volume requirement. These pumps are capable of delivering up to 200 psi (1400 kPa) pressure if necessary, but for most applications they are usually sized to produce 100 psi (700 kPa).

On some single-suction pump units, the impeller is mounted on the shaft of the motor driving the pump, with the motor bearings supporting the impeller (Figure 7-14A). These pumps are called CLOSE COUPLED. Single-suction pumps are also available with the impeller mounted on a separate shaft, which is connected to a motor with a coupling (Figure 7-14B). In this design, which is referred to as FRAME MOUNTED, the impeller shaft is supported by bearings placed in a separate housing, independent of the pump housing.

The casing for a single-suction pump is manufactured in two or three sections or pieces. All housings are made with a removable inlet-side plate or cover. This side plate is held in place by a row of bolts located near the outer edge of the volute. Removing the side plate provides access to the impeller. The pump does not have to be removed from its base in order to remove the side plate; however, all suction piping must be removed to provide sufficient access.

Some manufacturers cast the volute and the back of the pump as a single unit. Other manufacturers cast them as two separate pieces, which are connected by a row of bolts, similar to the inlet side plate. Units having separate backs permit the removal of the impeller and drive unit from the pump without disturbing any piping connections.

Double-suction pumps.   In DOUBLE-SUCTION PUMPS (Figure 7-15), water enters the impeller from two sides and discharges from the middle of the pump, off of the impeller vane tips. Although water enters the impeller from each side, it enters the housing at one location (usually on the opposite side of the discharge opening). Internal passages in the pump guide the water to the impeller suction and control the discharge-water flow. The double-suction pump is easily

A. Close-Coupled Design

B. Frame-Mounted Design

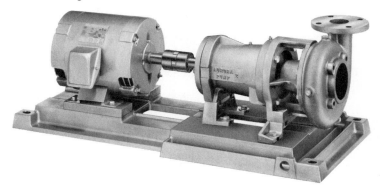

Courtesy of Aurora Pump

Figure 7-14.   Single-Suction Pumps

identified because of its casing shape (Figure 7-16). The motor is connected to the pump through a coupling, and the pump shaft is supported by ball or roller bearings mounted external to the pump casing.

The double-suction pump is usually referred to as a horizontal split-case pump. The term horizontal does not indicate the position of the pump, but refers to the fact that the housing is split along the centerline of the pump shaft, which is normally set in the horizontal position. However, some horizontal split-case pumps are designed to be mounted in the vertical position with the drive motor placed on top. Double-suction pumps can pump over 10,000 gpm (38,000 L/min), with heads up to 350 ft (100 m) and are widely used in large systems.

Removing the bolts that hold the two halves of the double-suction casing together permits the top half to be removed. Most manufacturers place two dowel pins in the bottom half of the casing to ensure the proper alignment between the halves when they are reassembled. It is important that the machined surfaces not be damaged when separating the halves.

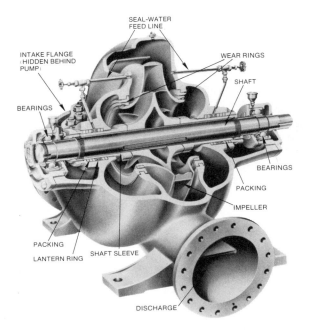

Figure 7-15.   Double-Suction Pump

Figure 7-16.   Double-Suction Pump

A. Semiopen

B. Closed

Courtesy of Goulds Pumps, Inc.

Figure 7-17.   Impellers

Impeller. Almost all pump impellers for water-utility use are made of bronze, although a number of manufacturers offer cast iron or stainless steel as alternative materials. The overall impeller diameter, impeller width, inlet area, vane curvature, and speed affect impeller performance and can be modified by the manufacturer to attain required operating characteristics. Impellers for single-suction pumps may be of the open, semiopen, or closed design, as shown in Figure 7-17. Most single-suction pumps in the water industry use impellers of the closed design, although a few have semiopen impellers. Double-suction pumps use only impellers of the closed design.

Wear rings. In all centrifugal pumps having impellers of the closed design, a flow restriction must exist between the impeller discharge and suction areas, to prevent excessive circulation of water between the two. This restriction is accomplished by WEAR RINGS. In some pumps, only one wear ring is used, mounted in the case. In others, two wear rings are used, one mounted in the case and the other on the impeller (Figure 7-15). In inexpensive pumps, wear rings are not used, but both the impeller and the case are machined to perform the same function.

The rotating impeller wear ring (or the impeller itself) and the stationary case wear ring (or the case itself) are machined so that the running clearance between the two effectively restricts leakage from the impeller discharge to the pump suction. The clearance is usually 0.010–0.020 in. (0.25 mm–0.50 mm). Rings are normally machined from bronze or cast iron, but stainless-steel rings are available. The machined surfaces will eventually wear to the point that leakage occurs, decreasing pump efficiency. At this point, the rings need to be replaced or the wearing surfaces of the case and impeller need to be remachined.

Shaft. The impeller is rotated by a pump SHAFT, usually machined of steel or stainless steel to provide sufficient strength and stiffness. There are several methods used to secure the impeller to the shaft. A common method for double-suction pumps is with a KEY and light INTERFERENCE FIT (also called a shrink fit) of 0.002–0.003 in. (0.050–0.075 mm). This method offers simplicity of design with reasonable ease of maintenance. An ARBOR PRESS or gear puller will be required to remove an impeller secured in this manner.

In pumps of the end-suction design, the impeller is mounted on the end of the shaft and held in place by a nut. The end of the shaft may be machined straight or with a slight taper; however, removing the impeller will usually not require a press. Several other methods are also used for mounting impellers.

Shaft sleeves.   Most manufacturers provide pump shafts with replaceable sleeves for the packing rings (discussed in the next paragraph) to bear against. If sleeves were not used, the continual rubbing of the packing would eventually cause the shaft to wear and require replacement. A shaft could be ruined almost immediately by overtightening the packing gland. Where shaft sleeves are used, a damaged surface can be repaired by replacing the sleeve, a procedure considerably less costly than replacement of the entire shaft. The sleeves are usually made of bronze alloy, which is much more resistant to the corrosive effects of water than steel. Stainless-steel sleeves are usually available for use where the water contains abrasive elements.

Packing rings.   There is a potential for leakage at the point where the shaft protrudes through the case. Either PACKING RINGS or a MECHANICAL SEAL (discussed below) is used to seal the space between the shaft and the case, preventing most of this leakage.

Packing consists of one or more (usually no more than six) separate rings of graphite-impregnated cotton, flax, or synthetic materials placed on the shaft or shaft sleeves (Figure 7-18). (Asbestos material, once common for packing, is no longer used as a packing material on potable-water systems.) The section of the case in which the packing (or mechanical seal) is mounted is called the STUFFING BOX. The adjustable PACKING GLAND maintains the rings under slight pressure against the shaft, stopping air from leaking in or water from leaking out.

To reduce the friction of the packing rings against the pump shaft, the packing material is impregnated with graphite or polytetrafluoroethylene to provide a small measure of lubrication. It is important that packing be installed and adjusted properly, as discussed in more detail later in this section.

Lantern rings.   When a pump operates under SUCTION LIFT, pressure at the impeller inlet is below atmospheric, and air will enter the water stream along the shaft if the packing does not provide an effective seal. It may be impossible to tighten the packing sufficiently without causing excessive heat and wear on the packing and shaft or shaft sleeve. To solve this problem, a LANTERN RING (Figure 7-19) is placed in the stuffing box. Pump discharge water is fed into the ring and flows out of it through a series of holes leading to the shaft side of the packing. From there, water flows both towards the pump suction and out from the packing gland. This water acts as a seal, preventing air from entering the water stream. It also provides lubrication for the packing.

Mechanical seals.   If the pump must operate under a high suction head (60 psig [400 kPa (gauge)] or more) the suction pressure itself will compress the packing rings regardless of the operator's care, and packing will require frequent replacement. Most manufacturers will recommend use of a mechanical seal under these conditions, and many manufacturers use mechanical seals for suction lift conditions as well. The mechanical seal (Figure 7-20) is provided by two machined and polished surfaces—one is attached to and rotates with the

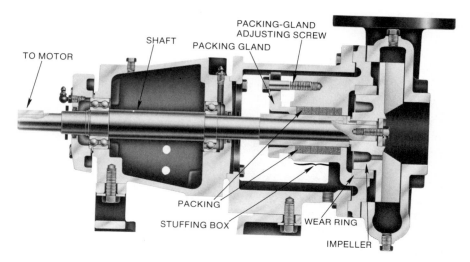

Figure 7-18.   Packing

Courtesy of Aurora Pump

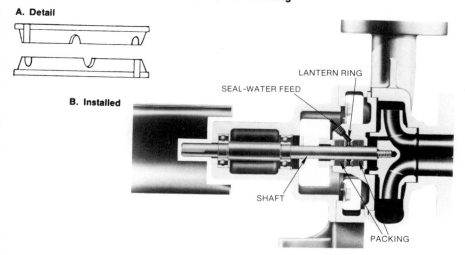

Figure 7-19.   Lantern Ring

Courtesy of Aurora Pump

shaft; the other is attached to the case. Contact between the seal surfaces is maintained by spring pressure.

The mechanical seal is designed so that it can be hydraulically balanced, so that the wearing force between the ground surfaces does not vary regardless of the suction head. Most seals have a 5000–20,000-hour operating life. In addition, there is little or no leakage from a mechanical seal—a leaky mechanical seal indicates problems that should be investigated and repaired. A major disadvantage of mechanical seals is the relative difficulty of replacement

A. Detail

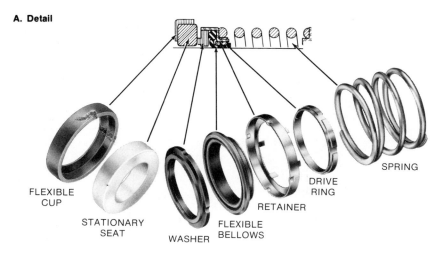

FLEXIBLE
CUP

STATIONARY
SEAT

WASHER

FLEXIBLE
BELLOWS

RETAINER

DRIVE
RING

SPRING

B. Installed

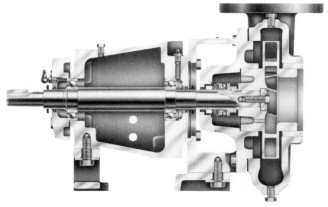

Courtesy of Aurora Pump

Figure 7-20.   Mechanical Seal

compared with packing rings. Replacement of the mechanical seal on many pumps requires removal of the shaft and impeller from the case. Another disadvantage is the fact that failure of a mechanical seal is usually sudden, accompanied by excessive leakage. Packing rings, by contrast, normally wear gradually, and the wear in most cases can be detected long before leakage becomes objectionable.

Bearings.   Most modern pumps are equipped with ball-type radial and thrust BEARINGS. These bearings are offered with either grease or oil lubrication and provide good service in most water-utility applications. They are reasonably easy to maintain when manufacturer's recommendations are followed, and new parts are readily available if replacement is required. A ball bearing will usually give audible warning of impending failure so that time can be provided for a planned shutdown.

Couplings.   The pump and motor may be directly coupled (close coupled), with the impeller actually mounted on the motor shaft, or they may have separate shafts connected by a separate COUPLING. In addition to their primary function of transmitting the rotary motion of the motor to the pump shaft, couplings are also designed to allow slight misalignment between the pump and motor, and to absorb the start-up shock when the pump motor is switched on. Although the coupling is designed to accept misalignment, the more accurately the two shafts are aligned, the longer the coupling life will be, and the more efficiently the unit will operate.

Various designs of couplings are supplied by pump manufacturers. Couplings may be installed dry or may be lubricated. The majority of the couplings that are furnished are of the lubricated style and require periodic maintenance, usually lubrication at 6-month or annual intervals. Dry couplings using rubber or elastomer membranes do not require any maintenance except for periodic visual inspection to make sure they are not cracking or wearing out. The rubber or elastomer used for the membrane must be carefully selected for the pump, because the corrosive chemicals used in water treatment plants can affect the life and operation of the coupling.

Inspection and Maintenance

A well-defined schedule of inspection and maintenance is necessary to ensure long and reliable pump service and to preserve warranty rights on new units. It is important to maintain complete records of inspections and any service performed. Often pump condition can best be evaluated by comparing the pump's current performance to its performance when first installed, so complete testing immediately following installation is important. When problems are encountered, the Pump Trouble-Shooting Chart given in Appendix F will be useful in locating the problem. Periodic checks of the following factors will ensure maximum operating efficiency and minimum maintenance expenditures:

- Priming
- Packing and seals
- Bearings
- Vibration
- Alignment
- Sensors and controls
- Head (pressure gauges).

Priming.   The pump must be checked to be sure it is primed before start-up. With a partially-primed pump, capacity will be reduced and water-lubricated internal wear rings may be damaged. If a pump is primed from an overhead tank or other gravity supply, it may be started as soon as water shows at the top vent cocks. If it is primed by a vacuum pump, the action of the device itself will indicate when the casing is filled with water. Be sure that all valves in the suction line are fully open. Petcocks on top of the pump case should be opened routinely during operations to bleed off any air that might collect there. Continuous bleeds or air-release valves may be necessary for some applications (for example, automatically controlled pumps).

Packing.   A stuffing box that has been correctly packed and adjusted should be trouble-free. Packing should be inspected annually if a pump is run on a regular basis. The easiest way to accomplish this is to remove the top of the casing or the packing gland. When the packing wears or is compressed to the point where it is impossible to tighten the gland further, a new set of rings should be installed. It is generally not considered good practice to add new rings on top of old ones in order to make up for wear and compression, although the addition of one more ring may sometimes be allowed by manufacturer's maintenance specifications.

To achieve proper life, new packing must be installed with care. The following guidelines should be observed:

- Keep parts arranged in the order in which they were removed, to ensure that they are reinstalled correctly.

- Clean the stuffing box.

- Check the shaft sleeve; replace if badly worn.

- During disassembly, record the order in which parts are removed, including the number of rings before and after the lantern ring.

- Use packing and sleeve material of the proper size and of a material that is compatible with the expected service. For a severe abrasive or corrosive service, consult the pump manufacturer. Although precut packing is available, most operators purchase bulk packing and cut what they need for each job.

- Replace parts in the same order that they were removed.

- Replace all packing rings. The ends may be cut diagonally or square. They should be carefully butted, with joints staggered at least 90°. Four or more rings are usually placed with their ends at 90° intervals. Be sure to replace lantern ring in its original sequence.

- Never overtighten the packing. Draw each ring down firmly. After all of the rings are installed, back off on the gland nuts about one full turn before the pump is started. Before its initial start-up with new packing, the pump should rotate fairly easily by hand unless it is very large. If it does not turn over, the packing should be loosened until it does, provided there are no other problems.

- At start-up, the packing should be allowed to leak freely until conditions stabilize and the operator has determined that there are no hot spots in the area of the packing or bearings. The packing gland can then be tightened very gradually until the packing just drips.

- While the pump is running, adjust the gland by tightening the gland-bolt nuts. Tighten the nuts evenly, and tighten each nut no more than $1/6$ of a turn every 20–30 min.

- Never tighten packing glands to the point where there is no leakage, since this will cause premature packing wear and scored shaft sleeves.

- After the initial installation and adjustment, check the packing regularly and adjust the gland whenever leakage increases. The leakage rate should be checked daily if possible.

- If cooling or sealing water is injected into the box, the flow pressure should be set to 15–20 psi (100–140 kPa) more than the pressure on the inboard end of the stuffing box. Excessive pressure will cause increased wear of the packing and sleeve.

Mechanical seals. The operating temperature of a mechanical seal should never exceed 160° F (71° C). It may be necessary to use water for cooling. The water can be supplied by the pump discharge or from an external source of the same (or better) quality as the water being pumped. The water must not contain dirt, grit, or other abrasive materials that could damage the seal.

Check to be certain that the mechanical seal is designed for the stuffing-box pressure at which it will operate. A pump that develops a partial vacuum on the suction side must be fitted with a close-clearance bushing between the seal and the suction passage of the pump, in order to maintain lubrication, cooling, and flushing fluid at the seal.

Bearings. Regular inspection and lubrication of bearings is essential to efficient pump and motor operation. Lubrication points should be checked and lubricated at the intervals prescribed by the manufacturer. The following checks are important:

- Oil level in bearing housings

- Free movement and proper operation of oil rings

- Proper oil flow for pressure-feed systems

- Proper type of grease for grease-lubricated bearings

- Proper amount of grease in the housing

- Bearing temperature.

Oil-lubricated bearings. The bearing housing of oil-lubricated bearings should be filled with a good grade of filtered mineral oil. The oil should be changed after the first month of operation. After the first oil change, future oil changes should be performed every 6 to 12 months, depending on operating frequency and environmental conditions. Whenever oil is changed, and especially after the first oil change, the oil should be inspected for signs of bearing wear or excessive dirt. The following viscosities are recommended for use with various antifriction bearings:

- Ball and cylindrical roller bearings—70 ssu (Saybolt Standard Units) oil rated at operating temperature

- Spherical roller bearings—100 SSU oil rated at operating temperature

- Spherical roller and thrust bearings—150 SSU oil rated at operating temperature.

It is important that the bearing housing not be overfilled with oil. Most housings have the correct oil level indicated on a sight glass.

Grease-lubricated bearings.   For grease-lubricated bearings, similar maintenance is required. During the initial run-in period, bearing temperatures must be closely monitored. Bearings that are running within an acceptable temperature range can be touched with the bare hand. After about one month of initial full-service operation, all bearings should be regreased.

Bearings should not be overgreased. Because of the internal friction caused by the churning grease, a bearing will run hotter if the grease pocket is packed too tightly.

The grease used for lubricating antifriction pump bearings should be a sodium soap-base type meeting Anti-Friction Bearing Manufacturer's Association group number 1 or 2 classifications. Bearings should be regreased every 3 to 6 months using the following procedures:

- Open grease drain plug.

- Fill the bearing with new grease until it flows from the drain plug.

- Run the pump with the drain plug open until the grease is warm and does not flow from the drain.

- Close the drain plug.

It is not necessary to add grease to bearings between intervals unless the bearing seals are bad and grease has been lost. If bearings run hot, the grease drain plug should be opened and some grease allowed to drain out. If the bearing has been disassembled, cleaned, and flushed, then the housing should be refilled to one third of its capacity.

Bearing replacement.   The life of a ball bearing will vary with the conditions of load and speed. Most pump bearings have a minimum operating life of 15,000 hour (about 1 1/2 years). As a general rule, if a pump has been in operation for one or two years of continuous service, then the bearings should be replaced when the pump is taken out of service for repairs. It is more economical to replace bearings while other work is being performed than to wait until excessive heat or noise warns that the bearings are about to fail.

Considerable care is needed when bearings are removed or installed. Appropriate bearing pullers and hydraulic presses should be used, especially if there is a chance that bearings will have to be reused. It is often a good policy to leave serviceable bearings mounted until replacements have been obtained. Proper installation procedures should be followed to prevent damage to new bearings during installation.

Vibration monitoring.   On high-speed or large pump units, instruments may be used to periodically monitor vibration in the vertical, lateral, and axial planes, in order to get an early indication of possible future problems. When problems occur, an experienced person will usually be able to detect undesirable vibration merely by touching the unit. In general, on units where periodic checks of vibration will be performed, a test of vibration should be made immediately after installation to establish a baseline condition. This should be followed by periodic measurements of vibration at intervals recommended by the manufacturer. For critical equipment operated continuously, monthly checks may be required. Less critical equipment may be checked annually.

Sensors may be installed on the pumping unit, or portable equipment may be used for the measurements. Vibration readings should be taken on the shafts in or near the bearings. Any change from the baseline measurements of vibration magnitude or frequency indicates potential problems.

When vibration or overheating of a bearing or coupling is observed, the pump rotating member should be inspected immediately for possible imbalance. The imbalance could be a result of the presence of foreign material, including the accumulation of scale, loss of material due to corrosion, CAVITATION, or mechanical breakage.

Alignment.   Excess vibration may also be caused by misalignment. Pumping units should be given an initial alignment check after they are first installed and brought up to operating temperature (often called a "hot alignment" check). This is particularly important for frame-mounted pumps. A record should be made of the initial dial-indicator readings, using the pump coupler key as a reference point. Periodically, alignment should be checked again, to ensure that the initial readings have not changed. Vibration due to misalignment is a common cause of bearing failures.

Sensors and controls.   Operating personnel should know how and why each part of the automatic control system functions and understand its effect on the operation of the system. The normal sequence and timing of each operation should be determined and any deviations identified for correction. Technicians familiar with each type of equipment should repair and adjust all of the more complex control devices. Pressure and float sensors may require adjustment seasonally or with changes in water demand.

Head.   Pump discharge and suction heads should be checked and compared to baseline performance figures for the pump when it was first installed. Some wear will necessarily reduce performance, but major reductions in capacity should be identified and corrected before the inefficient unit begins to waste large amounts of energy. Strip or circular chart recordings of pump performance can be helpful in identifying reduced pump capacity.

An installed centrifugal pump can be checked for wear by closing the discharge valve and then reading the discharge pressure. The observed discharge pressure can be compared with the original pump characteristics after making appropriate deductions for suction pressure. If the shut-off head is close to the original value, the pump is not greatly worn.

Cavitation.   Under certain circumstances, a pump can pull water so hard that some of the water turns to small bubbles of vapor. This is called CAVITATION. The bubbles explode against the impeller, eroding the metal. To avoid cavitation, a pump should never be operated continuously above its design conditions. In addition, the suction requirements for the pump must be met, and turbine and submersible pumps must be provided with a deep enough sump to keep them submerged. Intake screens should be routinely cleaned to avoid suction restrictions.

Major repairs. Major repair jobs that may be required for pumps at infrequent intervals include replacement of bearings, wearing rings, shaft sleeves, and impellers. These jobs all require that the pump be removed from service for some length of time, and planning is essential to ensure that the work goes smoothly and quickly. In many cases, the pump will be returned to the manufacturer or the utility shop for repair. If the operator will be performing the repair in the field, it is especially important to have all parts on hand before the pump is removed from service.

Care should be taken to ensure that the correct replacement parts are used. Bearings, for example, may be available that look identical to the original part and install with no trouble, but that are not adequate to handle the pump's full load. The parts themselves or accurate specifications can be obtained from the pump manufacturer before the repairs are begun.

In addition to replacement parts, any necessary special tools should be on hand before the job is started. For example, bearing removal will usually require a bearing puller or a hydraulic press, possibly fitted with special collars.

7-4. Types of Electric Motors and Controls

Electric motors are used as PRIME MOVERS to power more than 95 percent of the pumps used in municipal water-supply operations. Electric motors are available in a wide range of types, speeds, and power capabilities. Their inherent characteristics of smooth power output and high starting torque make them well suited for direct connection to centrifugal pumps. Most electric motors used by water utilities are powered by alternating current (AC)[2] electricity supplied by power utilities. Direct current (DC) electric motors are not covered in this module.

Basic Principles

Any electric motor, AC or DC, uses an arrangement of electrical circuits and magnetic fields in such a way that a rotating magnetic field is established within the stationary windings (the STATOR).[3] The rotating magnetic field pulls the iron of the ROTOR (the rotating member of the motor) around with it. The magnetic field has a number of north and south poles. The number of poles in a motor must be at least two, but it can be more, depending on how the stator is made and how the windings (current-carrying coils of wire) are wired and connected.

Speed. The speed at which the magnetic field rotates is called the motor's SYNCHRONOUS SPEED and is expressed in revolutions per minute (rpm). For a motor operating on an electric power system having a frequency of 60 Hz[4] (60

---

[2]*Basic Science Concepts and Applications,* Electricity; Electricity, Magnetism, and Electrical Measurements.

[3]*Basic Science Concepts and Applications,* Electricity; Electricity, Magnetism, and Electrical Measurements.

[4]*Basic Science Concepts and Applications,* Electricity; Electricity, Magnetism, and Electrical Measurements (Alternating Current).

cycles per second), the maximum synchronous speed is 3600 rpm, or 60 revolutions per second, which is achieved with a 2-pole motor. Motors may be designed to run at fractions of 3600 rpm by increasing the number of poles in the stator—a 4-pole motor has a synchronous speed of 1800 rpm; a 6-pole has a synchronous speed of 1200 rpm. Depending on the type of motor (synchronous or induction), it may run at exactly its synchronous speed or somewhat less (for example, 3450, 1750, or 1150 rpm). Electronic motor controllers are now available that can vary the speed of some types of motors over a wide range.

Starting current.   It takes more current to start most motors than it does to keep them running. The current drawn by the motor the instant the motor is connected to the power system, while the motor rotor is still at rest, is called the LOCKED-ROTOR CURRENT, MOTOR-STARTING CURRENT, or INRUSH CURRENT. The locked-rotor current is often from 5 to 10 times the normal full-load current of the motor. When the motor is started, the locked-rotor current starts out at its maximum value, then decreases to the motor's ordinary current draw as the motor reaches full speed.

The time required for a motor to start from rest and reach full speed varies with the load and type of motor, but it generally will be from 1 to 8 sec. During this time, the excessive current drawn by the motor causes an abnormal voltage drop in the power system. Excessive voltage drops during starting can cause dimming of lights and other problems in the system, such as the drop-out of relays, solenoids, or other motor starters.

The size of the motor has a considerable influence on the motor-starting problem and its effects. Most electric utilities have a DEMAND CHARGE for power, based in part on a motor's starting current, in addition to the usual charge for kilowatt-hours of power used.

Single-Phase and Polyphase Motors

The alternating current electricity used to power electric motors in utilities is supplied from power utilities in the form of THREE-PHASE current.[5] This may be reduced to SINGLE-PHASE current within the plant. Light-duty applications use single-phase electricity driving single-phase motors. This type of power and motor is the same as commonly used for 110-V household electrical systems and appliances; 240-V single-phase power is also available. Most utility pumping operations require the greater power that the three-phase electrical system can supply to industrial-type three-phase motors.

Single-phase motors.   Single-phase motors are generally fractional-horsepower units, but they can be supplied with ratings up to 10 hp (7.5 kW) at 120 or 240 V. A motor with a straight single-phase winding has no starting torque, but if it is started and brought up to speed by some outside device, it will run and continue to carry its load. Thus, some means must be provided to start the motor turning. The starting mechanism is usually built into the motor in the form of specially wired additional windings.

---

[5] *Basic Science Concepts and Applications,* Electricity, Electrical Quantities and Terms (Phase).

Single-phase motors used as prime movers are usually one of three types:

- Split phase—the motor rotor is a plain squirrel cage requiring no windings; it has a comparatively low starting torque and requires a comparatively high starting current.

- Repulsion-induction—compared to the split-phase motor, this motor is more complex and expensive; it has a high starting current.

- Capacitor start—this motor has a high starting torque and a high starting current, but it is limited to applications in which the load can be brought up to speed very quickly and infrequent starting is required.

Three-phase motors. Most motors ½ hp (4 kW) and larger in water distribution systems are three-phase, rated at 230, 460, 2300, or 4000 V. There are three general classes of three-phase motors:

- Squirrel-cage induction
- Synchronous
- Wound-rotor induction.

Squirrel-cage induction. The SQUIRREL-CAGE INDUCTION MOTOR is the simplest of all AC motors. The rotor windings consist of a series of bars placed in slots in the rotor, connected together at each end—this gives the rotor the appearance of a squirrel cage. The stator windings, located in the frame, are connected to the power supply. Current flowing through the stator windings sets up a rotating magnetic field within the motor. This rotating magnetic field induces a current flow in the rotor windings, which generates an opposing magnetic field. The force between the two magnetic fields causes the rotor to turn.

No elaborate controls are required for squirrel-cage motors. Simple starting controls are adequate for most of the normal- and high-starting-torque applications for which these motors are used.

Synchronous. The power-supply leads of a SYNCHRONOUS MOTOR connect to the stationary windings in such a way that a revolving magnetic field is established, rotating at synchronous speed. The rotor is constructed with poles to match the poles of the stator, and the rotor pole pieces are supplied with direct current. This requires that a slip-ring assembly, called a COMMUTATOR, and BRUSHES be used to connect the direct current to the rotating parts of the motor. Thus, in the synchronous motor, alternating current is supplied to the stator and direct current is supplied to the rotor. Where the consumer of purchased power pays a penalty for low POWER-FACTOR[6] conditions, the synchronous motor is often a logical choice, since it has a power factor of 1.0. Where speed must be held constant, the synchronous-type motor is necessary.

Wound-rotor induction. The WOUND-ROTOR INDUCTION MOTOR has a stator similar to that of the squirrel-cage induction motor, with a number of pairs of

---

[6] *Basic Science Concepts and Applications,* Electricity, Electrical Quantities and Terms (Power Factor).

poles and their associated stator windings connected to the three-phase power supply. The rotor has the same number of poles as the stator, and the rotor windings are wired out through slip rings. This allows the resistance of the rotor circuit to be controlled while the motor is running, which varies the motor's speed and torque characteristics.

The wound-rotor motor offers ease of starting in addition to variable-speed operation. The starting current required by a wound-rotor motor is seldom greater than normal full-load current. In contrast, squirrel-cage and synchronous motors may have starting currents from 5 to 10 times their normal full-load currents.

Electric Motor Construction

Industrial electric motors are used for a wide range of loads, environmental conditions, and mounting configurations. To meet these conditions, motor manufacturers have standardized most motors and motor modifications in sizes up to 200 hp (1500 kW). Larger motors are not necessarily standardized; however, they have certain standard construction features.

Temperature.   Motors convert electrical energy into mechanical energy and heat. The heat given off by a motor can be readily calculated from its horsepower (watt) rating and its efficiency. For example, if a 100-hp (750-kW) motor is 95-percent efficient, 95 percent of the energy supplied to it generates useful work—the remaining 5 percent (5 percent of 750 kW = 37.5 kW) is lost as heat. This heat must be removed from the motor as fast as it is produced to prevent the motor temperature from rising too high. The construction of a motor is influenced greatly by its cooling and venting requirements.

Motors are designed to operate with a maximum AMBIENT (surrounding) temperature of 104° F (40° C). However, the internal temperature a motor can operate under will be higher than the ambient temperature, depending on the type of insulation used for the motor windings. The air supplied to ventilate a motor should be at the lowest temperature possible and must not exceed 104° F (40° C). The useful life of any motor is extended by having plenty of cool ventilating air available.

Mechanical protection.   Various motor designs provide different protection to the windings and internal motor parts from falling, airborne, or wind-driven particles and from various atmospheric and surrounding conditions. Motor designs commonly available include open, drip-proof, splash-proof, guarded, totally enclosed, totally enclosed with fan-cooling, explosion-proof, dust-proof, etc. In most cases, the name clearly describes the unique features of each type of construction. Note that internal explosions could occur in explosion-proof motors; however, the explosions would be contained within the motor housing, preventing damage or ignition of the surrounding area.

Motor-Control Equipment

When electric motors were first used, simple manual switches were used to start and stop the motors. Each starting and speed-control operation had to be performed by hand, with the operator moving a manual switching device from one position to another. Control of large motors required great physical effort.

The development of magnetic starters allowed the operator to use push-button stations and control switches requiring practically no effort to operate the starter. This also permitted the location of push-button stations and control switches at various places away from the starting equipment, which, in turn, gave rise to centralized automatic control.

Motor-control equipment includes not only the motor starter but all of the miscellaneous devices that prevent the motor from being started, all of the devices that cause the motor to stop, and all of the devices required to protect the motor and the motor-driven equipment during operation. The functions of motor control fall into two general categories: (1) those functions that are for the protection of the motor and the motor feeder cables, and (2) those functions that determine when and how the motor is to operate.

Motor-control systems. Automatic controls to start and stop pumps eliminate the need for constant station attendance and in some cases provide closer control of operations. Control switches are usually operated by changes in discharge-line pressure or in the water level in elevated storage tanks. Other control methods include time clocks and computer control based on demand changes.

Full-voltage and reduced-voltage controllers. A motor controller will be one of two types, FULL VOLTAGE or REDUCED VOLTAGE. A full-voltage (across-the-line) controller uses the full line voltage to start the pump motor, and the starting current is drawn directly from the line. A reduced-voltage controller is used when the starting current of the motor is of such magnitude that it may cause damage to the electrical system. The controller uses a reduced voltage and current to start the pump motor. When the motor gains full speed, full voltage is applied. Reduced-voltage starting is an effective means to eliminate damage to the system.

Location of control point. Pump operation is often controlled based on pressure changes measured at the elevated storage tank. If the control switch is operated by pressure changes close to the pumping station, erratic operation (RACKING) can be a problem. Racking occurs when the pressure surge resulting from pump start-up immediately causes the pump-control pressure switch to shut the pump off again. Another reason for locating the control point near the elevated tank is that the discharge pressure measured at the pump station can be affected by system demand as well as the level of water in the elevated tank. The primary function of the pump may be to maintain a specified level of water in the tank, and that level cannot be consistently determined based on the pressure at the pump.

Control methods. Pump-control systems may monitor pressure or water level. No matter where in the system a pump-control pressure sensor is located, there is some chance that it will be improperly activated or deactivated by sudden pressure surges that do not accurately reflect the actual system pressure. The effect of surge can be reduced by mounting the pressure switch at the top of an air chamber that is connected to the discharge header by a pipe having a small orifice. This arrangement provides a substantial time delay between pressure change and activation of the switch. Similar delay may be obtained by using an electrical time-delay relay in the control circuit. Where altitude valves are installed at elevated storage tanks, the rise in pressure that occurs as the altitude valve slowly closes activates the pressure switch, which stops the pump.

Control by water level, in elevated storage tanks without altitude valves, is accomplished either by a float switch, by electrodes within the tank, or by a mercury U-tube with electrical contacts at the base of the tank. In some cases, it is more economical to lay a ¾-in. (19-mm) pipe from the station and use pressure switches, rather than installing electrical circuits. To ensure immediate replenishment of storage in case of sudden system draw caused by fire flow, dual control is often provided by an additional pressure switch that starts the pump immediately when system pressure lowers significantly.

Multiple-pump controls.   Automatic controls may be provided to operate any combination of pumps simultaneously or in sequence to match pumping capacity with demand. The pumps may be actuated by individual pressure switches or through a sequence matrix panel (a series of switches that can be programmed to change the pump starting sequence) reading a level in a storage tank. Switchboard circuits can be set to stop one or more smaller pumps when larger pumps are in operation. For full flexibility, a single control device with multiple contact points is used to control all pumps.

Accessory controls.   Pumping stations with automatic controls for starting and stopping the pumps require only occasional attendance during the day to ensure proper functioning. To increase reliability and further reduce the cost of attendance, additional controls are provided to guard against mechanical or electrical defects and abnormal hydraulic conditions. Complex controls are required to ensure satisfactory operation. Reliability is increased by installing alarm equipment to notify operators immediately in case of improper functioning. Pressure- and flow-recording instruments are used to provide a full record of operation. (These topics are discussed in greater detail in Module 9, Instrumentation and Control.)

Motor protection equipment.   Some of the automatic relays used to control and protect motors are described in the following paragraphs.

Starters.   Starters are equipped with thermal-overload relays to prevent the pump motor from burning out if abnormal operating conditions increase the load on the motor to more than its design capacity. When current is excessive, the relays (sometimes called heaters) placed on each phase of the power supply open the control circuit and stop the motor. These relays are normally set to stop the motor when current exceeds the design load by 25 percent.

To protect against short circuits, there is a fuse in each line to the motor. Fuses are normally located in the safety switch just ahead of the starter. In COMBINATION STARTERS, the safety switch is combined with the motor starter. These units save space and field wiring; however, they may cost slightly more than separate installations.

Overcurrent relays.   Overcurrent relays are often referred to as overload relays. Their purpose is to sense current surges in the power supply and to disconnect the motor if a surge occurs. They can be fitted with time-delay or preset thermal-overload mechanisms. The overload relays may reset automatically or they may require manual resetting after they are tripped.

Lightning surge arresters.   Pump motors can be seriously damaged by high voltage surges (spikes) caused by lightning striking the power lines. To prevent damage to electrical equipment, a device known as a SURGE ARRESTER may be installed, usually in the service entrance between the power lines and the natural ground. The arrester acts as an insulator to normal voltage, but automatically becomes a low resistance conductor to ground when the line voltage exceeds a predetermined amount. An arrester may be able to conduct up to 10,000 A to ground harmlessly, and is so constructed that it can protect against repeated surges without any maintenance.

Voltage relays.   Undervoltage relays are frequently used to indicate loss of power and to initiate a switch-over to alternate power sources. Undervoltage relays are also used to shut down motors if voltage drops too low. Voltage relays with various timing mechanisms are used for time-delay functions, either on loss of voltage or after restoration of voltage to the system.

Frequency relays.   Frequency relays respond to changes in cycles per second of an AC power supply. Usually they are found on large stations or on smaller stations where local generation of power is involved. Frequency relays are also used on synchronous motor starters to sense when the motor has reached synchronizing speed.

Phase-reversal relays.   If, for some reason, a condition occurs in which any two of the three lines of a three-phase power system are interchanged, the phase sequence will be reversed and all motors will run backward. A phase-reversal relay senses this change and opens the control circuit to disconnect the motors. A phase reversal can be particularly serious on deep-well pump applications in which the pump shafting can become unscrewed, allowing the pump to fall.

Differential relays.   Differential relays are frequently used on large equipment or switchgear. The differential relay checks whether all of the current entering a system comes back out of the system. If it does not, the relay closes a contact that shuts down the equipment. Shutdown of equipment by a differential relay indicates major trouble, so an electrician should always be called.

Reverse-current relay.   A reverse-current relay senses a change in normal direction of current or power flow and activates an alarm for the operator. It can also open circuits to isolate the faulted portion of the system.

Time-delay relays.   Quite often it is intended that some condition last for a specified length of time before some other action is begun. Time-delay relays are used in such cases. For example, if a pump motor must be given sufficient time to reach full speed before the discharge valve opens, a time-delay relay can be energized at the same time that the motor is started. The relay is then set to close its contacts several seconds later to activate the valve-opening control circuits.

Bearing-temperature sensors.   Sensors to monitor bearing temperature can be placed on any pump or motor. The sensors incorporate a contact that opens or closes at a predetermined temperature. This action can be used to provide both an alarm and emergency electrical-control shutdown.

Speed sensor. As for pumps, centrifugal-speed switches and tachometer generators are used in combination with relays to generate overspeed alarms and to trigger emergency shutdown. They are also used on variable-speed motors.

7-5. Improving Efficiency of Electrically Driven Pumps

The power consumed by electrically driven pumps is one of the most costly items in many water-utility budgets. Power costs can be reduced in three major ways:

- Total power use can be reduced by increasing system efficiency.

- Peak power use can be reduced by spreading the pumping load more evenly throughout the day.

- Power-factor charges can be reduced.

Although most major reductions in power cost require replacement of equipment, some pumping-schedule changes can reduce costs with little or no investment.

Reducing Total Electric Power Usage

In addition to the obvious steps of maintaining motors and pumps in peak condition, running the pumps at the peak of the efficiency curve will reduce power usage. This can be readily accomplished in systems that have multiple pumps.

Where system pressure is higher than necessary for fire and other demands, pressure reduction can reduce the load on pumping facilities. Pressure reduction can often be achieved by the use of high-service pumps in areas where fire protection is critical or by having two-speed motors that are pressure activated. A reduction in system pressure from 85 to 75 psi (590 to 520 kPa) could result in energy savings of 5 to 25 percent, depending on the system flow and pump arrangement.

Reducing Peak-Demand Charges

Peak demand. Water utilities and electric utilities share a common problem—satisfying peak demand (See Volume 1, *Introduction to Water Sources and Transmission*, Module 2, Water Use). Unfortunately, water systems and electric systems experience peak demands during the same periods—early mornings, late afternoons, and the hottest afternoons and evenings of the summer season. The extra capacity installed by electric utilities to provide for peak-demand periods is paid for by demand charges, which high-demand customers pay in addition to the usual rate per kilowatt-hour of electricity.

The demand charge for a billing period is generally calculated based on the daily peak kilowatt demand during the billing period. Demand charges can be minimized by using electricity at a minimal, constant rate—that is, by keeping the peak demand as low as possible. Demand charges will also be lower if the water utility's peak demand occurs during a period that is off-peak for the electric utility.

The methods used to reduce total energy usage will reduce peak demands somewhat. Certain specific methods that can be used to reduce demand charges include:

- Use of gravity-feed storage
- Use of engine-driven pumps
- Reduction of peak water usage
- Changing the time of demand.

Gravity-feed storage.   Gravity-feed storage is an effective way to smooth demand, as well as minimize friction head and reduce overall energy use. For a water utility, the energy conservation of gravity storage in meeting peak demand is very close to 100 percent. By providing sufficient storage to minimize variation of pumping rate, peak electrical demand and corresponding demand charges can be minimized.

Engine-driven pumps.   Engine-driven pumping units may be used instead of electric motor-driven units during periods of peak demand in order to reduce electricity usage and demand charges. This may be more energy efficient than standby electrical generation, transmission, and distribution for meeting peak demands. In some areas, natural gas could be used economically to power the engine-driven pumps, since summer would be off-season usage for the natural gas.

Reduction of peak water usage.   Most sources agree that reduction of peak residential water usage can be accomplished to some degree by lowering service pressure, using flow restrictors, or instituting pricing policies. Experience in this area is still very limited. However, it is felt that some peak usage reduction could be accomplished by lowering service pressure. The use of flow restrictors in fixtures seems to have little effect on irrigation demand. Preliminary data on pricing structures that increase the price of water during summer months suggest that a peak-usage reduction of 12 percent may be feasible. More data is needed, however, before peak usage reduction by these methods can be estimated with any degree of confidence.

Changing the time of demand.   The methods discussed for reducing peak electrical demand also apply to changing the time of demand to the time of day when off-peak electrical rates are available. Additional gravity-feed storage may be used to permit increased off-peak pumping. A switch to engine-driven pumping units or a reduction of peak water usage could be used to reduce storage requirements during periods of peak water usage. Excess storage would then be available for use during off-peak pumping. Large electric motors should never be operated unnecessarily during peak-demand periods.

Power-Factor Improvement

The POWER FACTOR[7] of a motor indicates how effective it is at using the power available to it. A power factor of less than 1.0, which is the highest possible, does not indicate that the motor is inefficient, only that it requires larger power lines and transformers than are needed for the power it actually uses. For example, a 50 kW motor with a power factor of 1.0 draws 50 kW·h of power during an hour, and that is all the power it must be supplied with. A 50-kW motor with a 0.8 power factor must be supplied with $50/0.8 = 62.5$ kW of power to operate properly. The electric utility imposes a charge for the lower power factor, since it must size its facilities to produce and supply the full 62.5 kW. Power factors vary with the type of motor and the load imposed on the motor. They can also be affected by other components in the motor circuit. Methods of improving power factors include:

- Change in motor type
- Change in motor loading
- Use of capacitors.

Change in motor type. Premium-efficiency motors have higher power factors than standard motors and this may result in added cost savings.

Change in motor loading. Although induction-motor efficiencies remain relatively high between 50 and 100 percent of rated horsepower, the power factor decreases substantially and continuously as the rated load falls below 100 percent. For example, at 100 percent of rated load, the power factor of an induction motor may be 80 percent; at 50 percent of rated load, it may be 65 percent; and at 25 percent of rated load, it may drop to 50 percent. Therefore, although reducing the load on motors may decrease power usage somewhat, this advantage is offset to some extent by the reduction in power factor.

Use of capacitors. The power factor of an induction-motor circuit may be improved by installing a capacitor in parallel with the motor. The power-factor improvement resulting from a certain capacitance depends on the motor characteristics. Power-factor improvement beyond a certain point may cause problems to an electrical system. The motor manufacturer should be consulted to determine the capacitance that may be used with a particular motor.

7-6. Maintenance of Electric Motors

A preventive-maintenance schedule should be developed and followed for all motors in the system. In general, motors should be inspected at regular intervals, usually once a month. Under severe conditions, more frequent inspections are advisable. Log cards (discussed further under section 7-9, Record Keeping)

---

[7] *Basic Science Concepts and Applications,* Electricity, Electrical Quantities and Terms (Power Factor).

should be kept for each motor, recording all details of inspections carried out, maintenance required, and general conditions of operation. This allows the motor's condition to be checked at a glance, and any maintenance required should be immediately apparent. Regular (annual or semiannual) in-depth inspections are better than frequent casual checks. Items that should be checked during an inspection include:

- Housekeeping
- Alignment and balance
- Lubrication
- Brushes
- Connections, switches, and circuitry
- Slip rings.

Housekeeping

Routine housekeeping includes keeping the motor free from dirt or moisture, keeping operating space free from articles that may obstruct air circulation, checking for oil or grease leakage, and routine cleaning. If the correct motor enclosure has been chosen to suit the conditions under which the motor operates, cleaning should only be necessary at infrequent intervals. Dust, dirt, oil, or grease in the motor will choke ventilation ducts and deteriorate insulation. Dirt or moisture will shorten bearing life.

Alignment and Balance

Procedures for checking the alignment and balance on new motors or motors that have been removed are covered in Appendix E. During regular inspections, alignment should be checked semiannually if the motor is mechanically coupled to the driven unit. The check is performed according to manufacturer's instructions, using thickness gauges, a straightedge, and/or dial indicators.

The most likely causes of vibration in existing installations are imbalance of the rotating elements, bad bearings, and misalignment resulting from shifts in the underlying foundation. Because of this, vibration should always be viewed as an indicator that can be monitored to reveal other problems. Both magnitude and frequency of vibration should be measured and compared to the results of measurements when the unit was first installed.

Lubrication

The oil level in sleeve bearings should be checked and replenished as needed at approximately six-month intervals. The correct type and grade of oil must be used. The oil ring, if fitted, should be checked to ensure that it operates freely. Oil wells should be filled until the level is approximately $1/8$ in. (3 mm) below the top of the overflow. Overfilling should be avoided.

Grease in ball or roller bearings should be checked and replenished when necessary (usually about every 3 months) with grease recommended by the manufacturer. Bearings should be prepared for grease according to manufacturer's instructions.

New motors fitted with ball or ball-and-roller bearings should be supplied with the bearing housing correctly filled with grease. The bearings should be flushed and regreased at approximately 12- to 24-month intervals, using the type of grease recommended by the motor manufacturer. If the motor operates in an environment with a high ambient temperature, then a grease with a high melting point should be selected. Where the ambient temperature varies considerably throughout the day, or with climatic changes, the manufacturer should be asked for specific advice on the type of grease to use.

It is usually convenient to regrease the bearings, after removing the old grease and cleaning, by taking off the outside bearing covers when the motor is dismantled for periodic cleaning. The bearing housing should not be overfilled, and the operator should take care that no grit, moisture, or other foreign matter enters the dismantled bearings or the housing. A grease gun can be used for motors fitted with grease plugs or nipples, without removing the bearing cap.

Oil or grease should not be allowed to come into contact with motor windings, since this could cause deterioration of the insulation. In all motors fitted with ball-and-roller bearings, the bearing housings are fitted with glands to keep in the grease and keep out the dirt. However, over-lubrication of the bearings will result in grease leakage. An overgreased bearing tends to run hot, and the grease melts and creeps through the gland and along the shaft to the windings. Over-lubricated sleeve bearings can also contaminate motor windings.

Brushes and Slip Rings

Brushes should be checked for wear quarterly and replaced before they wear beyond the manufacturer's recommended specification. For the three-phase motors commonly used by water utilities, brushes should generally not be allowed to wear below $1/2$ in. (12 mm).

Brushes. New brushes should be bedded in to fit the curvature of the slip rings by placing a strip of sandpaper between the slip rings and brushes, with the rough surface towards the brushes. (Emery cloth should not be used because the material is electrically conductive and may contaminate components.) Normal brush-spring pressure will hold the brushes against the sandpaper. The sandpaper should be moved under the surface of the brushes until the full width of each brush contacts the surface of the slip ring. Where the motor rotates in one direction only, it is best to move the sandpaper in this direction only, then release the brush-spring pressure when the sandpaper is moved back for another stroke.

All carbon dust should be removed after bedding in, including dust that may have worked between the brushes and brush holders. Brush-spring tension should be set to 2.5–3 lb (1.1–1.4 kg). A small spring balance is useful for checking spring tension. The brush must slide freely in the brush holder, and the brush spring should bear squarely on the top of the brush.

Slip rings. When properly maintained, slip rings acquire a dark glossy surface with complete freedom from sparking between slip rings and brushes. Sparking will quickly destroy this surface and rapid brush and slip-ring wear will result. Sparking may result from excessive load (starting or running), vibration (for example, from driven machine), worn brushes, sticking brushes, worn slip rings, or incorrect grade of brushes. Periodic inspection and checking will ensure that trouble is reduced to a minimum. If inspection reveals slip rings with a rough surface, the rotor must be inspected for wear and horizontal movement, then repaired as necessary.

Insulation

Motor-winding insulation should be checked periodically to make sure it remains uncontaminated with oil, grease, or moisture. In general, insulation should exhibit at least 1 MΩ (1 megohm) of resistance.

Connections, Switches, and Circuitry

NOTE: All electrical circuits must be de-energized and "locked out" before any inspection or maintenance is performed.

Wiring should be checked semiannually or annually. Wires should be examined and all connections should be checked to ensure that they have not worked loose. Continuous vibration and frequent starting will gradually loosen screws and nuts. Vibration can also break a wire at the point where it is secured to a terminal.

Circuit-breaker and contactor contacts should be examined and cleaned regularly. Slightly pitted and roughened contact surfaces that fit together when closed should not be filed—the burned-in surfaces provide a far better contact than surfaces that have been filed down, no matter how carefully. Any beads of metal that prevent complete closure of the contacts should be removed. Contactor retracting springs should be examined and tested to ensure that they retain sufficient tension to provide satisfactory service.

Moving parts should be operated by hand to check the mating of main, auxiliary, and interlocking contacts. Faces of the contactor armature and holding electromagnet should be clean and fit together snugly. Anything on the faces that prevents an overall contact when the contactor is closed is likely to lead to chatter and excessive vibration of the mechanism, which increases wear and the possibility of a failure.

Oil-immersed gears must also be inspected occasionally, even if it is necessary to remove the oil. In any case, the oil must be examined for condition, since it becomes carbonized with time due to the arcing between contacts when the gears operate. The rate of deterioration of the oil depends on how often the contacts operate and the amount of current involved. It will be necessary to change oil at appropriate intervals.

Protective devices and fuses should be checked to be sure they have not tripped or blown and to be sure they are in proper working order. If a fuse has blown or a protective device tripped, the cause of the trouble should be identified before the circuit is reset and restored to service.

Moving parts must be kept clean to avoid any tendency to stick. Check the settings to make sure that the adjustments have not been moved by vibration or altered by unauthorized persons. Oil dashpots used with magnetic current relays may need replenishing with oil of the correct grade in order to maintain the required triggering time delay.

Phase Imbalance

Phase imbalance is caused by defective circuitry in the electrical service. It is the principal cause of motor failures in three-phase pumps. The effect of phase imbalance is to produce a large current imbalance between each phase or leg of a three-phase service. This imbalance, in turn, will produce a reduction in motor starting torque, excessive and uneven heating, and vibration. The heat and vibration will eventually result in failure of the motor windings and bearings. Some phase imbalance will occur in any system, but it generally should not exceed 5 percent. Each phase imbalance should be checked monthly with a volt–ammeter for balance, and the values should be logged. If excessive imbalance is found, an electrician or the power utility should be called to isolate and repair the problem.

When a fuse blows on one leg of a three-phase circuit, a phase-imbalance situation known as SINGLE PHASING occurs. The fuse may have blown because of an insulation failure in the motor or because of problems with the power line. The motor will continue to run on the two remaining phases. However, the windings that are still under power will be forced to carry a greater load than they were designed for, which will cause them to overheat and eventually fail.

7-7.   Types of Combustion Engines

Internal-combustion engines, and occasionally external-combustion (steam) engines, are used by water utilities to power pumps for emergency back-up, for portable dewatering applications, and in some remote installations where electric power is not economically available. They are also used as a means to reduce peak-demand electrical charges.

Internal-combustion engines commonly used by utilities include gasoline engines (which can also be run on propane, natural gas, or methane with minor modifications) and oil engines (diesels). External-combustion engines include reciprocating and turbine-type steam engines.

Gasoline Engines

Gasoline engines are generally used as standby or emergency units. Their initial cost is low, but fuel and maintenance costs are relatively high. Small gasoline engines may be used to power portable dewatering pumps. Gasoline engines, available in sizes from one to several hundred horsepower, generally operate at variable speeds between 600 to 1800 rpm. They can be connected directly to centrifugal pumps. A common practice is to install an electric motor at one end of a pump shaft and a gasoline engine at the other, where it can be easily coupled on if the current or motor fails.

Diesel Engines

Diesel engines run at constant, low speed. They are high in initial cost but reliable and generally economical to operate. Diesels are often used to power pumps where electricity is unavailable, expensive, or unreliable. They also frequently function as emergency generator drives, furnishing power to an entire station that is normally supplied with power from an electric utility. Diesel engines are available in sizes from 50 to 10,000 horsepower, with V or in-line blocks and 1 to 16 cylinders. The operation and maintenance of modern diesels requires considerable training and skill.

Steam Engines

Reciprocating, or direct-acting, steam engines are rarely used for water pumping today. A number of these units have been preserved as historical civil engineering works. The more modern STEAM TURBINE may be a viable alternative pump drive where steam is readily available or where steam-pressure reduction is required. This engine may be especially valuable where electric power is intermittent or exceptionally costly.

7-8.   Operation and Maintenance of Internal Combustion Engines

Proper engine starting and operating procedures should be posted near the engine. The following sections describe general service procedures to be performed before starting an engine, during initial and continued operation, and after shutdown. A more detailed general check list is given in Table 7-3.

Before-Operation Service

Before putting the engine in operation, a general inspection of the engine and all attached components should be performed. Items to check include all fluid levels (fuel, oil, coolant), all rubber components (belts, hoses), and any regularly adjusted items, such as V-belts. Any leaks or loose components should be repaired or, if minor, noted for future repair.

If the engine has a separate cooling-water source, make sure all of the valves are turned on before starting the engine. Also, make sure that any clutches are disengaged.

Initial-Operation Service

As soon as the engine starts, check idle speed, oil pressure, ammeter, water temperature, and any other gauges that are attached. If any unusual gauge readings are observed, shut the engine down and correct the problem.

If idle-speed checks show no problems, allow the engine to warm up according to manufacturer's recommendations, then engage the clutch or other transmission device and put the engine under load. With the engine under load, keep a close watch on the engine and its gauges to make sure it is running properly. Watch for leaks and loose components, and listen for unusual noises. If the engine will not come up to speed or will not hold speed under the load, shut it down and correct the problem immediately. A fouled fuel filter or a sticking governor may be reducing engine performance, or the engine may require major maintenance.

Table 7-3. Operation of Internal Combustion Engines

Before-Operation Service

Check fire extinguishers for ease of removal and tight mountings, full charge, and closed valves. See that valves and nozzles are not corroded or damaged.

Look for signs of tampering, damage, or injury, such as loosened or removed accessories or drive belts.

Check the amount of fuel in tanks, note signs of leaks or tampering. Add fuel if necessary. Fuel should be protected against freezing by use of appropriate additives.

Check the oil level and add oil if necessary. Check the level and condition of the coolant. During the season when antifreeze is used, test the coolant with a hydrometer; add antifreeze and/or water if necessary. Do not use alkali water. Check for any appreciable change in fuel or water level since the last after-operation service.

Check all attachments such as carburetor, turbocharger, compressor unit, generator, regulator, starter, fan, fan shroud, and water pump for loose connections or mountings.

Look for signs of fuel, oil, water, or gear-oil leaks. Check the cooling system for leaks, especially the radiator core and connectng hose. Check for leaks from the engine crankcase, oil filters, oil tanks, oil coolers, and lines. Check the fuel system for indications of leaks. Trace all leaks to their source and correct them.

Check the electrolyte level in the batteries. Make sure that the battery is not leaking and the battery, cables, and vent caps are clean and secure. Check the voltmeter to see that it registers at least a nominal battery rating.

Set the choke or operate the primer if necessary.

Activate the starter. Note whether the starter has adequate cranking speed and engages and disengages properly without unusual noise.

Initial-Operation Service

When the engine starts, adjust the throttle to normal (fast idle) warmup speed and continue the servicing procedure. CAUTION: The engine may be damaged or its service life appreciably reduced by placing it under load before it reaches normal operating temperature.

Observe the operation of the oil-pressure gauge or light indicator. If these instruments do not operate properly within 30 sec, stop the engine immediately and determine the cause. Do not operate the unit if the oil pressure drops below indicated normal range at normal operating speed.

As the engine warms up, reset the choke to prevent over-choking and dilution of engine oil.

The ammeter may show a high charging rate for the first few minutes after starting, until current used in starting is restored to the battery; after this period, the ammeter should register zero or a slight positive charge with accessories turned off and engine operating at fast idle; investigate any unusual drop or rise in reading; an extended high reading may indicate a dangerously low battery or faulty generator regulator.

Note whether the tachometer indicates the engine rpm and varies with engine speed through the entire speed range.

(continues on next page)

Table 7-3.   Operation of Internal Combustion Engines *(continued)*

───────────────────── Initial-Operation Service (cont.) ─────────────────────

Check voltmeter for proper operation; it should register at least nominal battery voltage and will register slightly higher if the generator or alternator is working properly.

Check the temperature gauge to be sure engine temperature increases gradually during the warm-up period; if temperature remains extremely low after a reasonable warm-up period, the engine cooling system or temperature control device needs attention; engines should not be operated for an extended time at temperatures above or below the normal operating range; all stationary gasoline engines will be equipped with thermostatically controlled coolant regulators recommended by the manufacturer.

───────────────────── During-Operation Service ─────────────────────

The clutch or engine drive train should not grab, chatter, or squeal during engagement or slip when fully engaged. If the clutch lever does not have sufficient free travel before the clutch begins to disengage, the clutch may slip under load. Too much free travel may keep the clutch from disengaging fully, causing clashing gears and damage when shifting.

The transmission gears should shift smoothly, operate quietly, and not creep out of mesh during operation. Gears jumping out of mesh indicate wear in the shifting mechanism or gear teeth, or incorrect alignment of transmission or clutch housing.

Note poor engine performance such as a lack of usual power, misfiring, unusual noise, stalling, overheating, or excessive exhaust smoke. See that the engine responds to controls satisfactorily and that the controls are correctly adjusted and are not too tight nor too loose.

Check the instruments regularly during operation. As a general rule, do not let the oil pressure drop below the indicated normal range at normal operating speed.

───────────────────── Short-Stop Service ─────────────────────

Short-stop service is performed whenever the engine is stopped for a brief period. This consists of correcting any defects noted during operation and making the following inspections:

If fuel tank is engine mounted, check fuel supply and, if operated continuously, fill tank. When refueling, use safety precautions for grounding static electricity. Allow space in filler neck for expansion. See that filler-vent caps are open and pressure-cap valves are free. Replace caps securely.

Check crankcase oil level and add oil if necessary. Make sure oil drain is closed tightly. If there is evidence of loss or seepage, determine the cause.

Remove radiator filler cap, being careful of steam, especially if a pressure cap is used. See that coolant is at proper level and replenish as necessary. Do not fill to overflowing, but leave enough space for expansion. If engine is hot, add coolant slowly while engine is running at fast idle speed.

Check for leaks.

Check accessories and belts; make sure fan, water pump, and generator are secure, their drive belts properly adjusted and undamaged.

Check air cleaners; inspect air cleaners and breather caps to see that they are delivering clean air; clean and service if necessary.

Table 7-3.   Operation of Internal Combustion Engines *(continued)*

——————————————————— After-Operation Service ———————————————————

Check fuel, oil, and water.

Check engine operation.

Check instruments.

Check battery and voltmeter.

Check accessories and belts. Inspect the carburetor, generator, regulator, starter, fan, shroud, water pump, and other accessories for loose connections or mountings. Inspect the fan and accessory drive belts and adjust if necessary. Belts should deflect. Replace any damaged or unserviceable belts.

Check electrical wiring. Make sure that all ignition wiring is securely mounted and connected, clean, and not damaged. Repair or replace as necessary.

Check air cleaner and breather caps. Oil in the air cleaner must be kept at the correct level. Rub a drop of oil between fingers; if it feels dirty, drain and refill with fresh oil. If operating in a dusty area, remove and clean the air cleaners and breather caps whenever necessary. Check all fuel filters for leaks.

Check engine controls. Check linkage for worn or damaged joints and connections. Correct or report any defective linkage.

Check for leaks. Inspect for any fuel, oil, or water leaks and correct or report any found.

Check gear-oil levels. Check the lubricant level in drives and transmissions after they have cooled enough to be touched by hand. If the lubricant is hot and foamy, check to determine the level of the liquid below the foam.

During-Operation Service

Once the engine is placed into normal service, gauges and overall operation should be checked periodically to ensure no problems have occurred. Fuel level should be monitored, especially for diesels, since running out of fuel can cause major damage to diesel fuel-injection systems.

After-Operation Service

After the engine is shut down and disengaged from the pump, any minor problems that were noticed during operation should be corrected. This is also a good time to check again the lubricant levels in the engine and gear cases. Lubricant should be checked for foaming, which may indicate contamination by water.

Regular Maintenance

The traditional use of internal-combustion engines has been to provide backup and emergency power for pumping during periods of peak demand or when there are electric-power outages. In these applications, it is essential that the engine function perfectly on very short notice. A program of regular inspection and maintenance is necessary to ensure this reliability. As part of this program, all

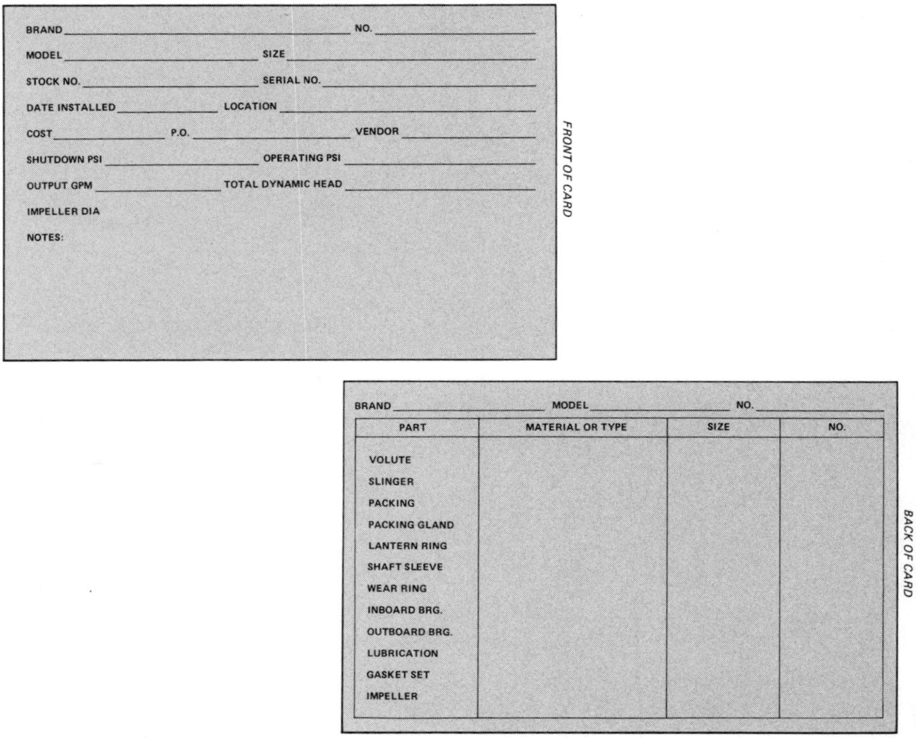

<table>
<tr><td>BRAND</td><td></td><td colspan="2">NO.</td><td rowspan="9">FRONT OF CARD</td></tr>
<tr><td>MODEL</td><td colspan="2">SIZE</td><td></td></tr>
<tr><td>STOCK NO.</td><td colspan="2">SERIAL NO.</td><td></td></tr>
<tr><td>DATE INSTALLED</td><td colspan="2">LOCATION</td><td></td></tr>
<tr><td>COST</td><td>P.O.</td><td>VENDOR</td><td></td></tr>
<tr><td>SHUTDOWN PSI</td><td colspan="2">OPERATING PSI</td><td></td></tr>
<tr><td>OUTPUT GPM</td><td colspan="2">TOTAL DYNAMIC HEAD</td><td></td></tr>
<tr><td>IMPELLER DIA</td><td></td><td></td><td></td></tr>
<tr><td>NOTES:</td><td></td><td></td><td></td></tr>
</table>

BRAND		MODEL		NO.	
PART	MATERIAL OR TYPE		SIZE	NO.	BACK OF CARD
VOLUTE					
SLINGER					
PACKING					
PACKING GLAND					
LANTERN RING					
SHAFT SLEEVE					
WEAR RING					
INBOARD BRG.					
OUTBOARD BRG.					
LUBRICATION					
GASKET SET					
IMPELLER					

Figure 7-21. Pump Record Card (Front and Back)

engines should be started and operated continuously for at least 15 min each week. Running the engine under load is advisable, since this will be more likely to bring out any problems that have developed.

Regular maintenance checks of specific items should be scheduled according to manufacturer's recommendations. These include oil and other fluid draining and replacement, filter and belt changes, inspection and replacement of hoses and other rubber parts, and more major operations as necessary. Where antifreeze is used, its strength should be checked every six months, more often if leakage has required frequent coolant replenishment. Battery electrolyte level should be checked at least quarterly, more often in hot climates. A schedule of battery testing should be established, especially for standby units that may require maximum battery efficiency to start during winter months.

Services recommended for diesel engines are essentially the same as listed in Table 7-3. One item of particular importance for diesels is fuel filtering. The injection-type fuel system used in diesels is a close-tolerance mechanical system that can be easily fouled or damaged by improperly filtered fuel. Only qualified personnel should attempt repairs of a diesel fuel-injection system. Diesel engines require large volumes of air, so the air cleaners must be well maintained.

7-9. Record Keeping

An adequate equipment and maintenance record system should assist the operator in scheduling inspections and needed service work, evaluating pump equipment, and assigning personnel. An appropriate system could be based on data cards like those in Figure 7-21. Each card should list the make, model, capacity, type, date and location installed, and other information for both driver and driven unit. The remarks section should include the serial or part numbers of special components (such as bearings) that are likely to require eventual replacement parts.

A separate operating log should be kept listing all pumping units along with a record of the operating hours. This record is an essential feature of any reasonable periodic service or maintenance schedule. In addition, a daily work record should be kept on each piece of equipment. The log sheet shown in Figure 7-22 is generally appropriate for pumps and motors used in water supply.

7-10. Safety

Specific areas where the need for safety precautions must be recognized and observed when dealing with pumps and prime movers include:

- Material handling (refer to Module 1)
- Personal-protection equipment (refer to Module 1)
- Hand tools (refer to Module 1)
- Electrical devices
- Moving machinery
- Fire safety.

Electrical Devices

There is no safety tool that will protect absolutely against electrical shock. Plastic hard hats, rubber gloves, rubber floor mats, and insulated tools should be used when working around electrical equipment. However, these insulating devices do not guarantee protection, and the operator using them should not be lulled into a false sense of security.

Electrical shocks from sensors are possible in many facilities, such as pumping stations, because many instruments do not have a power switch disconnect. It is important to tag such instruments with the number of their circuit breaker so that it can be quickly identified. After the circuit breaker has been shut off, an operator should tag or lock the breaker so other employees will not reenergize the circuit while repairs are being performed. Even after disconnecting a circuit, it is good practice to use a voltmeter to be certain that all electrical power has been removed.

```
                                                      DAILY OPERATION LOG
                                  DATE _____  TIME _____ INITIALS _____
                                                       Page 3 of 3

PUMP STATION
Pump #1:  On_____  Off_____  Run Time _____-_____=_____hrs
          Lube Leak_____  Water Leak_____  Noise_____  Temp_____  Vibra_____  Other_____
          All OK_____  Pressure _____psi
Pump #2:  On_____  Off_____  Run Time _____-_____=_____hrs
          Lube Leak_____  Water Leak_____  Noise_____  Temp_____  Vibra_____  Other_____
          All OK_____  Pressure _____psi
Pump #3:  On_____  Off_____  Run Time _____-_____=_____hrs
          Lube Leak_____  Water Leak_____  Noise_____  Temp_____  Vibra_____  Other_____
          All OK_____  Pressure _____psi
Pump #4:  On_____  Off_____  Run Time _____-_____=_____hrs
          Lube Leak_____  Water Leak_____  Noise_____  Temp_____  Vibra_____  Other_____
          All OK_____  Pressure _____psi
B. W. Meter:  Totalizer _____-_____=_____gals x 100
          Discharge Pressure _____psi

MISCELLANEOUS
Auxiliary Power:  Run Time _____-_____=_____hrs
          Fuel Level _____  Battery Charger: On_____ Off_____
Control Panel Battery:  On_____  Off_____
Temp _____°F  Precipitation _____ins
P.S. Meter _____-_____=
Demand _____
Chlorine Cylinders:  Full _____  Empty _____
Alarms On _____           Res. level _____ft.
          _____                    _____gals.
          _____

COMMENTS
_____
_____
```

Courtesy of Highlands Ranch Water and Sanitation District
Figure 7-22. Pump Log Sheet

It is very easy to damage an electrical or electronic instrument by inadvertently shorting a circuit while making adjustments. Insulated screwdrivers should be used for electronic adjustments to reduce the chance for damage.

Blown fuses are usually an indication of something more than a temporary or transient condition. The cause of the overload should be identified. Never replace a blown fuse with a fuse of an amperage higher than the circuit's designed rating.

Broken wires should be replaced instead of repaired. Also, secure terminal connectors should be used on all wires at all points of connection. Charred insulation on a wire is a warning of a serious problem.

Extreme care should be taken in working around transformer installations. Maintenance activities should be performed only by employees of the power company. Power to a transformer cannot be locked out except by the power company, so operators should always assume that all transformers are energized and observe full safety precautions.

Electric switchboards should be located and constructed in a manner that will reduce the fire hazard to a minimum. They should be located where they will not be exposed to moisture or corrosive gases, and their location should allow a clear working space on all sides.

Adequate illumination should be provided for the front, and back if necessary, of all switchboards that have parts or equipment requiring operation, adjustment, replacement, or repair. All electric equipment, including switchboard frames, should be well grounded. Insulating mats should be placed on the floor at all switchboards.

Open switchboards should be accessible only to qualified and authorized personnel and should be properly guarded or screened. Permanent and conspicuous warning signs should be installed for panels carrying more than 600 V. Areas screened off because of high voltage should be provided with locks that open from the inside without keys.

Switches should be locked open and properly tagged when personnel are working on equipment. Fully enclosed, shockproof panels should be used when possible. Such equipment should be provided with interlocks so that it cannot be opened while the power is on.

Moving Machinery

Machinery should always be stopped before cleaning, oiling, or adjusting it. The controlling switchgear should be locked out before beginning any work, so that the machinery cannot be started by another person. A conspicuous tag should be posted on or over the control panel, giving notice that the equipment is under repair and should not be restarted, as well as the name of the person who locked it out. Before a machine is restarted, the operator should check to be sure that all personnel are clear of danger and that working parts are free to move without damage.

When disconnecting equipment, be familiar with the manufacturer's instructions for disconnecting and securing drive and rotating equipment. Place the shift or drive-belt clutch into the neutral position before starting.

Guards should be adequately secured in place in order to shield, fence, enclose, or otherwise guard prime movers, power transmission equipment, and machines. Guards should be provided with hinged or movable sections in places where it is necessary to change belts, make adjustments, or admit lubricants. When a guard or enclosure is within 4 in. (100 mm) of a moving part, the maximum opening in the screen should not exceed ½ in. (13 mm) across. Guards placed more than 4 in. (100 mm), but less than 15 in. (380 mm) from a moving part can have openings no more than 2 in. (50 mm) across. The guard should be strong enough to provide real safety, and guard structures should be constructed so that they cannot be pushed or bent against moving parts. Guards should be removed and replaced for maintenance only when the machinery is not in operation.

Fire Safety

Suitable fire extinguishers should be kept near at hand and ready for use. The location and operation of fire extinguishers should be familiar to all employees. Water or soda-acid extinguishers should never be used on electrical fires or in the vicinity of live conductors; carbon dioxide or dry-powder extinguishers are recommended for these situations.

Detailed fire safety requirements vary considerably from one installation to another and from one locality to another. Therefore, it is recommended that the advice and, where necessary, the approval of one or more of the following organizations be secured: local fire department, state fire marshal's office, fire insurance carrier, or local and state building and fire prevention bureaus. Generally, the fire insurance carrier and one of the fire-protection agencies will provide all of the advice that is needed.

Selected Supplementary Readings

American National Standard for Vertical Turbine Pumps—Line Shaft and Submersible Types. ANSI/AWWA Standard E101-77. AWWA, Denver, Colo. (reaffirmed 1982).

Anderson, Donald. Selecting Lubricants for Plant Equipment. *OpFlow*, 10:12:5 (Dec. 1984).

Booster Pumps Pose Potential Health Hazards. *OpFlow*, 8:5:3 (May 1982).

Care of Motors. *OpFlow*, 3:10:1 (Oct. 1977).

Emmerling, C.R. Hints to Help Operators Keep Pumps Running. *OpFlow*, 2:12:1 (Dec. 1976).

Gierer, Bill. Pumps—Mechanical 'Hearts' of the Water Industry. *OpFlow*, 8:4:1 (Apr. 1982).

Gros, W.F.H. Save Energy: Use This Method to Check Pump Efficiency. *OpFlow*, 2:9:1 (Sept. 1976).

Gros, W.F.H. Pump Rehabilitation Improves Efficiency, Saves Dollars. *OpFlow*, 2:10:1 (Oct. 1976).

Gros, W.F.H. Check Pump Efficiency. *OpFlow*, 3:5:5 (May 1977).

Gros, W.F.H. How to Check Pump Efficiency. *OpFlow*, 6:11:1 (Nov. 1980).

Hitt, M.J. Choosing and Applying Non-Centrifugal Pumps. *OpFlow*, 5:8:3 (Aug. 1979).

How to Keep Well Pumps Operating. *OpFlow*, 1:1:3 (Jan. 1975).

Hydraulic Institute Standards for Centrifugal, Rotary and Reciprocating Pumps. Hydraulic Institute, Cleveland, Ohio. (14th ed., 1983).

Improving Well and Pump Efficiency. AWWA, Denver, Colo. (1983).

Magney, H.C. More Efficient and Prolonged Life From Your Submersible. Proc. AWWA DSS, Milwaukee, Wis. (Mar. 1982).

Maintenance of Pumps and Motors. *OpFlow*, 2:1:3 (Jan. 1976).

Price, B.C. Performance Curves and Evaluation. *OpFlow*, 1:2:3 (Feb. 1975).

Pumps—Lubricating Bearings. *OpFlow*, 4:3:4 (Mar. 1978).

Shubert, W.M. Preventive Maintenance of Controls. *OpFlow*, 5:3:3 (Mar. 1979).

Glossary Terms Introduced in Module 7

(Terms are defined in the Glossary at the back of the book.)

Ambient
Arbor press
Axial-flow pump
Bearing
Brushes
Casing
Cavitation
Centrifugal pump
Close coupled
Combination starter
Commutator
Coupling
Demand charge
Diffuser bowl
Diffuser vanes
Double-suction pump
Foot valve
Frame mounted
Full voltage
Head
Impeller
Inrush current
Interference fit
Internal backflow
Jet pump
Key
Kinetic pump
Lantern ring
Locked-rotor current
Mechanical seal
Mixed-flow pump
Motor-starting current
Packing gland
Packing rings

Positive-displacement pump
Power factor
Prime mover
Priming
Racking
Radial-flow pump
Reciprocating pump
Reduced voltage
Rotary pump
Rotor
Shaft
Single phase
Single phasing
Single-suction pump
Slip
Squirrel-cage induction motor
Stator
Steam turbine
SSU (Saybolt Standard Units)
Stuffing box
Submersible pump
Suction lift
Surge arrester
Synchronous motor
Synchronous speed
Three phase
Throttling
Velocity pump
Venturi
Vertical turbine pump
Volute
Water hammer
Wear ring
Wound-rotor induction motor

Review Questions

(Answers to the Review Questions are given at the back of the book.)

1. How do the two basic parts of a velocity pump operate?

2. What are the two designs used to change high velocity to high pressure in a pump?

3. In what type of pump are centrifugal force and the lifting action of the impeller vanes combined to develop the total dynamic head?

4. Identify one unique safety advantage that velocity pumps have over positive-displacement pumps.

5. What is a multistage centrifugal pump? What effect does the design have on discharge pressure and flow volume?

6. What are two types of vertical turbine pump, as distinguished by pump and motor arrangement, that are commonly used to pump ground water from wells?

7. What type of vertical turbine pump is commonly used as an inline booster pump?

8. Describe the two main parts of a jet pump.

9. What is the most common use of positive-displacement pumps in water plants today?

10. What is the purpose of the foot valve on a centrifugal pump?

11. How is the casing of a double-suction pump disassembled?

12. What is the function of wear rings in centrifugal pumps of the closed-impeller design? What is the function of lantern rings?

13. Describe the two common types of seals used to control leakage between the pump shaft and the casing.

14. What feature distinguishes a close-coupled pump and motor?

15. What is the value of listening to a pump or laying a hand on the unit as it operates?

16. Define the term "racking" as applied to pump and motor control.

17. When do most electric motors take the most current?

18. What are three major ways of reducing power costs where electric motors are used?

19. What effect could overlubrication of motor bearings have?

20. Why should emery cloth not be used around electrical machines?

21. What are the most likely causes of vibration in an existing pump installation?

22. What can happen when a fuse blows on a single leg of a three-phase circuit?

23. Name at least three common fuels for internal-combustion engines.

24. List the type of information that should be recorded on a basic data card for pumping equipment.

25. What is the first rule of safety when repairing electrical devices?

Study Problems and Exercises

1. Outline the basic steps that should be considered when installing a vertical turbine pump that will pump from a wet well.

2. Outline a basic scheme for the periodic inspection and maintenance program of a pump house with four motors and pumps and a diesel auxiliary power plant.

Module 8

Water Storage

Storage is an essential part of any water system. It is becoming increasingly important as continued growth, expanding service areas, and additional uses increase the demand for water. DISTRIBUTION STORAGE refers to the storage of water ready for distribution; this does not include the impoundment of untreated water.

After completing this module you should be able to

- Identify the principal reasons for storing water in the distribution system.

- Explain the difference between operating storage and emergency storage requirements.

- Identify and explain the common construction features and accessory devices for water storage tanks.

- Specify key factors that should be considered in selecting the proper size and location for a reservoir.

- Describe what maintenance procedures are required for water storage facilities; develop a maintenance schedule to reflect routine inspection intervals; and recommend items of information that should be recorded during the course of operating, inspecting, and maintaining distribution storage tanks and reservoirs.

- Recommend safety procedures that should be observed when operating, inspecting, and maintaining water storage facilities.

8-1. Water Storage Requirements

The type and capacity of water storage required in a distribution system varies from small, pneumatic tanks used to supply a single customer to large tanks and reservoirs used to provide reserve capacity for the entire system.

Purpose of Water Storage

Water storage in the distribution system is required for the following reasons:

- Equalizing supply and demand
- Increasing operating convenience
- Leveling out pumping requirements
- Providing water during source or pump failure
- Providing water to meet fire demands
- Providing surge relief
- Increasing detention times
- Blending water sources.

Equalizing supply and demand.   The demand for water normally changes throughout the day and night (Figure 8-1). If treated water was not available

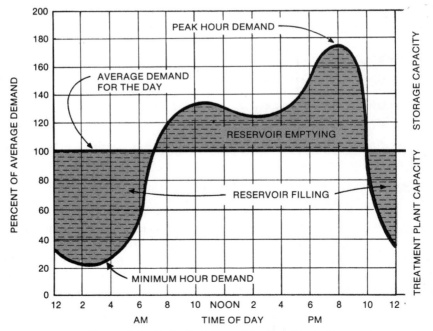

Figure 8-1. Daily Variation of System Demand

from storage, the water treatment plant would have to be large enough to meet the PEAK HOUR DEMAND. In Figure 8-1, the peak hour demand is approximately 175 percent of the average demand for the day illustrated. If stored water was not available, the plant capacity would have to be almost double the size needed to meet the average demand.

With adequate storage, water can be treated and pumped into the distribution system at a rate equaling the average demand. Figure 8-1 shows that from 10 p.m. to 7 a.m. the demand on the system is below the average rate, so during this time the RESERVOIRS are being filled. From 7 a.m. to 10 p.m. the demand is greater than the supply, so the reservoirs are being used to feed water back into the system.

Increasing operating convenience. The previous paragraphs indicate that storage allows smaller treatment plants to be built and operated 24 hours a day at the average demand rate. In some situations, however, it may be more practical and economical to build a larger plant and only operate it for 8 or 16 hours a day, thereby limiting personnel duty periods to one or two shifts. In this situation, water treated during this period would be stored for use when the plant was not operating.

Leveling out pumping requirements. As discussed, demands for water change from hour to hour and day to day. Without storage, pumps would have to match the changing demand by frequently turning on and off. This cycling of the pumps causes increased wear on pump controls and motors, as well as increased electrical costs.

Storage also allows pumping costs to be reduced by taking advantage of rate structures for electrical power. Many electric power utilities have "time of use" rates that are lower during off-peak hours. These lower rates make it more economical to fill the reservoir at night. Storage can also be used to avoid the use of additional pump stations, intended only to meet occasional peak demands, thereby saving the electrical demand charges on these standby units.

Providing water during source or pump failure. There are times when all pumps may not be available when needed, because of power failure, mechanical breakdown, or preventive maintenance. Sudden increases in demand can be caused by main breaks, broken hydrants, or similar problems. In addition, the maximum output of water from the treatment plant may not be available at all times. All or part of the water treatment plant may be out of operation for the same reasons as for pumps. The plant may have to shut down because raw-water quality has become temporarily unacceptable for reasons such as turbidity after a storm or a chemical spill. As much as 25 percent of storage capacity may be reserved to meet such situations.

Providing water to meet fire demands. One of the major purposes for distribution system storage is meeting FIRE DEMANDS. Although fire demands may not occur very often or last very long, they can be much greater than consumer peak demands. Water systems, including storage, are usually designed to meet fire demand in addition to normal customer needs. Fire demand can account for as much as 50 percent of the total capacity of a storage system.

Providing surge relief.   As pumps turn on and off, and valves are opened and closed, tremendous pressure changes, known as water hammer, can surge through the distribution system. Water hammer can seriously damage pipes and appurtenances. Reservoirs tend to lessen the shocks of valve and pump operation, which helps to maintain a reasonably constant pressure in the system.

Increasing detention times.   The added detention time provided by having water storage has two major advantages:

- Disinfection continues even at low chlorine levels
- Sand, floc, or precipitated solids settle out before reaching the mains and customers.

Blending water sources.   Some water systems use water from different sources that can vary in quality. Blending these different sources together in a reservoir will often improve the quality of marginally acceptable water. It also provides a more uniform quality of water to the customers, rather than water that changes in taste and composition from day to day.

Capacity Requirements

The capacity of distribution storage is based on the maximum water demands in the different parts of the system. Capacity varies for different systems and can only be determined by qualified engineers after a careful analysis and study of a particular system. Storage capacity needed for fire protection should be based on the recommendations of fire underwriters' organizations. Because there are so many variables involved, operators should contact the Insurance Services Organization Office or the Fire Insurance Rating Office in their state to obtain any available information.

Additional storage capacity may be necessary to meet emergencies such as pump failure, source failure, or transmission-line break. The need for emergency storage should be based on the reliability of the supply and pumping equipment and the availability of backup equipment and standby power sources.

8-2.   Types of Treated-Water Storage Facilities

Water storage tanks can be classified in several ways:

- Type of service
- Configuration
- Type of construction material.

Type of Service

Reservoirs can be used for either OPERATING STORAGE or EMERGENCY STORAGE. Operating storage generally "floats" on the system, as shown in Figure 8-1. The storage reservoir fills when demand is low and empties when demand exceeds supply. Emergency storage is designed to be used only in exceptional situations, such as high-demand fires or failure of the supply or of a transmission main. An example of emergency storage would be the ELEVATED STORAGE many companies have for use in their own fire-protection sprinkler systems. Because emergency storage water is not constantly used, its quality can become questionable.

Configuration

Distribution storage facilities can be either ground level or elevated. These facilities may be referred to as TANKS, STANDPIPES, or RESERVOIRS. The term tank generally refers to any structure used for containing water. A standpipe is a type of tank resting on the ground that has a height greater than its diameter (Figure 8-2). A general disadvantage of a standpipe is that only the upper portion of the water in the standpipe has enough pressure to be used in the distribution system. The lower portion can only be used for emergency storage. The term reservoir is more commonly used for very large storage facilities. Reservoirs are generally ponds, lakes, or basins that are either naturally formed or constructed in part using surrounding natural geographic features. (Most large reservoirs are used for raw-water storage, not for storage of treated water to be used in the distribution system.)

Courtesy of Seattle Water Dept.

Figure 8-2.   Standpipe

Courtesy of Tnemec Company, Inc.

Figure 8-3.   Elevated Tank

ELEVATED TANKS are usually supported by a steel or concrete tower (Figure 8-3). This is done to gain elevation and maintain pressure in the system.[1] GROUND-LEVEL TANKS (Figure 8-4) are usually less expensive to build and maintain than elevated tanks, but they may require a booster pump to provide adequate pressure.

---

[1] *Basic Science Concepts and Applications,* Hydraulics Section, Head.

Figure 8-4.   Ground-Level Tank

In almost any system, there are very high or remote areas with only a few customers. Such areas can be served quite adequately with a pneumatic steel pressure tank (Figure 8-5) filled partially with pressurized air. This is called a HYDROPNEUMATIC SYSTEM. The compressed air on top of the water provides the necessary pressure for the area. The pump is located in a small booster system connected to the tank. Pump cycling frequency is reduced because the pump is set to automatically turn on only when the pressure in the tank drops below a certain point. The maximum amount of water that may be withdrawn during a cycle is $^1/_3$ of the total tank volume.

Type of Construction Material

Over the years, reservoirs have been constructed from a variety of materials. Early reservoirs were constructed with earth-embankment techniques. Today, steel and concrete are the most widely used construction materials.

Earth-embankment reservoirs.   Probably the first type of distribution storage in common use was the earth-embankment reservoir. These reservoirs were generally built on higher ground, in or near the service area. The early earth-embankment reservoirs were generally constructed using partial excavation and partial embankment; they were paved with stone riprap, brick, or concrete on the slopes and bottoms; and they were almost always open at the top.

Although these installations were effective and valuable in system operation, they are seldom built now, and few, if any, remain in service. Problems that can occur with uncovered reservoirs include:

- Water loss through leakage and evaporation
- Freezing
- Loss of chlorine from exposure to air and light
- Contamination by birds, animals, humans, and airborne pollutants
- Increased maintenance costs and taste-and-odor problems because of algae growth.

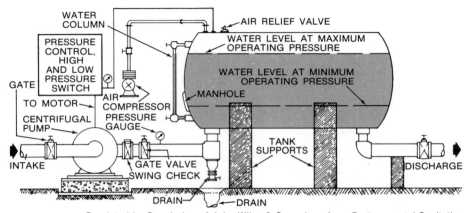

Figure 8-5.   Pneumatic Steel Pressure Tank

Water loss from leakage through the bottom and the embankment can be prevented by lining the reservoir with concrete, asphalt concrete, or non-permeable rubber or plastic membranes, but water is still lost to evaporation, and freezing of uncovered reservoirs is still a problem in cold climates. In addition, most state health authorities now require that distribution storage systems be covered because of potential water quality problems.

Steel tanks.   Most elevated tanks and standpipes, and many ground-level tanks, are constructed out of bolted or welded steel. Since steel is vulnerable to corrosion, it must be painted and may also require CATHODIC PROTECTION (refer to Module 1, Pipe Installation and Maintenance, for a discussion on cathodic protection). However, steel tanks often cost less than concrete tanks.

Concrete tanks and reservoirs.   The first concrete reservoirs were mostly open, but were later roofed with wood or concrete. Once a reservoir is covered and properly ventilated, the sanitation problem is almost solved. The major disadvantage of concrete reservoirs is their relatively high initial cost of construction.

Cast-in-place concrete tanks.   The shape of ordinary cast-in-place concrete reservoirs is determined by the forms or molds used to construct them. This limits their shape—usually they are square or rectangular. Other construction problems include the placing of the reinforcing steel and the fact that construction joints may leak.

Prestressed-concrete tanks.   The construction of PRESTRESSED-CONCRETE tanks is begun with an inner core wall that determines the reservoir's circular form. Steel prestressing wire (wire under high tension) is wrapped around the core wall (Figure 8-6). The finished reservoir wall is then covered with layers of HYDRAULICALLY APPLIED CONCRETE. A properly designed and con-structed prestressed-concrete tank has many advantages: it is competitively priced, it is relatively watertight, and it does not require a coating or cathodic protection. Because of their strength, prestressed tanks can be made thinner and

Courtesy of DYK-BBR Prestressed Tanks, Inc.

**Figure 8-6. Prestressed Concrete Tank Covered With Layers
of Hydraulically Applied Concrete**

with less reinforcing steel than cast-in-place concrete tanks. However, it is
important that such tanks are built only by qualified contractors.

Hydraulically-applied-concrete-lined reservoirs. Another way to build
reservoirs using hydraulically applied concrete is to cover earthen embankments
with hydraulically applied concrete, reinforcing wire mesh, and then more
hydraulically applied concrete. This is a relatively low cost method of building a
reservoir, but such reservoirs are usually shallow and hard to cover.

8-3. Location of Distribution Storage Systems

The location of distribution storage is closely associated with the system
hydraulics and the water demands, as well as their variations throughout the day
in different parts of the system.

Elevated Storage

Locations for placing elevated-storage tanks so that distribution main size is
minimized are illustrated in Figure 8-7. Elevated storage located at the source is
illustrated in Figure 8-7A. Since the part of the service area that is farthest from
the elevated tank has inadequate pressure due to head loss in the system,[2] main
size has to be increased in order to provide enough pressure. To avoid the cost of
increased main size, the elevated tank can be located beyond the service area
(Figure 8-7B), so that the pressure will be significantly improved with existing
main size. An even better solution is shown in Figure 8-7C, where the elevated
tank is located in that part of the service area that originally had the lowest
pressure, rather than beyond the surface area. In this case, not only is the pressure

---

[2] *Basic Science Concepts and Applications,* Hydraulics Section, Head Loss (Friction Head Loss).

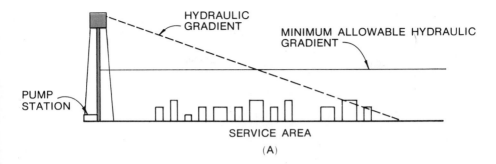

(A)

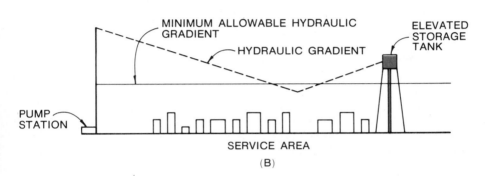

(B)

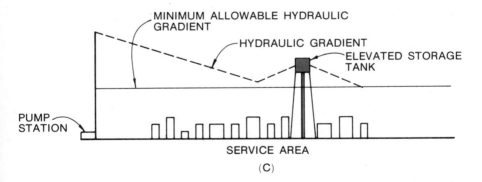

(C)

NOTE: Hydraulic Gradients Represent System During Maximum Hour Periods

Courtesy of Public Works Magazine

Figure 8-7.   Location of Elevated Storage

slightly improved over the situation before the tank was installed, but smaller mains can be connected to the tank since the flow from the tank is split into two directions.

Often, it is desirable to provide several smaller storage units in different parts of the system in place of a larger tank of equal capacity at a central location.

Smaller pipelines are required to service decentralized storage and, other things being equal, a lower flow-line elevation and pumping head result.

Minimizing pumping requirements.   The effect of storage tank location on pumping requirements was illustrated in Figure 8-7. If no storage is provided (Figure 8-8), the pump station must provide sufficient pumping head to overcome system friction losses at the demand rate and maintain a minimum head in the service area.

Environmental concerns.   A reservoir should be acceptable to the community. It should be unobtrusive and should blend, as much as possible, into the surroundings. Naturally, the first considerations in reservoir design and construction are economy and efficiency, but an aesthetically pleasing site will help promote good public relations by being an asset rather than an eyesore to the community.

Community-acceptance factors (which must be balanced against site availability, reliability, and cost) include aesthetics, landscaping, control of noise and television interference, and multiple-use of both site and structure.

Ground Storage

Where topography or community acceptance does not permit the economical location of elevated storage tanks, ground-level storage and an accompanying pump station may be the best method for providing adequate storage. If the distribution system has several pressure zones, ground-level storage and booster pumping are often located at the boundaries of these pressure zones. Water from the lower pressure zone flows into the reservoir and is pumped to the higher pressure zone.

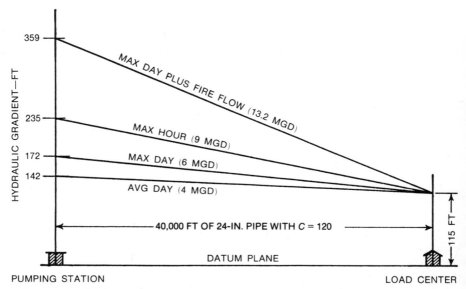

Figure 8-8. Sufficient Pumping Head Must Be Provided When There Is No Storage

In environmentally sensitive areas, ground-level storage facilities are much less obtrusive than elevated storage tanks. The roofs of ground storage reservoirs can be designed for multiple uses, such as parking lots, tennis courts, etc.

8-4. Accessory Equipment

Typical storage tanks, with their parts labeled, are illustrated in Figures 8-9 (elevated storage tank) and 8-10 (ground-level storage tank). A variety of accessory devices may be installed on storage tanks, depending on need.

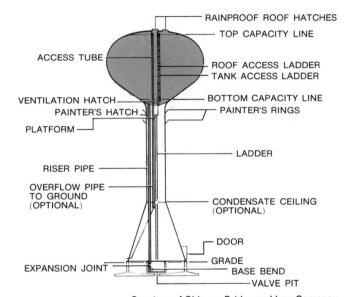

Courtesy of Chicago Bridge and Iron Company

Figure 8-9.   Elevated Storage Tank

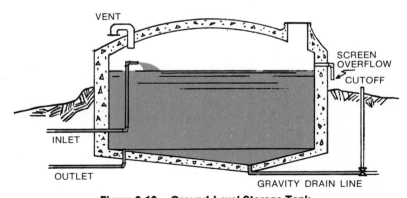

Figure 8-10.   Ground-Level Storage Tank

Inlet and outlet pipes.   In an elevated tank, the same pipe is generally used as both inlet and outlet pipe. This pipe is called a RISER. Elevated tanks with risers over 8 in. (200 mm) in diameter should have protective bars over the riser opening inside the tank to prevent someone from falling through the opening during the tank-cleaning operation. In cold climates, the riser is usually 36 in. (910 mm) in diameter or larger, to allow for expansion of ice during freezing temperatures, and it may be insulated or heated. In locations where freezing temperatures do not occur, a small riser pipe may be used.

Ground-level storage tanks generally have separate inlet and outlet pipes. This tends to improve the circulation of water within the tank so that the quality of the water leaving the tank remains uniform. A SILT STOP on the outlet pipe is used to keep sediment from reaching the customer. Often the silt stop is removable so that the sediment can be easily removed during cleaning operations.

Overflow pipe.   An overflow pipe is necessary in case the water-level controls fail. This pipe should be brought down to within 1 ft (0.3 m) of the ground surface and should discharge to a splash plate or drainage inlet structure to prevent washing away of soil around the tank base. The overflow pipe should never be connected to a sewer or storm drain without an air gap.

Drain pipe.   The drain pipe used to empty the tank should be installed with the same precautions as the overflow pipe.

Monitoring devices.   The water level in the tank can be measured either by a pressure sensor at the base of the tank or a level sensor inside the tank. For these sensors, some means should be provided to indicate the water level at the site and transmit that information to a central location (see Module 9, Instrumentation and Control). In many systems, the water-level sensors control the pumps as well as the alarms used for indicating high and low water levels.

Valving.   Altitude and isolation valves associated with water storage facilities are discussed in Module 3, Valves.

Air vents.   Air vents must be installed to allow air to enter and leave the tank as the water level falls and rises. Vents should be screened to keep out birds and animals that might contaminate the water. A screen with ½-in. (13-mm) mesh openings is usually satisfactory. Insects seldom cause contamination of stored water and insect screening is not recommended because it may become clogged or covered with ice. This could cause the tank to collapse if water is withdrawn when air cannot enter through the vent.

Access hatches.   Water storage tanks should be built with convenient access for cleaning and maintenance. Hatches should be installed so that surface runoff cannot get into the tank. All hatches and manholes should be locked when not in use to prevent vandalism. When existing tanks are repainted it may be necessary to install additional hatches so that adequate ventilation can be provided to remove evaporating paint solvents.

Ladders   Three types of ladders should be provided on a tank to allow safe and convenient access to the different parts of the tank and access hatches—a tower ladder, an outside tank ladder, and a roof ladder.

The tower ladder, with side rails, should extend from a point 8 ft (2.5 m) above the ground up to and connecting with either the horizontal balcony girder or the tank ladder if there is no balcony. Having the tower ladder begin several feet above the ground reduces the chances of vandalism and unauthorized use; however, it requires the use of a lift truck or portable ladder for access.

In all cases, an outside ladder with side rails should be provided on the outside of the tank shell, connecting either with the balcony or with the tower ladder if there is no balcony. The outside tank ladder may be attached to the roof ladder.

Access should also be provided to all roof hatches and vents. The roof ladder should be such that it can be reached from either the outside tank ladder or the riser ladder on pedestal tanks.

Safety devices that should be provided with the ladders include a safety cage, a rest platform, or roof-ladder handrails. The necessity for these is governed by federal or local regulations.

Ladders may also be provided within the tank. In such instances, access is through a normally locked manhole in the sidewall of the tank or through a roof hatch for access to the exterior on top.

Coatings. All steel tanks and some concrete tanks are protected from corrosion with some type of paint or coating. Because of the variety of coatings available (alkyd, epoxy, vinyl, petroleum wax, etc.), it is important that the coating selected will be effective, safe, and not cause taste or odor problems. It is recommended that a reliable manufacturer be consulted in selecting coating systems that meet AWWA standards (AWWA D102, Standard for Painting Steel Water-Storage Tanks) and are approved by the state health agency for potable water use.

Cathodic protection. In addition to a good inside and outside coating system, corrosion of the interior walls below the water line of steel tanks is often minimized through cathodic protection. Cathodic protection is a system that reverses the flow of current that tends to dissolve iron from the surface, causing rust and corrosion. By placing electrodes in the water and connecting them to the steel tank, an electrical current can be run through the tank so the electrodes will corrode instead of the tank (Figure 8-11). Because of the obvious difficulty of emptying storage reservoirs and repainting them, cathodic protection may be desirable as an added protection against corrosion.

Lighting. Depending on the tank's height and location, the Federal Aviation Administration (FAA) may require that obstruction lights be placed on the tank to warn aircraft in the vicinity. In particularly hazardous locations, orange and white checkerboard painting may also be required.

8-5.   Operation and Maintenance of Water Storage Facilities

Operation of storage facilities on a day-to-day basis is usually a matter of monitoring automated systems. Regular maintenance and inspection are required to ensure that facilities remain in efficient and sanitary operational condition and to provide for maximum service life.

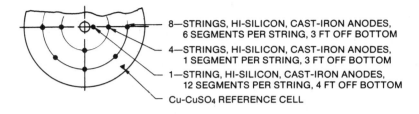

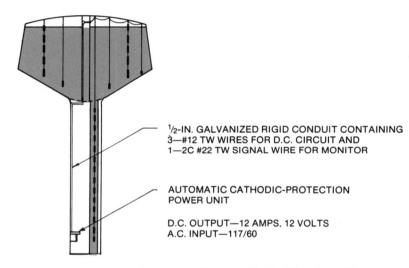

Courtesy of Public Works Journal Corporation

Figure 8-11.   Cathodic Protection

Operation

Operation of storage facilities is generally automated, with the operation being designed to level out supply and demand. Tanks fill when supply exceeds demand and empty into the system when demand exceeds supply. Ordinarily, distribution systems should be operated to keep the pressures throughout the system between 35 and 100 psi (240 and 690 kPa).

Cold-weather operation.   In cold climates, freezing in tanks can be a serious problem. Freezing is more likely in systems using surface-water sources in which the raw water is just above freezing. However, ground-water supplies, which provide relatively warm water, can also experience icing problems in storage tanks during very cold weather.

If ice forms inside the tank, it may damage the paint, cathodic-protection system, ladders, and overflow pipes inside the tank. As the water freezes, it expands, causing forces that could damage the tank walls or riser pipe. If a layer of ice forms all across the surface of the water, it may fall as the water level drops, also causing damage.

Since water demands are usually lower in the winter, water is not replaced in the tank as fast as it is in the summer. To help keep the water that remains in the tank too long from freezing, the operator can increase the variation in the water level, either by adjusting the automatic controls or by using manual controls. The operator may also be able to lower the maximum water level of the tank; however, the fire department should be consulted before doing this. If ice buildup is noticed in a storage tank, it should be melted using a high-pressure hose or a steam generator.

Maintenance

Water storage tanks should be drained, cleaned, painted (if necessary), and disinfected annually. The surfaces of the walls and floor should be cleaned thoroughly with a high-pressure water jet, sweeping, scrubbing, or other methods. All water and dirt should be flushed from the tank. Following the cleaning operation, the vent screen should be checked.

Successful and economical painting of tanks depends on proper surface preparation; selection of suitable paints; good workmanship; adequate drying and curing; and proper maintenance through periodic inspection and spot, partial, or complete removal of old paint, followed by repainting where necessary. Painting materials and methods should conform to AWWA D102, Standard for Painting Steel Water-Storage Tanks. Proper safety precautions should be stringently observed to avoid death or injury caused by paint fumes accumulated in a tank. Only reputable firms with well established experience should be used for tank maintenance and painting.

After cleaning and/or painting, water storage tanks must be disinfected before being placed in service. Liquid chlorine, sodium hypochlorite solution, or calcium hypochlorite granules or tablets may be used. Three alternate methods of chlorination are recommended by AWWA C652, Standard for Disinfection of Water Storage Facilities. The state water supply agency should be consulted for any specific requirements. In the first method, the volume of the entire tank is chlorinated, so that the water will have a free chlorine residual of at least 10 mg/L after the proper detention time. The detention time is 6 hours if the disinfecting water is chlorinated before entering the tank, and 24 hours if the water is mixed with hypochlorite in the tank. The second method involves spraying or painting all interior tank surfaces with a solution of 200 mg/L available chlorine. This is a hazardous procedure and should only be done by trained, experienced, and properly equipped personnel. In the last method, 6 percent of the tank volume is filled with a solution of 50 mg/L available chlorine for at least 6 hours. Then the tank is completely filled and the solution held for 24 hours.

The amount of chlorine needed can be determined from Table 8-1. If the chlorine solution must be neutralized before being discharged, Table 8-2 indicates the amount of chemical required, depending on the tank size and chlorine residual. Check with your local sewer department before discharging into a sanitary sewer and with the state health agency before discharging highly chlorinated water elsewhere.

Table 8-1. Amounts of Chemicals Required to Give Various Chlorine Concentrations in 100,000 gal of Water

Desired* Chlorine Concentration in Water mg/L	Pounds of Liquid Chlorine Required	Gallons of Sodium Hypochlorite Required			Pounds of Calcium Hypochlorite Required
		5% Available Chlorine	10% Available Chlorine	15% Available Chlorine	65% Available Chlorine
2	1.7	3.9	2.0	1.3	2.6
10	8.3	19.4	9.9	6.7	12.8
50	42.0	97.0	49.6	33.4	64.0

*Amounts of chemicals are for initial concentrations of available chlorine. Allowance may need to be made for chlorine depletion where low concentrations are held for extended time periods.

Table 8-2. Amounts of Chemicals Required to Neutralize Various Residual Chlorine Concentrations in 100,000 gal of Water

Residual Chlorine Concentration mg/L	Pounds of Chemical Required			
	Sulfur Dioxide (SO_2)	Sodium Bisulfite $(NaHSO_3)$	Sodium Sulfite (Na_2SO_3)	Sodium Thiosulfate $(Na_2S_2O_3 \cdot 5H_2O)$
1	0.8	1.2	1.4	1.2
2	1.7	2.5	2.9	2.4
10	8.3	12.5	14.6	12.0
50	41.7	62.6	73.0	60.0

After the chlorination procedure is completed, and before the tank is placed in service, water from the full tank must be tested for bacteriological safety (coliform test). If the tests show the sample to be bacteriologically safe according to the requirements of the state health agency and the Safe Drinking Water Act (see Volume 4, *Water Quality Analyses*, Module 1, Drinking Water Standards), then the tank can be placed in service. If the sample tests unsafe, then further disinfection must be performed and repeat samples taken until two consecutive samples test safe.

Inspection

Periodic inspection of water storage facilities is necessary for the facilities' continued operation. (See sample checklist, Figure 8-12.) Cracks or holes in reservoirs can cause a loss of water, and flaws in ground storage tanks can cause contamination of the water supply.

Tanks should be inspected for corrosion both inside and outside. This requires draining the tank to check the surfaces and the operation of the cathodic protection equipment, if used. Overflows and vents should be examined to make sure that they are not blocked and that screens are clean and in place, especially after routine maintenance has been performed. The altitude valve should be checked to make sure it is allowing water into the tank and stopping the flow when the tank is full. Level sensors and pressure gauges should be checked by

Item Inspected	Action	Frequency
Foundation	Check for settling, cracks, spalling, and exposed reinforcing	Semiannual
Concrete tanks	Check exterior for seepage; mark spots	Semiannual
	Check exterior for cracks, leaks, spalling, etc.	Annual (spring)
	Check floor expansion joints (if any) for leakage; check roof expansion joints (if any); check for missing sealant	Every five years
	Check roof condition; check hatches and screens on openings	Semiannual
Steel tanks	Check for ice damage	Annual (spring)
	Check interior and exterior walls for rust, corrosion, loose scale, leaky seams and rivets, and for condition of paint	Annual
Tower structures	Check tower structure (if applicable) for corrosion; loose, missing, bowed, bent, or broken members; loose sway bracing; misalignment of tower legs, evidence of instability	Semiannual
	Check surface of lattice bars, anchor bolts, boxed channel columns	Semiannual
Covers	Check obstruction (navigation) lights, hoods, shields, receptacles, and fittings for missing or damaged parts	Monthly
Risers	Check riser pipe insulation and heating system, if any	Annual—two months before freezing weather
Cathodic protection	Check rectifier according to manufacturer's instructions	Monthly
	Check anode condition and connections	Monthly
	Perform internal tank-to-water potential survey	Annual (spring)
Appurtenances	Check ladders, walkways, guardrails, etc., for rust, corrosion, damage, or deterioration	Semiannual

Figure 8-12.   Routine Inspection Checklist (Sample)

changing the water level in the tank. Controls should be examined to see if they are operating at the desired pressures.

Water-storage tanks should be checked frequently for vandalism and signs of forced entry. Access doors and their locks should be kept in good repair. Ladders must provide safe access for authorized personnel, but should have the access entry locked.

Aviation warning lights, if installed, should be maintained so they will provide adequate warning to aircraft. Lights should be checked regularly to make sure they are working and clean enough that they do not have reduced light output.

FAA regulations require that the bulbs be replaced before they reach 75 percent of their normal life expectancy.

Records

Because most water systems have a relatively small number of distribution storage tanks, the record-keeping system does not have to be extensive. Basic information to be recorded should include the location, the dates of inspection, conditions noted during the inspection, the dates maintenance was performed, and notes on the maintenance performed. Examples of storage tank records are shown in Figures 8-13 and 8-14. If maintenance is performed by outside contractors, it is a good practice to keep copies of specifications and contracts for future use. The phone numbers of equipment suppliers' service departments should also be available.

PITTING REPORT

LOCATION (BY PLATES NUMBERED FROM ROOF DOWN)	PERCENT OF AREA AFFECTED (APPROXI- MATELY)	MAX. DEPTH OF PITTING FOUND	TYPE OF PITTING	PLATE THICKNESS
1				
2				
3				
bottom				

Figure 8-13.   Sample Pitting Report

8-6.   Safety

Only trained and experienced operators should be allowed to work on elevated and ground storage tanks and standpipes. This is hazardous work and dangerous for untrained operators. Special precautions are needed while working on or in tanks because they are confined working areas and also provide the hazard of falling from considerable heights.

Specific areas where the need for safety precautions must be recognized and observed when dealing with water storage facilities include:

- Personal protection equipment (refer to Module 1)
- Chemicals (refer to Module 1)
- Hand tools (refer to Module 1)
- Portable power tools (refer to Module 1)
- Confined spaces (refer to Module 3).

WATER TANK SERVICE REPORT

FOR: _____

CUSTOMER P.O. # _____

JOB NO.: _____
CONTRACT TYPE: _____
ANNIVERSARY DATE: _____
TANK TYPE: _____
TANK LOCATION: _____
HEATED TANK: YES _____ NO _____

ON ARRIVAL, RECTIFIER UNIT # _____ , MANUFACTURED BY _____
_____ RATED AT _____ D.C. VOLTS, AND _____
D.C. AMPS, TYPE: MANUAL □ AUTOMATIC POTENTIAL CONTROL □ AND/OR
T.A.S.C. AUTO CONTROLLED # _____ , WAS FOUND OPERATING
AT _____ AMPS TO THE BOWL AND _____ AMPS TO THE RISER AT
_____ D.C. VOLTS WITH TAP SETTING _____ AND/OR _____
VOLTS POTENTIAL.

ALL ANODES, SUSPENSION, WIRING, ETC., WERE INSPECTED
AND THE FOLLOWING REPAIRS OR REPLACEMENTS WERE MADE.

ANODES	NO.	TYPE	ALUM. AND LENGTH OR NO. SECTIONS AND SPACING (HISILICON CAST IRON)
RING #1			
RING #2			
RING #3			
RING #4			
RECTIFIER			
WIRING			
SUSPENSION			
OTHER			

SYSTEM LEFT OPERATING AT:

MANUAL RECTIFIER
TAP SETTING _____ D.C. VOLTS _____ D.C. AMPS # 1 _____
D.C. AMPS # 2 _____ WATER LEVEL % _____ TANK-TO-WATER _____
POTENTIAL PROVIDED WITH CuCuSO4 REFERENCE ELECTRODE ON _____
OFF _____

AUTOMATIC RECTIFIER
TAP SETTING _____ D.C. VOLTS _____ D.C. AMPS # 1 _____
D.C. AMPS # 2 _____ WATER LEVEL % _____ CONTROLLER _____
INTERNAL POTENTIAL SET POINT _____ NATURAL POTENTIAL _____

ENERGIZED POTENTIAL _____
OTHER CuCuSO4 TANK-TO-WATER POTENTIAL
LOCATION _____ OFF POTENTIAL _____ ON POTENTIAL _____
LOCATION _____ OFF POTENTIAL _____ ON POTENTIAL _____
LOCATION _____ OFF POTENTIAL _____ ON POTENTIAL _____

TO INSURE CONTINUOUS CATHODIC PROTECTION, OF MANUALLY
OPERATED SYSTEMS, THE RECTIFIER OUTPUT SHOULD BE READ ONCE A
MONTH AND ADJUSTMENTS MADE TO MAINTAIN THE BOWL CURRENT
BETWEEN _____ AND _____ AMPS AND THE RISER
CURRENT BETWEEN _____ AND _____ AMPS WHEN THE
TANK IS FULL. THESE READINGS SHOULD BE NOTED ON THE POST
CARDS FURNISHED AND MAILED. SHOULD ANY PROBLEMS ARISE,
PLEASE WRITE OR CALL THE WATER TANK SERVICE DEPT. **DO NOT
ADJUST AUTOMATIC SYSTEMS AS THEY ARE PRE-SET TO OBTAIN
OPTIMUM CATHODIC PROTECTION UNDER ALL CONDITIONS.** FOR
OPTIMUM EFFECT THE POTENTIAL LEVEL METER SHOULD READ
MV. ± 25 MV.

FOR YOUR CONVENIENCE IN MAINTAINING RECORDS OF THE
MONTHLY READINGS ON THIS CATHODIC PROTECTION SYSTEM,
THE FOLLOWING FORM IS PROVIDED.

	TAP/ POTENTIAL SETTING	UNIT DC VOLTAGE	BOWL CURRENT	RISER CURRENT	WATER LEVEL
JANUARY					
FEBRUARY					
MARCH					
APRIL					
MAY					
JUNE					
JULY					
AUGUST					
SEPTEMBER					
OCTOBER					
NOVEMBER					
DECEMBER					

THE FOLLOWING REPAIRS (IF ANY) ARE REQUIRED BUT ARE NOT COVERED
BY YOUR SERVICE CONTRACT. A PROPOSAL FOR THIS WORK WILL
BE SENT FOR YOUR CONSIDERATION.

DATE OF INSPECTION _____ CREW ARRIVED _____ CREW LEFT _____

SIGNED: _____ CREW FOREMEN _____ SIGNED BY: _____ BY THE CUSTOMER

Courtesy of HARCO Corporation

Figure 8-14. Cathodic-Protection Report Form (Front and Back)

Selected Supplementary Readings

Antos, Lois. Application and Maintenance of Altitude Valves. *Op Flow*, 10:3:1 (Mar. 1984).

Brotsky, Bob. Interior Maintenance of Elevated Storage Tanks. *Jour. A WWA*, 69:9:506 (Sept. 1977).

Disinfection of Water Storage Facilities. AWWA Standard C652-86. AWWA, Denver, Colo. (1986).

Factory-Coated Bolted Steel Tanks for Water Storage. AWWA Standard D103-80. AWWA, Denver, Colo. (1980).

Graffiti 1980—The Good, The Bad, and The Ugly. *Op Flow*, 6:12:3 (Dec. 1980).

Hudson, W.D. Elevated Storage vs Ground Storage. *Op Fow*, 4:7:3 (July 1978).

Inspecting and Repairing Steel Water Tanks, Standpipes, Reservoirs, and Elevated Tanks, for Water Storage. AWWA Standard D101-53 (R1986). AWWA, Denver, Colo. (reaffirmed, 1986).

Larson, L.A. Cold-Weather Operation of Elevated Tanks. *Jour. A WWA*, 68:1:17 (Jan. 1976).

Leary, C.J. Water Towers and Ice Build Up. *Op Flow*, 4:1:1 (Jan. 1978).

Miller, A.L. Cleaning and Recoating Your Storage Tanks. *Op Flow*, 4:10:1 (Oct. 1978).

Miller, A.L. How to Cope With Freeze Damage and Ice Buildup on Outside Storage Tanks. *Op Flow*, 6:1:4 (Jan. 1980).

Miller, A.L. Sr. Cleaning, Recoating Water Storage Tanks. *Op Flow*, 6:8:1 (Aug. 1980).

Muller, K.C. Recoating Storage Tanks as Part of Tank Maintenance—Part I. *Op Flow*, 3:3:1 (Mar. 1977).

Muller, K.C. Recoating Storage Tanks as Part of Tank Maintenance—Part II. *Op Flow*, 3:4:1 (Apr. 1977).

Obstruction Marking and Lighting. Federal Aviation Administration, Washington, D.C. AC 70/7460-1F. (Sept. 1978).

Painting Steel Water-Storage Tanks. AWWA Standard D102-78. AWWA, Denver, Colo. (1978).

Proudfit, D.P. Adequate Storage Capacity is Vital for Peak Demands. *Op Flow*, 3:12:7 (Dec. 1977).

Thermosetting Fiberglass-Reinforced Plastic Tanks. AWWA Standard D120-84. AWWA, Denver, Colo. (1984).

Two Men Die—Overcome by Toxic Paint Fumes. *Op Flow*, 7:6:3 (June 1981).

Welded Steel Tanks for Water Storage. AWWA Standard D100-84. AWWA, Denver, Colo. (1985).

Glossary Terms Introduced in Module 8

(Terms are defined in the Glossary at the back of the book.)

Cathodic protection	Operating storage
Distribution storage	Peak hour demand
Elevated storage	Prestressed concrete
Elevated tank	Reservoir
Emergency storage	Riser
Fire demand	Silt stop
Ground-level tanks	Standpipe
Hydraulically applied concrete	Tank
Hydropneumatic system	

Review Questions

(Answers to Review Questions are given at the back of the book.)

1. Identify at least four reasons for providing water storage within a distribution system.

2. In terms of total storage capacity, what is the difference between operating storage and emergency storage?

3. What are the advantages and disadvantages of elevated tanks, ground-level tanks, and standpipes.

4. Describe the operation and intended application of hydropneumatic storage.

5. Identify two factors that significantly influence the location of water storage facilities.

6. List three types of piping found on every tank.

7. Describe what annual maintenance should be performed on water storage tanks.

8. After disinfection, what must be done before a water storage tank can be placed back in service?

9. At a minimum, explain what information should be recorded on water storage facilities for future reference.

10. Special safety equipment and procedures are required when performing maintenance on water storage facilities, but are unnecessary when making periodic inspections. Do you agree?

Study Problems and Exercises

1. On a map of your city, locate all or a minimum of four distribution storage tanks. Prepare a report stating briefly the advantages and disadvantages of the storage tank locations.

2. Select one storage tank from the above and determine:
   a. total capacity
   b. maximum water level
   c. minimum water level
   d. maximum and minimum water levels during normal operation
   e. pressure at the nearest customer outlet when the tank is full.

3. Discuss measures to prevent cold weather problems with elevated storage tanks or reservoirs.

4. Prepare a maintenance schedule and check list for inspecting storage tanks.

5. Prepare a list of safety measures to be used when inspecting or performing maintenance on a water storage tank.

Module 9

Instrumentation and Control

Much of the day-to-day operation of a large water-distribution system is performed from a central location. INSTRUMENTATION allows operators in the control center to monitor flow rates, pressures, levels, and other important information from all parts of the distribution network. CONTROL EQUIPMENT allows the same centrally located personnel to change valve settings, turn pumps on and off, and otherwise adjust the system for efficient operation.

In many modern utilities, instrument readings are fed to one or more computers, which evaluate distribution-system needs and adjust controls as necessary, maintaining system balance with only minimal need for operator intervention. Most systems continue to maintain simpler, on-site instrumentation at remote sites, for testing purposes and as a backup to the main system. Some small utilities depend on similar direct-acting equipment for daily operations.

After completing this module you should be able to

- Identify the major components of an instrument.

- Distinguish between analog and digital systems.

- Explain the operation of several common sensors.

- Explain the operation of a basic telemetry system.

- Identify common systems of manual and automatic control.

- State general requirements for maintenance of instrumentation and control equipment.

- Observe safety precautions when dealing with electric and pneumatic instruments and control facilities.

9-1. Instrumentation

INSTRUMENTS are used to measure, display, and record the conditions and changes in a water-distribution system. Some instruments indicate what is happening at a given instant. Others guard against equipment overload and failure. Still others yield permanent records that are used to determine operating efficiency.

These functions are similar to the functions of an automobile instrument panel. The speedometer indicates the current vehicle speed; the oil-pressure light warns of impending engine failure; and the mileage indicated on the odometer can be used to calculate gas mileage and determine the need for regular maintenance. In both the distribution system and the automobile, the instruments do not replace the operator. They simplify the operator's work and help to improve the performance, safety, and reliability of the equipment.

Basic Instrument Components

The simplest instruments have only two parts: a SENSOR and an INDICATOR. The sensor responds to the physical condition (the PARAMETER) being measured, converting it to a signal that can activate the indicator. The signal may be simple physical motion, or it may be electrical current or a change in PNEUMATIC (air) pressure. The indicator may display the result immediately, or it may be replaced or supplemented by RECORDERS or TOTALIZERS to monitor conditions over a period of time.

In a simple instrument, the sensor and indicator are directly connected, so that when the parameter being measured changes, the indicator immediately displays the sensor's response. An example of a simple instrument is the bellows-type pressure gauge shown in Figure 9-1. The bellows acts as a sensor, responding to changes in pressure by expanding or contracting. The motion of the bellows causes the attached dial-gauge indicator to display the pressure.

Where the parameter being measured is some distance from the indicator, the sensor and indicator are not directly connected. Instead, the sensor unit includes or connects to a TRANSMITTER, also called a TRANSDUCER. The transmitter changes the sensor's output to a standard signal, either air pressure, electric current, or electric voltage. That signal is then carried through a TRANSMISSION CHANNEL (pneumatic lines or electric wiring) to a RECEIVER, which includes or activates an indicator at the operator's station.

To monitor conditions at very distant locations, such as a remote pump station or reservoir, a TELEMETRY system may be used. The components of the telemetry system are essentially the same as those for any remote-indicating instrument. A sensor is connected to a transmitter, which sends a signal over a transmission channel to a receiver/indicator combination. However, the signal used is designed to maintain its accuracy over a long distance, and usually consists of audio tones or electrical pulses. The transmission channel is either special interference-resistant wiring, a leased telephone line, or a radio or microwave transmission. The receiver converts the signal to standard electrical values or digital signals to operate the indicator.

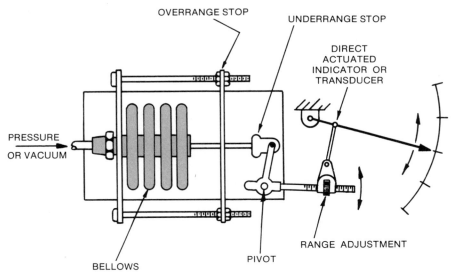

Figure 9-1.   Bellows-Type Pressure Gauge

Analog and Digital Systems

The terms ANALOG and DIGITAL are commonly used in describing various components and functions of instrumentation. Analog values range smoothly from the minimum to the maximum value of a given range. The varying voltage from a rheostat circuit is an analog signal, a dial indicator is an analog display. Digital values, on the other hand, take on only a fixed number of values within a range. The pulses sent out by a dial telephone are a digital signal, as are the 12 distinct tones sent out by a touch-tone phone; and a digital watch has a digital display. An important use of digital signals today is in the internal circuitry of computers.

Most parameters measured in water distribution are continuous in nature, like an analog signal or display. However, analog-to-digital converters allow continuous values to be displayed on digital indicators or transmitted over digital transmission channels, and digital-to-analog converters allow the reverse conversion. Digital signals are less susceptible to error and interference than analog signals, and digital displays are easier to read accurately. Analog equipment has advantages where simplicity of indication and operation is required.

Sensors

SENSORS that respond to all important parameters in a distribution system are the major components of every type of instrumentation, whether it is mechanical, pneumatic, electric, or electronic.

Pressure sensors.   Pressure sensors are used to determine suction and discharge pressures at pumps, pressure regulators, and selected points in a distribution grid. They also determine pressures of plant water, eductors, storage tanks, and air compressors. The four most common pressure sensors are:

- Strain gauge
- Bellows (low pressure)
- Helical element (medium pressure)
- Bourdon tube (high pressure).

The diaphragm STRAIN GAUGE sensor is the type used most widely in modern instrumentation. As shown in Figure 9-2A, it consists of a section of wire fastened to a diaphragm. As the variable being measured changes, the diaphragm moves, changing the length of the wire, thus increasing or decreasing its resistance. This changing resistance can be measured and transmitted by standardized electrical circuitry.

The remaining three types of sensors were once widely used but are seldom installed today. The BELLOWS SENSOR (Figure 9-2B) is a flexible copper can. The sensor expands and contracts with changes in pressure. The HELICAL SENSOR (Figure 9-2C) is a spiral wound tubular element that coils and uncoils with changes in pressure. The BOURDON TUBE (Figure 9-2D) is a semicircular tube with an eliptical cross section that tends to assume a circular cross-sectional shape with changes in pressure, thereby causing the C-shape to open up.

Level sensors. Level sensors are used to determine the height of liquid in wet wells, storage reservoirs, and elevated tanks, as well as to determine the levels of stored chemicals. The three most common level sensors are:

- Float mechanism
- Diaphragm element
- Bubbler tube.

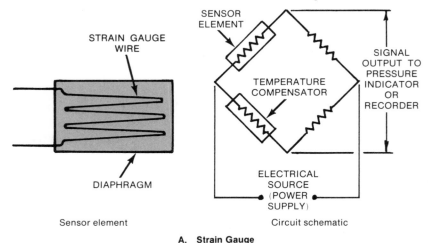

A. Strain Gauge

Figure 9-2. Types of Pressure Sensors

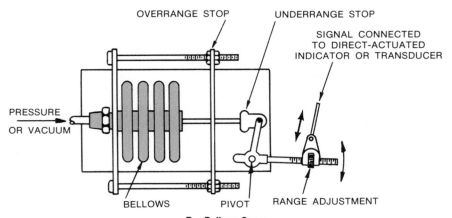

B. Bellows Sensor

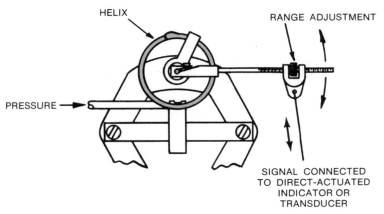

C. Helical Sensor

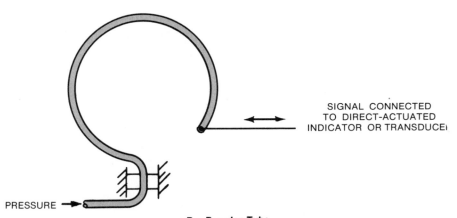

D. Bourdon Tube

Figure 9-2. (Cont.) Types of Pressure Sensors

A FLOAT MECHANISM (Figure 9-3A) operates on the displacement principle, with the float sized to generate whatever power is necessary to drive the transducer. A DIAPHRAGM ELEMENT (Figure 9-3B) operates on the principle that a confined volume of air at atmospheric pressure will compress to a new pressure equal to the head of water above the diaphragm. A BUBBLER TUBE (Figure 9-3C) operates on the principle that a constant volume of air flow to the bubbler will require a pressure approximately equal to the head above the tip of the tube. Both the diaphragm and the bubbler-tube sensors commonly use a strain gauge sensor to convert the physical variable into an electrical signal.

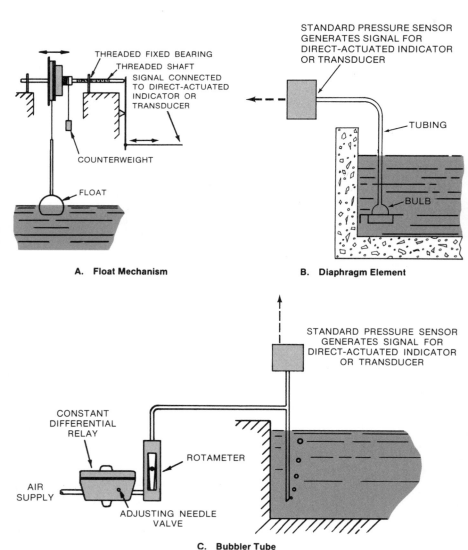

A. Float Mechanism

B. Diaphragm Element

C. Bubbler Tube

Figure 9-3. Types of Level Sensors

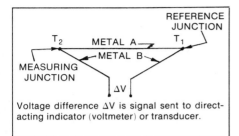

Voltage difference ΔV is signal sent to direct-acting indicator (voltmeter) or transducer.

A. Thermocouple

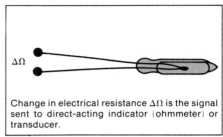

Change in electrical resistance ΔΩ is the signal sent to direct-acting indicator (ohmmeter) or transducer.

B. Thermister

Figure 9-4.   Types of Temperature Sensors

Temperature sensors. Temperature sensors determine the condition of pump bearings, switchgear, motors, building heat, and water. There are two temperature sensors commonly used in distribution systems:

• Thermocouple

• Thermister.

Schematics of both the thermocouple and the thermister are shown in Figures 9-4A and B. The THERMOCOUPLE uses two wires of different materials, joined together at two points—the sensing point and the reference junction. Temperature changes between the two points cause an electromotive force or voltage to be generated. This can be read out directly or amplified through a transducer. A THERMISTER uses a semiconductive material, such as cobalt oxide, that is compressed into a desired shape from the powder form. The material is then heat-treated to form crystals to which wires are attached. Temperature changes are reflected by a corresponding change in resistance of the thermister, as measured through the attached wires.

Flow sensors. Flow measurement is used to bill customers, check the efficiency of pumps, monitor for leaks, and help control or limit the delivery of water. The most common flow-measuring devices used in distribution systems are:

• Venturi meter

• Propeller meter

• Magnetic meter.

Construction of the three commonly used flow meters is shown in Figures 9-5A, B, and C.

A VENTURI METER operates on the principle that flow passing through a constriction (the meter's throat) will increase in velocity with an accompanying decrease in pressure. The difference in pressure (the PRESSURE DIFFERENTIAL) between the inlet and the throat will be proportional to the square of the flow velocity. The pressure differential is usually measured with a strain gauge. A PROPELLER METER operates on the principle that water colliding with the propeller blades, which are shaped to reduce slippage and drag, will cause the propeller to rotate at a speed proportional to flow. A MAGNETIC METER operates

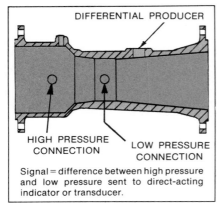

DIFFERENTIAL PRODUCER

HIGH PRESSURE
CONNECTION   LOW PRESSURE
CONNECTION

Signal = difference between high pressure
and low pressure sent to direct-acting
indicator or transducer.

A.   Venturi Meter

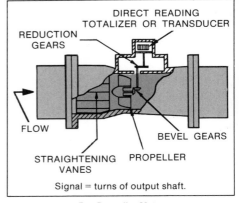

DIRECT READING
TOTALIZER OR TRANSDUCER

REDUCTION
GEARS

FLOW

BEVEL GEARS

STRAIGHTENING   PROPELLER
VANES

Signal = turns of output shaft.

B.   Propeller Meter

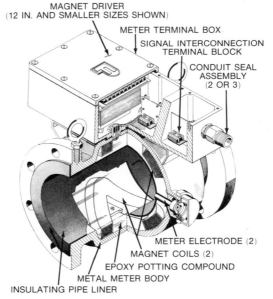

MAGNET DRIVER
(12 IN. AND SMALLER SIZES SHOWN)

METER TERMINAL BOX

SIGNAL INTERCONNECTION
TERMINAL BLOCK

CONDUIT SEAL
ASSEMBLY
(2 OR 3)

METER ELECTRODE (2)
MAGNET COILS (2)
EPOXY POTTING COMPOUND
METAL METER BODY
INSULATING PIPE LINER

Signal = voltage generated between electrodes, sent to direct-acting indicator (voltmeter) or transducer.

Courtesy of Fischer & Porter Company

C.   Magnetic Meter

Figure 9-5.   Types of Flow Sensors

on the principle that a conductive fluid (water) passing through the flux lines of an electromagnet will generate an electromotive force proportional to the flow.

Speed sensors.   A TACHOMETER GENERATOR is used to measure the speed of rotating equipment while it is operating. It is often necessary to measure the speed of a pump motor or chemical-feeder motor in order to be sure of its speed and operation. This is particularly true when operating equipment from a remote station.

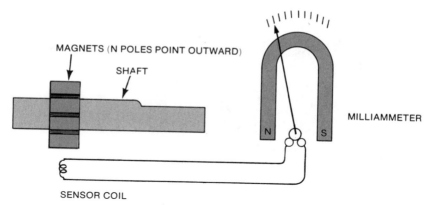

MAGNETS (N POLES POINT OUTWARD)

SHAFT

MILLIAMMETER

SENSOR COIL

Figure 9-6.   Construction and Circuitry of Tachometer Generator

Figure 9-6 illustrates the construction and circuitry of a tachometer generator. The instrument uses an electrical coil that produces a voltage when excited by magnets mounted on a spinning shaft. Various electrical devices can be wired to the output, which is proportional to speed.

Electrical sensors.   Four important electrical parameters are monitored for various pieces of equipment in the distribution system:[1]

- ELECTROMOTIVE FORCE, measured in VOLTS. The VOLTAGE in a circuit is analogous to the pressure in a distribution main.

- CURRENT, measured in AMPS. The current in a circuit is analogous to the flow rate in a main.

- RESISTANCE, measured in OHMS. The resistance of a circuit is analogous to the friction factor of a main.

- POWER, measured in WATTS. The power drawn by a circuit is directly related to the amount of work that the equipment fed by the circuit performs.

Because the equipment used to measure these parameters is itself electrical in nature, it is difficult to make clear distinctions as to what part of the equipment is the sensor and what part is the indicator. In fact, the measurements of volts, amps, and ohms are all made with the same unit, the D'ARSONVAL METER (Figure 9-7). Electric current passing through the meter's coil creates a magnetic field. The field reacts with the field of the permanent magnet surrounding it, causing the coil and the attached indicator needle to move [2]

By connecting a D'Arsonval meter into an electrical circuit, it can be made to measure electrical parameters without any special sensor. Which parameter is measured—volts, amps, or ohms—depends on how the meter is connected in the circuit, as shown in Figure 9-8. Depending on what it measures, the D'Arsonval

---

[1] *Basic Science Concepts and Applications*, Electricity Section, Electrical Measurements and Equipment.
[2] *Basic Science Concepts and Applications*, Electricity Section, Electricity and Magnetism.

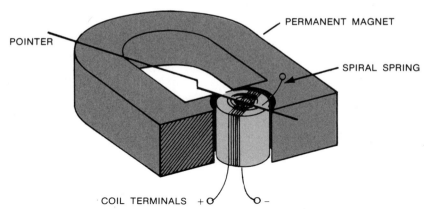

POINTER

PERMANENT MAGNET

SPIRAL SPRING

COIL TERMINALS   + O   O −

Reprinted by permission of John Wiley & Sons, Inc., New York, N.Y., from Basic Electric Circuits, *3rd ed., by Donald P. Leach. Copyright © 1984.*

Figure 9-7.   D'Arsonval Meter

meter circuit will be referred to as a VOLTMETER, an AMMETER, or an OHMMETER.

To measure power (watts), an instrument must combine the measurements of volts and amps, since watts = volts × amps (assuming power factor = 1).[3] The ELECTRODYNAMIC METER movement is used for this purpose. The electrodynamic meter is similar to a D'Arsonval meter, except that the permanent magnet in the D'Arsonval movement is replaced with a fixed coil. The interaction between the magnetic fields of the fixed coil and the coil attached to the indicator needle causes the needle to move. A circuit using an electrodynamic meter as a WATTMETER is shown in Figure 9-9. Electrodynamic meters can also be used to measure power factor.

The electric energy used by a utility, which determines its electric bill, is measured as kilowatt-hours (kW·h). One thousand watts (that is, 1 kW) drawn by a circuit for one hour results in an energy consumption of 1 kW·h. The meters used to determine kilowatt-hour usage are essentially totalizing wattmeters (Figure 9-10).

On most kilowatt-hour meters, a rotating disk can be seen through the glass front or through a small window in the front of the case. Revolutions of the disk can be counted by watching for the black mark on the disk. The speed of the disk indicates the kilowatt load. Checking the disk speed with a stop watch is a quick and accurate way of determining kilowatts used over a short period of time, using the following formula:

kilowatts = (disk watt-hours constant × revolutions × 3600)/(seconds × 1000)

Transmitters

Because most measured variables must be sensed at a point that is inconvenient to the operator, TRANSMITTERS, also called TRANSDUCERS, are used to convert the signal put out by the sensor into a standard signal that can be sent

---

[3] *Basic Science Concepts and Applications,* Electricity Section, Electrical Quantities and Terms.

VOLTS

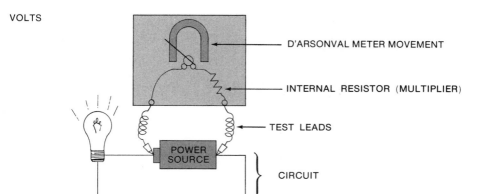

D'ARSONVAL METER MOVEMENT

INTERNAL RESISTOR (MULTIPLIER)

TEST LEADS

CIRCUIT

This setup measures voltage between the points where the test leads are connected.

A.   Measuring Potential (Voltmeter Circuit)

AMPS

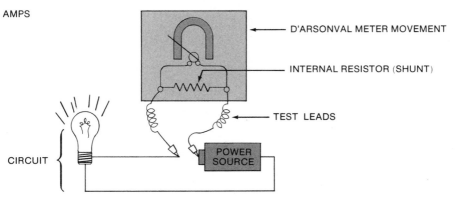

D'ARSONVAL METER MOVEMENT

INTERNAL RESISTOR (SHUNT)

TEST LEADS

CIRCUIT

This setup measures current (amps) passing through the circuit.

B.   Measuring Current (Ammeter Circuit)

OHMS

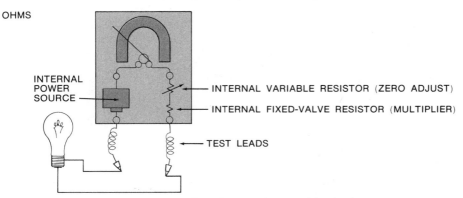

INTERNAL POWER SOURCE

INTERNAL VARIABLE RESISTOR (ZERO ADJUST)

INTERNAL FIXED-VALVE RESISTOR (MULTIPLIER)

TEST LEADS

This setup measures the resistance (ohms, Ω) of the circuit.

C.   Measuring Resistance (Ohmmeter Circuit)

**Figure 9-8.   Circuits Used to Measure Volts, Amps, and Ohms With
a D'Arsonval Meter Movement**

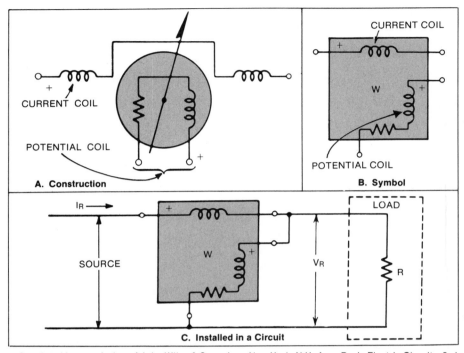

Figure 9-9. Wattmeter

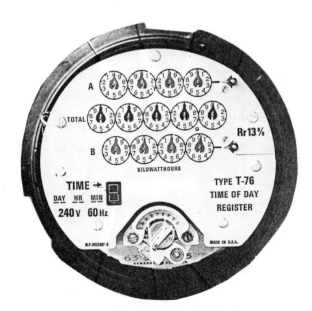

Figure 9-10. Totalizing Wattmeter

to a receiver/indicator in another location. The most common signal standards are:

- 3–15 psig-pneumatic (21–103 kPa)
- 15-sec electrical pulse duration modulation (PDM)
- Digital
- 4–20 mA DC
- 1–5 V DC.

Receivers and Indicators

RECEIVERS convert the signal sent by the transmitter to an indicator reading for the operator to monitor. The INDICATOR may be a direct-reading display that shows the current value of the parameter being monitored; it may be a record that preserves the values for later examination; it may be a totalizer that gives the total accumulated value since the instrument was last reset; or it may be some combination of these units.

Indicator displays and records are of two types: analog and digital. Analog indicators (Figure 9-11) include dial gauges and strip or circle charts. The indicated values on an analog display range smoothly from the display's zero to its maximum, allowing an infinite number of different readings. The D'Arsonval meter is commonly used as an analog indicator for electrical signals at standard voltage or current levels.

Digital indicators, like digital watches, display values as a decimal number, and the number of possible readings within a given range is limited by the number of digits displayed (Figure 9-12). Digital indicators are usually more accurate than analog indicators, because they are not subject to the errors associated with electromechanical or mechanical systems, and because they are easier to read correctly. However, analog indicators may be preferable for at-a-glance monitoring to ensure a value remains within a given range or to observe its rate of change.

Telemetry

TELEMETRY systems allow flow rate, pressure, and other distribution-system parameters to be sensed at one or more remote sites and indicated at a central

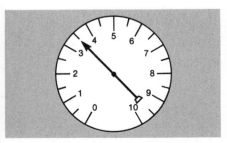

Figure 9-11.   Analog Indicator Display

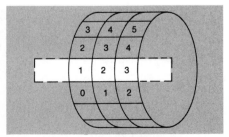

Figure 9-12.   Digital Indicator Display

location. Telemetry means metering at a distance. Every telemetry system has three basic components:

- A transmitter
- A transmission channel
- A receiver.

The transmitter takes in data from one or more sensors at the remote site (Figure 9-13). It converts the data to a signal that is sent to the receiver over the TRANSMISSION CHANNEL. The receiver changes the signal into standard electric values that are used to drive indicators and displays, recorders, or automatic control systems.

Telemetry transmission channels.   The transmission channel in a telemetry system may be a utility-owned cable for short distances, such as between two buildings on a common site. In most cases, however, the channel is either a leased telephone line, a radio channel, or a microwave system. The leased telephone line may be a dedicated metallic pair, which is relatively expensive but highly reliable and interference free, or it may be a standard voice-grade phone line. Most modern transmitters generate signals that are designed to be sent over voice-grade lines.

Radio channels can be in the vhf (very high frequency) or uhf (ultrahigh frequency) band. Both radio and microwave systems generally require a LINE-OF-SIGHT path, which is unobstructed by buildings or hills between the transmitter and the receiver. To bypass obstructions or to ensure signal strength over very long distances, RELAY STATIONS may be required.

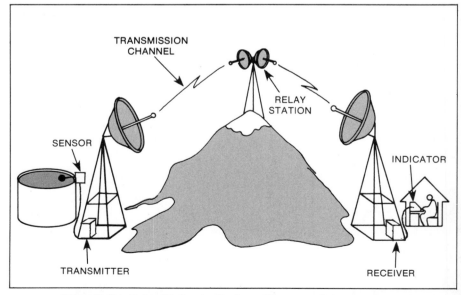

Figure 9-13.   Relay Station for Radio or Microwave Telemetry System

Telemetry signal types.   The signal sent over the transmission channel can be either analog or digital (Figure 9-14). Commonly used analog signal types include:

- CURRENT—the DC current generated by the transmitter is proportional to the measured parameter.
- VOLTAGE—the DC voltage generated by the transmitter is proportional to the measured parameter.
- PULSE-DURATION MODULATION (PDM)—the time period that a signal pulse is on is proportional to the value of the measured parameter.
- VARIABLE FREQUENCY—the frequency of the signal varies with the measured parameter.

The current and voltage signals can only be used for short-distance systems with utility-owned cable or a leased metallic pair. The pulse-duration-modulation and variable-frequency signals can be used for any distance over any type of channel.

Digital signal systems generally send BINARY CODE, in which the transmitter generates a series of on-off pulses that represent the exact numerical value of the measured parameter (for example, off-on-off-on represents 5). These signals can

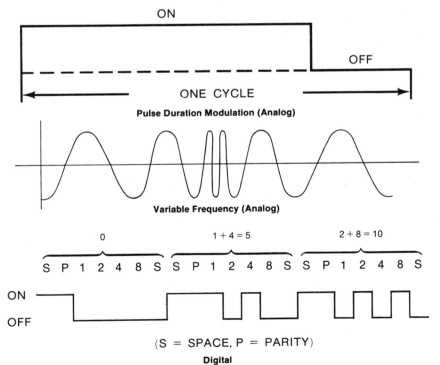

Figure 9-14.   Telemetry Signals—Analog or Digital Form

be used over long or short distances with any transmission channel. The binary-code signal is well adapted to computerized systems.

Multiplexing. MULTIPLEXING systems allow a single physical channel—a physical metallic pair, a leased phone line, or a radio channel—to carry several signals simultaneously. TONE-FREQUENCY MULTIPLEXING accomplishes this with tone-frequency generators in the transmitter and tone-frequency filters in the receiver. The signal representing each measured parameter is assigned a separate audio frequency. The transmitter sends data representing each signal only over its assigned frequency, and the filters in the receiver allow it to respond to each frequency separately. Up to 21 distinct frequencies can be sent over a single voice-grade phone line. Tone-frequency multiplexing can be used with all but current, voltage, and frequency signals, and it can be applied to any type of transmission channel.

Scanning. A second method of sending multiple signals over a single line or transmission channel is SCANNING. A scanner at the transmitter end checks and transmits the value of each of several parameters one at a time, in a set order. The receiver decodes the signal and displays each value in turn. Scanning can be used with all types of signals and with all types of transmission channels. Scanning and tone-frequency multiplexing can be combined to allow even more signals over a single line. A four-signal scanner combined with a 21-channel tone-frequency multiplexer would yield 84 distinct signal channels on a single line.

Polling. Another method for using a single line or channel is POLLING. In a polling system, each instrument has a unique address, or identifying number. A system controller unit sends messages over the line telling the instrument at a given address to transmit its data. The process of asking an instrument to send data is called polling. In many systems, the controller is programmed to poll instruments as often as necessary to monitor the system—some instruments may need checking more often than others. In more sophisticated systems, the controller regularly scans the status of each instrument to see whether there is new information to be transmitted. If the status indicates that new information exists, the controller instructs the instrument to send its data. In some systems, critical instruments also have the capability to interrupt a long transmission by another instrument in order to call the controller's attention to urgent new data. Since data is transmitted only when needed under the polling system, a single line or channel can handle more instruments than with a simple scanning system.

Duplexing. In many telemetry installations, the instrument signals received at the operator's central location must be acted on by sending control signals back to the remote site. Duplexing allows this to be accomplished with a single line.

- FULL DUPLEX allows signals to pass in both directions simultaneously.

- HALF DUPLEX allows signals to pass in both directions, but only in one direction at a time.

- SIMPLEX allows signals to pass in only a single direction.

Full-duplex systems usually make use of tone-frequency generators to divide the line into transmission and receiving channels. Half-duplex systems may use tone-frequency generators, or they may simply rely on timing signals (similar to scanners) or on signals indicating status, such as end-of-transmission or ready-to-receive.

9-2. Control

CONTROL EQUIPMENT is used to operate other equipment, such as motors and valves. Control equipment can be completely independent of instrumentation, as in manual control or remote manual control systems, or it may operate in direct response to instrument signals, as in automatic control, semiautomatic control, and most computer control systems. This module discusses control systems in general; specific information on valve-control systems and motor-control systems is included in Module 3, Valves, and in Module 7, Pumps and Prime Movers.

Manual and Remote Manual Control

In a LOCAL MANUAL CONTROL system, the operator operates switches or levers on the equipment to turn the equipment on or off or otherwise change its operating condition. A valve operated by a handwheel is a common manually controlled piece of equipment. Some older electrical equipment, such as motor starters and oil-filled circuit breakers, may have operating levers to allow direct manual operation.

REMOTE MANUAL CONTROL also requires an operator to turn the switch or push the button that activates and deactivates the equipment. However, the operator's controls may be located some distance from the equipment itself. When the operator activates the control switch, an electric RELAY or SOLENOID or motor is energized, which in turn activates the equipment. Power valve operators and magnetic motor starters are common examples of remote manual control devices.

The solenoids and relays used for remote control are common components of all types of control systems. A solenoid (Figure 9-15) is an electric coil with a moveable magnetic core. When an electric current is passed through the coil, a magnetic field is generated that pulls the core into the coil. The core can be attached to any piece of mechanical equipment that needs to be moved by

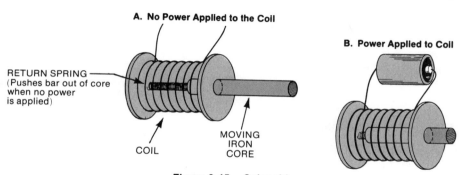

A. No Power Applied to the Coil

B. Power Applied to Coil

RETURN SPRING
(Pushes bar out of core
when no power
is applied)

COIL

MOVING
IRON
CORE

Figure 9-15.   Solenoid

remote control. For example, a solenoid might pull a lever to disengage the clutch between a diesel engine and a pump.

A relay is constructed of a solenoid that operates an electrical switch or a bank of switches. The most common use of a relay is to allow a relatively low-voltage, low-current control circuit to activate a high-voltage, high-current power circuit. Relays designed for this purpose are called POWER RELAYS or CONTROL RELAYS.

Automatic and Semiautomatic Control

An AUTOMATIC CONTROL system turns equipment on and off or adjusts its operating status in response to signals from instruments and sensors. The operator does not have to touch the controls under normal conditions. Automatic control systems are quite common, with simple examples including the thermostatic control for heating and air conditioning systems, automatic pump start/stop in response to tank water level, and automatic activation of security lighting systems at nightfall. More complicated automatic control systems may be installed to allow entire pump stations to operate unattended.

A number of MODES (logic patterns) of operation are available under automatic control. One common mode is ON-OFF DIFFERENTIAL CONTROL. In this mode, equipment is turned full on when a sensor indicates a preset value, then is turned full off when the sensor indicates a second preset value. The on-off differential control mode is commonly used to control pumping equipment in response to a tank water-level sensor. Another mode is PROPORTIONAL CONTROL, which opens a valve or increases a motor's speed slightly when the sensor shows a slight variation from its preset intended value, and opens the valve or increases the speed a great deal if the sensor shows a large error. This mode could be used in a treatment plant to control flow rate. Other, more complicated modes are availabe for special applications.

SEMIAUTOMATIC CONTROL combines manual or remote manual control with automatic control functions for a single piece of equipment. A circuit breaker, for example, may disconnect automatically in response to an overload, then require manual reset.

Computer Control

A computer control system uses a computer to generate signals that operate pumps, valves, and other distribution system equipment. The computer system may be fully automatic, requiring no operator intervention; this is called CLOSED-LOOP CONTROL. It may also be a sophisticated remote manual system, with the computer advising the operator of instrument status, then timing and sequencing valve and pump operation in response to the operator's commands. This type of system is called OPEN-LOOP CONTROL.

Direct-Wire and Supervisory Control

Within a plant or an attended pump station, equipment is usually connected to the central control panel through electrical wiring. This system is known as

DIRECT-WIRE CONTROL. Where a remote station is unattended, control of the station equipment from an operator's control location is accomplished using SUPERVISORY CONTROL EQUIPMENT, which transmits control signals over telemetry channels (discussed in the Instrumentation section of this module). The unattended site is sometimes called the OUTLYING STATION, and the operator's station may also be called the DISPATCHER'S STATION. Many large utilities have a large, central facility known as the DISTRIBUTION LOAD-CONTROL CENTER, where the entire distribution system is monitored and controlled (Figure 9-16).

9-3. Operation and Maintenance of Instrumentation

Instrumentation represents 7–8 percent of the capital investment in mechanical equipment for the average distribution system. Although this may not be a large sum, the proper operation of controls can make a substantial contribution to savings in the performance of the overall system. There are numerous routine maintenance tasks that an operator can perform to keep the instrumentation functioning properly. An instrument's useful life can be increased significantly by routinely checking to be sure it is not exposed to moisture, chemical gas or dust, excessive heat, vibration, or other damaging environmental factors.

Maintenance of Sensors and Transmitters

Routine maintenance that can be performed by an operator is often required for sensors. Special skills may be required to work on transmitters.

Pressure sensors.   Every sensor that in any way responds to liquid pressure will perform poorly if air enters the sensor. Air or gases may enter into the sensing element when released from the fluid line. A vent should be provided on the mechanism and activated on a regular basis. Eventually, experience will provide the basis for a maintenance schedule. Any pressure sensor located above

Figure 9-16.   Distribution Load-Control Center

the point of connection to the line being monitored is vulnerable to the entrance of air or gas. Wherever possible, the sensor should be lowered so that trapped gases flow upward toward the monitored line.

In cold climates, freezing is a possibility. Most pressure sensors either have a provision for a heater and thermostat or can be equipped with special protective cabinets. Heater tapes can also be used for sensing lines.

Pressure sensors, switches, and gauges require a small connection, usually ¼-in. NPT, to the pipeline. These small connections can be a continual source of trouble where dissimilar metals—cast iron and brass or bronze—are used together. A valve cock should be installed between the pipeline tap and the device to allow easy disconnection and service. At least twice a year, the sensor should be removed and the valve cock blown and rodded out if necessary. Corrosion of nipples should be checked and, if any weakness is apparent, the failed part should be replaced.

Primary flow elements.   There are several primary flow elements used in distribution systems, and each has its own particular maintenance requirements. A differential producer, such as a Venturi tube, orifice, flow nozzle, or equivalent device, has small ports that connect the process fluid to the transmitter mechanisms. These should be blown out periodically. Propeller meters are all-mechanical devices, and over a period of time, wear produces lower readings. The manufacturers' recommendations for lubrication should be carried out regularly, and only the recommended lubricant should be used. A magnetic meter is an electrical unit so regular checks should be made for corrosion or insulation breakdown around conduit connections or grounding straps.

Transmitters.   The operator without special training in instrument repair can generally do little more in the way of transmitter maintenance than to ensure a favorable environment for the equipment. In hot climates, it is important to protect electronic transmitters from exposure to high temperatures. Most electronic transmitters are rated for operation up to 130° F (55° C). Beyond this temperature, many electronic components may break down. Direct exposure to sunlight can contribute to serious heat buildup within units, and it is often necessary to provide a shield or fan if components are confined in a cabinet. Most transmitters are splash-proof and will function with occasional exposure to wetting. However, unless specified, most units will not withstand being submerged in water. Therefore, metering pits should have pumps and level alarms to warn of flooding.

Maintenance of Receivers and Indicators

Receiving and indicating devices usually require special skills for service and maintenance. However, the operator should attempt to maintain a favorable environment for the equipment. A dirty or damp atmosphere may damage receivers. Vibration, chemical dust (such as fluoride), and high chlorine content in the atmosphere should all be avoided. The high temperatures that affect transmitters also cause problems for receivers, and inking systems on recording indicators may perform poorly in cold areas.

Troubleshooting Guidelines

Some general guidelines for troubleshooting instrumentation systems are as follows:

- Never enter the facility with tools of any type—there is a tendency to start taking things apart without a thorough diagnosis.
- Make a complete diagnosis and attempt to confirm it by example.
- Consider ways to deal with the safety risks.
- Consider what will happen to the operating system while maintenance is being performed; in particular, where control valves are involved, consider the possibilities of water hammer.
- Always inform the proper authorities of planned actions beforehand.

Maintenance Shop

Plant designs usually provide an area for a repair shop appropriate for the specific type of plant and equipment. If an operator chooses to perform instrument maintenance, then the shop should contain the following equipment:

- Air compressor—the compressor should generate pressures equal to the maximum pressure rating of any transmitter.
- Precision pressure gauges—two are necessary so that differential pressures can be simulated for flow transmitters.
- Pressure regulators—two are necessary in order to adjust pressures to representative values for the transmitters.
- Volt-ohm-amp meter—this should be a portable unit for reading voltage (either AC or DC) in a circuit; reading the resistance in a circuit (such as telephone lines—3000 ohm maximum); detecting a short or open coil in a relay; and reading the current (amperes) in a circuit.
- Manometer—a 36-in. (910-mm) mercury manometer is sufficient to take a reading on any pressure-differential producer such as a Venturi tube; this reading then provides a basis for calibration of any transmitter used to measure flow.
- Dead weight lists—this device has a hydraulic system and a weight platform; gauges to be checked are threaded into a hydraulic connection and then weights are added for calibration of the pressure indication.
- Signal generators—the devices selected should be able to generate the standard signals of the instruments; these are usually 1–5 V DC, 4–20 A DC, pulse duration modulation, and pulse frequency.

Maintenance Records

The development of a maintenance file on instrumentation has many long-term benefits for any distribution system. These benefits include the following:

- Accurate accounting of the cost.

- Indication of areas where there may be a need for a change in the type of equipment.

- Projection of the life expectancy of each type of equipment.

- Maintenance of parts inventories at minimum cost based on operating experience.

- Proper parts inventories that can reduce emergency downtime substantially, which is another form of cost savings.

- Guidance for preventative maintenance steps.

Many facilities today have computers that print out maintenance schedules for each type of equipment as a guide to the preventative measures that need to be performed (Figure 9-17). Where computers are not used, reminders can be placed in an instrument chart supply so that the reminder comes up as the charts are rotated.

Maintenance records should include some form of index card for each instrument.

One of the most common problems for maintenance-program continuity is the loss of instruction manuals and calibration sheets. A file should be provided for these, and care should be taken to avoid their loss, especially when responsible personnel are transferred or terminated. Records should also be kept on services related to instruments not owned by the utility, such as telephone lines and power supplies to pump stations. Such information should be carefully documented so that there is concrete evidence in case it becomes necessary to register a complaint. System diagrams showing the routings for communication links are important. For telemetry systems it is useful to have a map showing channels or frequencies in each tone spectrum and the facilities to which they are connected.

D.B.W.C.	INSM-21C		INSPECTION MASTER FILE REPORT			DATE 06/21/84	PAGE 12
W RT C NR	INSP SEQ NR	LOC CODE	ITEM NUMBER	EQUIP CLASS	INSPEC FREQ AND DATES F DATE 1 1		APPLICATION
E 03	0	BL&C	966894	LTG	M	204	LIGHTING GENERAL
E 03	0	CD10	960156	LTG	M	194	LIGHTING GENERAL
E 03	0	CD10	960157	BLM	M	194	BLOWER MOTOR
E 03	0	CD10	960158	ACS	M	194	ACCUMMULATOR SYSTEM
E 03	0	CD10	960159	SMP	M	194	SUMP PUMP
E 03	0	CH&M	960103	LTG	M	194	LIGHTING GENERAL
E 03	0	JULN	960425	OHH	A	244	OVERHEAD HEATERS
E 03	0	KECT	960471	LTG	M	194	LIGHTING GENERAL

Figure 9-17. Computer Printout of Maintenance Schedule

9-4. Safety

Specific areas where the need for safety precautions must be recognized and observed when dealing with instrumentation and control include:

- Hand tools (refer to Module 1)
- Portable power tools (refer to Module 1)
- Electrical devices (refer to Module 7)
- Hydraulic and pneumatic devices (refer to Module 3).

Selected Supplementary Readings

Automation and Instrumentation. AWWA Manual M2. AWWA, Denver, Colo. (1983).

Basic Electrical Theory and Terminology. *OpFlow*, 8:7:4 (July 1982).

Computer-Based Automation in Water Systems—An AWWA Management Resource Book. AWWA, Denver, Colo. (1980).

Fragassi, P.A. Computers and Other Instrumentation Advances. Proc. Ann. Publ. Water Supply Engrg. Conf., Champaign, Ill. (April 1981).

Franson, D.C. Sizing Air Valves for Maximum Flow. Parker Hannifin Corp., Pneumatic Div., Otsego, Mich. (Feb. 1984).

Freeston, R.C. Maintenance of Instrumentation. AWWA Ann. Conf., Las Vegas, Nev. (June 1983).

Garrett, J.W. Automated Control Systems—The Operator's Viewpoint. *OpFlow*, 2:9:3 (Sept. 1976).

Johnson, J.L. et al. Do Utilities Really Need Operators? *OpFlow*, 1:5:1 (May 1975).

Lawton, J.; McHugh, J.F. Liquid Monitor. IBM Tech. Disclosure Bull., vol. 26, no. 6, Armonk, N.Y. (Nov. 1983).

Proceedings AWWA Water Plant Instrumentation and Automation Seminar. New Orleans, La. (June 1976).

Smith, G.R. Selecting a Telemetering System for Your Utility. *OpFlow*, 9:7:1 (July 1983).

Shubert, W.M. Preventive Maintenance of Controls. *OpFlow*, 5:3:3 (March 1979).

Von Rader, J.E. Telemetric Control of Water Distribution Systems. *Jour. AWWA*, 75:11:542 (Nov. 1983).

Glossary Terms Introduced in Module 9

(Terms are defined in the Glossary at the back of the book.)

Ammeter
Amp
Analog
Automatic control
Bellows sensor
Binary code
Bourdon tube
Bubbler tube
Closed-loop control
Control equipment
Control relay
Current
D'Arsonval meter
Diaphragm element
Digital
Direct-wire control
Dispatcher station
Distribution load-control center
Electrodynamic meter
Electromotive force
Float mechanism
Full duplex
Half duplex
Helical sensor
Indicator
Instrument
Instrumentation
Line-of-sight
Local manual control
Magnetic meter
Mode
Multiplexing
Ohm
Ohmmeter
On-off differential control
Open-loop control
Outlying station
Parameter
Pneumatic

Polling
Power
Power relay
Pressure differential
Propeller meter
Proportional control
Pulse-duration modulation (PDM)
Receiver
Recorder
Relay
Relay station
Remote manual control
Resistance
Scanning
Semiautomatic control
Sensor
Simplex
Solenoid
Strain gauge
Supervisory control equipment
Tachometer generator
Telemetry
Thermister
Thermocouple
Tone-frequency multiplexing
Totalizer
Transducer
Transmission channel
Transmitter
Variable frequency
Venturi meter
Volt
Voltage
Voltmeter
Watt
Wattmeter

Review Questions

(Answers to Review Questions are given at the back of the book.)

1. What are three basic functions of instruments?

2. What are the two basic parts of a simple instrument?

3. If the parameter being measured is located some distance from the operator who is monitoring the system, three additional components are needed in the instrument system; what are those components?

4. Where is a telemetry system required?

5. List at least four general types of sensors commonly found in a water distribution instrumentation system.

6. What type of pressure sensor is most commonly used in modern instrumentation? How does it operate?

7. Electrical measurements are commonly explained by comparing them to similar variables used to describe hydraulic conditions. Explain how electromotive force, current, and resistance are similar to conditions measured in a water main. What units are used to measure each of these electrical variables?

8. What is the fourth commonly measured electrical variable, not discussed in the previous question? What units is it measured in? What does it indicate?

9. What is the function of a transmitter?

10. What is the function of a receiver?

11. What is the difference between an analog and a digital indicator?

12. What transmission channels are commonly used for telemetry systems?

13. Much of the complexity of a telemetry system is the result of the need to monitor or control more than one system over a single transmission channel. List several ways of doing this.

14. Explain the operation of one of the techniques listed as an answer to the previous question.

15. Distinguish between manual, automatic, and semiautomatic control systems.

16. Distinguish between direct-wire control and supervisory control.

17. What is the simplest routine inspection procedure an operator can perform to significantly increase the life of all instrument components?

18. What specific maintenance procedure should be performed periodically on the pressure-differential producer type of primary flow element (such as a Venturi tube, orifice, or flow nozzle)?

19. Are rubber gloves or insulated tools the proper precaution to observe when working with energized electrical power circuits?

20. In general, which can cause the most rapid and widespread damage or injury—a faulty hydraulic device or a faulty pneumatic device?

Study Problems and Exercises

1. In a distribution control station or a pumping station, trace the signal path for a single simple instrument, perhaps a pressure or flow sensor. Identify each of the components of the instrument and record the model name and manufacturer. Using the utility's repair manuals or manufacturer's literature, investigate the internal construction of each component and identify the function of each of the parts.

2. Determine what types of telemetry systems a local utility uses, and for what functions. Is the equipment and the transmission channel owned or leased by the utility? Who is responsible for repairs? If a radio or microwave system is used, what repairs can be performed by a technician who does not hold FCC (Federal Communications Commission) certification as a radio repairman?

3. At the central control point for the distribution system, identify which systems are automatic, manual, and semiautomatic. What alarms are installed in case of system failure? What backup systems are available if the central control station or telemetry lines fail?

4. Use a VOM (volt-ohm-milliamp meter), a flashlight battery, and a flashlight bulb to construct circuits similar to those shown in Figure 9-8 of this module. Measure similar values in other low-voltage circuits. Try connecting several light bulbs or small resistors in various configurations and measuring the changes in voltage, current, and resistance.

Module 10

Maps, Drawings, and Records

Maps, drawings, and records provide a method for recording information necessary for the efficient operation and maintenance of a water distribution system. Since most of the distribution system is underground and out of sight, maps and records free water suppliers from dependence on the memory of longtime employees and help prevent the loss of valuable information due to an employee's retirement or death. Maps, drawings, and records also provide vital information to other utilities that may be working below ground in a specific area. They aid in troubleshooting, planning, and systematically developing the distribution system, and they furnish water distribution personnel with an accurate, comprehensive record of the physical assets of the system.

Maps and records must be updated regularly to include system changes or new construction. Out-of-date maps and records not only hinder daily system operation, but also pose additional problems during emergency situations.

After completing this module you should be able to

- State the importance of maintaining system maps, drawings, and records.

- Give examples of maps, drawings, and records commonly used in distribution system operations, and explain their purposes.

- Interpret standardized map symbols from water distribution system maps and drawings, and be familiar with construction plan terminology.

- Describe the intersection method and the plat and list method used for mapped records.

- Recommend the frequency for updating mapped records.

- Cite examples of card records and explain their use in relation to mapped records.

10-1.   Purpose and Types of Records

Maps, drawings, and records serve a variety of functions for both public and private water distribution systems. In addition to providing information on main, valve, hydrant, and service locations, they provide a permanent record of the system's assets. Permanent records of physical facilities and appurtenances (assets) are essential to determine system development changes and to document the system's underground investments. Aboveground assets can be easily inventoried, but underground assets, which often constitute a major portion of the distribution system, can only be inventoried through maps and records.

Maintaining up-to-date maps, drawings, and records should not be considered a secondary function to construction or other service work. Complete and accurate records help promote the efficient operation and maintenance of the distribution system. They allow personnel to determine which sections are undersupplied and why, whether hydrants are too close together or too far apart, and whether mains in a certain area are capable of handling a major fire. Maps, drawings, and records also aid in troubleshooting and systematically developing the system.

Some states have statutes or regulations requiring the maintenance of drawings and records in order to respond to emergencies, personnel injuries, or damage to underground utility services during excavations. Distribution system maps, drawings, and records provide essential information to other utilities, such as gas, electric, or telephone, whose services are also below ground.

10-2.   Basic Distribution System

There are a number of records needed for the efficient operation and maintenance of a water distribution system. Distribution system records can generally be divided into the following four groups:

- Mapped records
- Plan and profile drawings
- Card records
- Statistical records.

There are three principal mapped records: (1) a COMPREHENSIVE MAP outlining the entire system, (2) SECTIONAL MAPS providing a more detailed picture of the system on a larger scale, and (3) VALVE AND HYDRANT MAPS pinpointing valve and hydrant locations throughout the system. Large systems may also have ARTERIAL MAPS, valve closure maps, LEAK SURVEY MAPS, and PRESSURE ZONE MAPS.

PLAN AND PROFILE DRAWINGS are engineering records showing pipe depth, pipe location, and correct location from a starting point.

CARD RECORDS contain information that cannot be incorporated on mapped records. Typical card records include valve cards, hydrant cards, and service cards. These cards are cross-referenced to mapped records and provide information as to the location, description (type, direction to open, number of turns to open, etc.), and history of each valve, hydrant, and service within the system.

STATISTICAL RECORDS, or summary records, are used for reporting construction progress and growth. Statistical records are maintained for mains, valves, hydrants, and services. Statistical records for mains provide the mileage for each size of main laid or retired. Valve and hydrant statistical records summarize the number of valves and hydrants installed for given sizes. Service statistical records document the number of feet of each size of service pipe laid, the number and size of corporation stops installed, and the number and size of curb stops installed. (If services are installed and owned by a household, service statistical records need only summarize the number added or abandoned.) Many water suppliers graph their statistical records for easy comparison.

Mapped Records

Comprehensive maps. A comprehensive map (Figure 10-1) provides a clear picture of the entire distribution system. The comprehensive map (sometimes referred to as the wall map) is primarily used by the system manager or

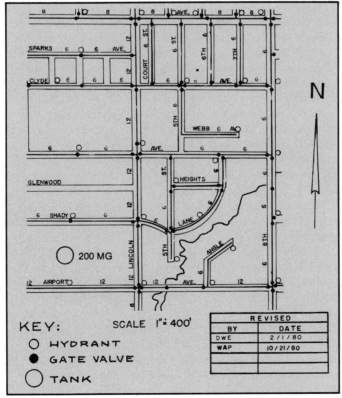

Courtesy of Water Works and Lighting Comm., Wisconsin Rapids, Wis.

Figure 10-1.   Part of a Comprehensive Map

Table 10-1.   Important Elements of a Comprehensive Map

Street names	Water source
Water mains	Scale
Sizes of mains	Orientation arrow
Fire hydrants	Date last corrected
Valves	Pressure zone limits*
Reservoirs and tanks	Closed valves at pressure zone limits
Pump stations (discharging into the distribution system)	

*Often done on a tracing of the comprehensive map because pressure zones may change from time to time.

engineering department to identify sections of the system that either need improvement or are not yet fully developed. The map points out places where short extensions will eliminate dead ends, where mains are inadequate to supply hydrants in case of large fires, or where additional valves or hydrants are needed. A comprehensive map also serves to document the financial value of physical assets within the distribution system. These costs are used to determine system development charges, which are used in setting water rates.

The comprehensive map should not be cluttered with distracting information. Important information to include on a comprehensive map is listed in Table 10-1. Since the primary purpose of the comprehensive map is to provide an overall picture, the scale of the map should be as large as possible. Typical comprehensive maps have scales ranging between 500 and 1000 ft to an inch (150 and 305 m to a millimetre) with some smaller communities using scales less than 500 ft to an inch (150 m to a millimetre). It is preferable that a scale be selected that allows the map to be drawn on one sheet, although some larger systems may have to use two sheets attached together. Some systems also have an overlay with CONTOUR LINES (topographic lines) for use with the comprehensive map.

Water suppliers should try to reference their comprehensive map to the state's COORDINATE SYSTEM or range and section lines. Referencing to such systems will facilitate easy additions whenever necessary. Ordinary commercial maps available in some cities must be checked carefully for accuracy prior to their use as a BASE MAP for a comprehensive map. Agencies such as highway departments and the US Department of the Interior, Geological Survey (USGS) are excellent sources for base maps. Once an accurate base map has been selected, photostatic enlargements or reductions must be done carefully so that the base map does not become distorted. The actual map should be drawn in ink on a durable material such as heavy tracing cloth. Prints can then be easily made from the master for regular use.

Copies of the comprehensive map are provided to the system manager and engineering department, and copies are also often provided to the mayor, city manager, city fire stations, and the water distribution maintenance department.

Comprehensive maps must be updated periodically. The normal practice is to either update them once or twice a year or immediately after the completion of major system extensions.

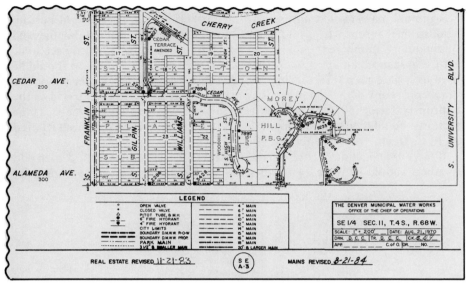

Figure 10-2. Section Map

Sectional maps.   A sectional map or PLAT (Figure 10-2) provides a detailed picture of a section of the distribution system. Sectional maps are generally used for day-to-day operations. They reveal the location and valving of existing mains, the location of fire hydrants, and the location of active service lines; they also match account numbers with lots. Because detail is important on a sectional map, the scale used is generally much larger than that used on a comprehensive map. Therefore, several maps are required to cover the entire system. Typical scales range from 50 to 100 ft to an inch (15 to 30 m to a millimetre) for most communities, to 200 ft to an inch (60 m to a millimetre) for larger systems. Information to include on a sectional map is listed in Table 10-2.

Table 10-2.   Important Elements of a Sectional Map

Section designation or number	Water account numbers
Adjacent section numbers	Measurements to service lines
Street names	Distances—main to curb box
Mains and sizes	Distances to angle points
Materials of mains	Distances to fittings
Date of main installation	Dead ends and measurements
Distances from property line	Date last corrected
Fire hydrants and numbers	Orientation or north arrow
Valves and numbers	Scale
Valve sheet designation shown in margin	Closed valves at pressure zone limits
Intersection numbers (if valve intersection	Service limits
plats are used)	Tanks and reservoirs
Block numbers	Pump stations (discharging into
Lot numbers	distribution system)
House numbers	Pressure zone limits

Sectional maps should not overlap each other; instead they should butt up against one another to avoid confusion. Each section should be indexed (individually numbered) and show matching points for adjacent sections. INDEXING sectional maps is not difficult, and the comprehensive map should be used as a base map for indexing. North–south and east–west lines should be drawn on a copy of the comprehensive map dividing it into sections based on the appropriate scale (50 ft [15 m], 100 ft [30 m], etc.) for the sectional maps. These sections should then be numbered 1, 2, 3, 4, etc. from north to south and lettered A, B, C, D, etc. from east to west (Figure 10-3).

If there are not enough letters in the alphabet for east–west indexing, then odd numbers should be used for north–south designations and even numbers for

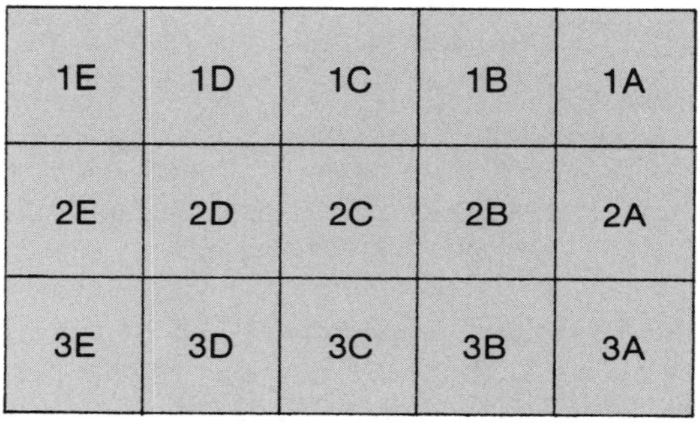

1E	1D	1C	1B	1A
2E	2D	2C	2B	2A
3E	3D	3C	3B	3A

Figure 10-3.   Dividing Small Comprehensive Map Into Sections

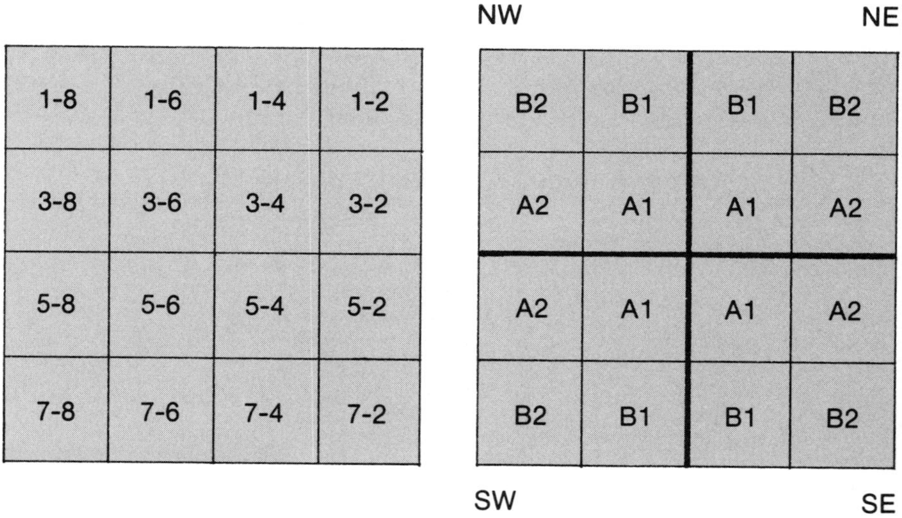

NW   NE

1-8	1-6	1-4	1-2
3-8	3-6	3-4	3-2
5-8	5-6	5-4	5-2
7-8	7-6	7-4	7-2

B2	B1	B1	B2
A2	A1	A1	A2
A2	A1	A1	A2
B2	B1	B1	B2

SW   SE

A. Odd for N–S, Even for E–W   **B. Sequential by Quadrants**

Figure 10-4.   Dividing Large Comprehensive Maps Into Sections

east–west designations (Figure 10-4A). Indexing should always begin some distance outside the present distribution system to allow for system growth.

Areas that experience growth in all directions may prefer a method in which the indexing starts in the middle of the mapped area and is expanded outward by quadrant (for example, NW, SE). Columns would be lettered sequentially from south to north in the NE and NW quadrants and north to south in the SE and SW quadrants. Rows would be numbered sequentially from west to east in the NE and SE quadrants and east to west in the NW and SW quadrants (Figure 10-4B).

Tax assessment maps, insurance maps, subdivision maps, or city engineering maps may be used as sectional base maps provided they are checked carefully for accuracy. Photostatic enlargements or reductions can be used to arrive at a desired scale. (Photostatic work may cause distortion if not performed carefully.) Many communities find that drawing an accurate base map at the desired scale is easier than searching for a suitable map and then checking its accuracy.

Sectional maps should be drawn in ink either on heavy tracing cloth (from which prints can be made) or other durable art material. Whether using tracing cloth or another material, a set of duplicate sectional maps should be kept in a safe location, preferably a fireproof vault, because of their importance. Microfilming sectional maps is also becoming a common, relatively low-cost method of maintaining a backup copy of maps and other system records.

Sectional maps must be updated more frequently than comprehensive maps. Many systems update working copies of their sectional maps daily as field reports come back to the main office. Original sectional map prints are then updated periodically (monthly or quarterly) from the working copies, and new copies with the corrected sections are distributed as often as necessary. In situations where major system extensions are added or major changes are made, the original print should be updated as soon as possible and a copy of the corrected section should be issued for each copy in use.

Valve and hydrant maps.   Valve and hydrant maps pinpoint valve and hydrant locations throughout the distribution system. Valve and hydrant maps are primarily used by field crews. They enable crews to quickly locate all valves and hydrants for general maintenance work or emergency situations when a section of main must be repaired. To be of value, valve and hydrant maps should be bound in books and then passed out to appropriate crews for field use.

Valve and hydrant maps should provide measurements from permanent reference points to each valve within the system, and they should pinpoint hydrant locations. (For communities with older flush-type hydrants at the surface level, measurements from permanent reference points may be needed for easy location.) Valve and hydrant maps should also provide appropriate information such as direction to open, number of turns to open, model, type, and date installed for hydrants and valves. Some maps also include the last date tested or repaired.

There are a number of methods for setting up valve and hydrant maps. The two most common of these are the PLAT AND LIST METHOD and the INTERSECTION METHOD. In the plat and list method, the plat (Figure 10-5) is the map or drawing

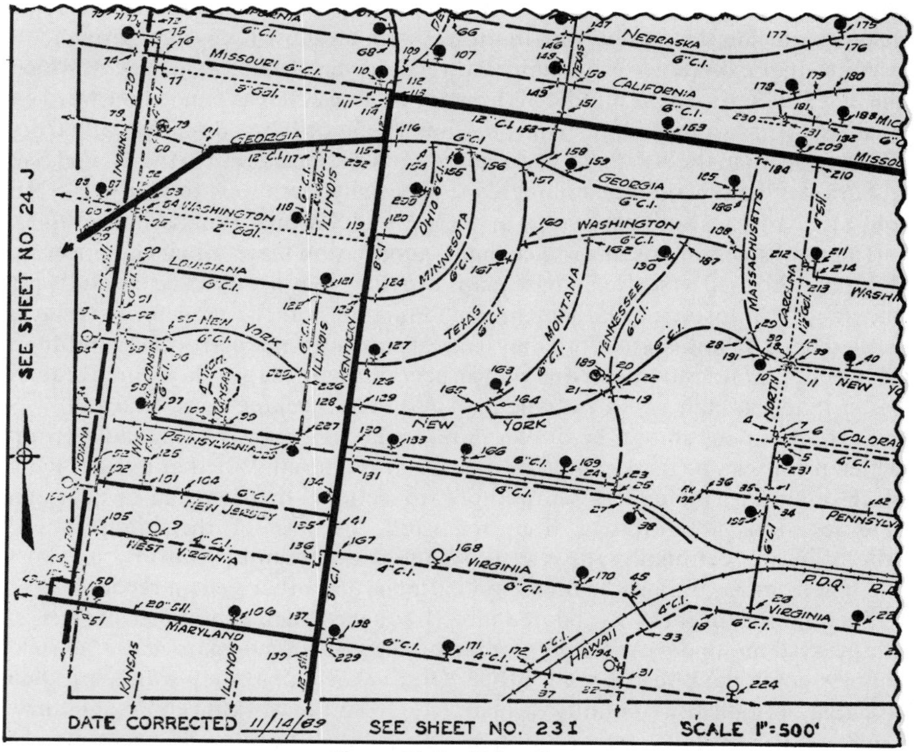

Figure 10-5. Valve Plat

SHEET NO. 24 I

VALVE NO.	PRINCIPAL STREET	DISTANCE FT.	DIR.	REFERENCE	INTERSECTING STREET	DISTANCE FT.	DIR.	REFERENCE	SIZE	MAKE	OPEN	TURNS	DATE SET	REMARKS
1	North Carolina	11	W	E.side E.walk	Pennsylvania	12	N	N.side N.walk	6	A.P.S.	L	13	5/22	
2	Pennsylvania	1½	S	N.⅊	Indiana	18	E	W.⅊	6	A.P.S.	L	16	7/38	Pat. Tap
3	Indiana	6½	E	W.⅊	New York	38¾	N	N.⅊	20	Pen.	L	62	1931	
4	North Carolina	11	W	E.side E.walk	Colorado	9	N	N.side N.walk	2	A.P.S.	-	T.Hd.	9/12	
5	Colorado	12	N	S.side S.walk	North Carolina	3	E	of E.side E.walk	4	Cha.	L	9	5/22	
6	Colorado	22	S	N.side N.walk	North Carolina	11	W	E.side E.walk	2	Cha.	-	T.Hd.	9/12	
7	Colorado	22	S	N.side N.walk	North Carolina	12¼	W	E.side E.walk	2	Cha.	-	T.Hd.	9/12	
8	New York	7¼	S	N.side N.walk	North Carolina	19½	W	E.⅊ priv. Rd.	2	P&C	-	T.Hd.	10/12	
9	West Virginia	7¾	S	N.side N.walk	Indiana	28½	E	W.⅊ 7615	6	P&C	L	19	10/16	6' S.W. of Hydrant
10	New York	15½	S	S.side S.walk	North Carolina	11½	W	of E.side E.walk	2	Mue.	-	T.Hd.	2/13	
11	New York	15¼	S	S.side S.walk	North Carolina	9¾	W	E.side E.walk	2	Mue.	-	T.Hd.	12/22	
12	North Carolina	19½	W	E.side E.walk	New York	10½	N	S.side S.walk	2	Eddy	-	T.Hd.	9/12	
13	Colorado	22⅞	S	N.side N.walk	Florida	5½	W	E.side E.walk	4	Ohio	L	11	3/14	
14	Colorado	11	N	S.side S.walk	Florida	8½	W	W.side W.walk	4	Ohio	L	12	3/14	
15	New York	16	S	N.⅊	Florida	14	W	W.⅊	6	Pen.	L	19½	9/35	Pat. Tap
16	New York	10½	S	S.side S.walk	Florida	2½	W	E.side E.walk	2	Ohio	-	T.Hd.	12/22	
17	Virginia	2¼	S	N.side N.walk	Florida	20¾	W	W.side W.walk	6	Fair.	L	13	5/15	
18	Virginia	8	N	S. side S.walk	North Carolina	3½	W	E side walk [illegible]	6	Fair.	L	13¼	11/23	
19	Tennessee	7⅞	E	W.side W.walk	New York	5½	N	N.side N.curb	6	Fair.	L	23	11/23	
20	Tennessee	7½	E	W.side W.walk	New York	6	N	N.side N.walk	6	M&H	L	19½	11/23	
21	New York	8	S	N.side N.walk	Tennessee	2½	E	E.side W.curb	6	M&H	L	22	11/23	
22	New York	6¾	S	N.side N.walk	Tennessee	12⅝	W	W.side E.curb	6	M&H	L	22	11/23	
23	Tennessee	2	E	W.⅊	Pennsylvania	8	N	N.⅊	6	Lud.	L	23	4/25	In Vault
24	Pennsylvania	7	S	N.⅊	Tennessee	8	E	W.⅊	6	Lud.	L	23	4/25	
25	Tennessee	6	W	W.⅊	Pennsylvania	7	S	N.⅊	6	Lud.	L	23	5/25	
26	Virginia	1½	N	N.curb	Drive to 7171 Virginia	5	W	℄	6	Eddy	L	20	7/27	
27	Pennsylvania	10	N	S.⅊	Tennessee	12	E	W.⅊	6	Eddy	L	20	5/28	
28	New York	7	S	N.side N.walk	Massachussetts	49½	W	W.side W.walk	6	Vogt.	L	13	11/23	
29	Massachusetts	13	E	W.side W.walk	New York	26	N	N.side N.walk	6	Vogt.	L	13	11/23	
30	New York	20	N	S.curb of park	Massachusetts	8½	E	W.curb park	6	Ken.	L	13½	11/23	
31	Hawaii	9	E	W.curb	Maryland	47	N	S.curb	4	Ken.	L	12	12/16	
32	Private Drive	40	E	W.curb Hawaii	Maryland	4	S	S.curb	4	Darl	L	13	1/23	
33	Virginia	4½	S	S.curb	Hawaii	323	E	E.curb	4	Darl	L	13	1/23	
34	Pennsylvania	4½	S	S.side N.walk	North Carolina	5	E	E.edge E.walk	6	Crane	L	13½	9/31	
35	Pennsylvania	12	S	N.⅊	North Carolina	10	W	W.⅊	6	Crane	L	13½	9/31	
36	Pennsylvania	12	S	N.⅊	North Carolina	233	W	W.⅊	6	R&C.	L	21	3/32	
37	Maryland	36	N	S.side S.walk	P.D.Q. R.R.	25	E	E.rail	4	Crane	L	10	12/18	
38	Pennsylvania	9	N	S.⅊	Tennessee	208	E	E.⅊	6	[illegible]	L	19½	10/23	1½' W. of Hydrant

Figure 10-6. Partial Valve List for Figure 10-5

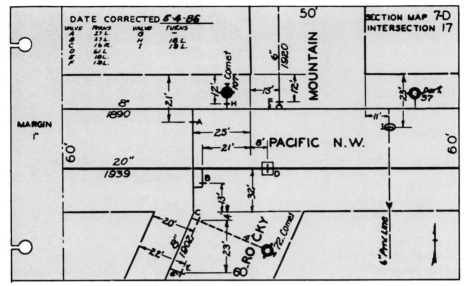

Figure 10-7.   Valve Intersection Plat

that shows street names, mains, main sizes, numbered valves, and numbered hydrants. Following the plat on the next page (or pages) is a LIST (Figure 10-6) providing appropriate information on the numbered valves and hydrants such as reference measurements, size and make, direction of operation and number of turns, date installed, date tested, and any remarks. Plats generally have a scale of around 500 ft to an inch (150 m to a millimetre). Plats can be indexed in the same manner or in a manner similar to sectional maps. Valves and hydrants on each plat should be assigned a number or designation. A reduced print of the comprehensive map or city map showing the coordinates for each plat may be used as a field index for valve and hydrant maps. These indexes need not be updated, unless extensions to the original system are added, since they are used only to show the general area each plat covers.

The alternative to the plat and list method is the intersection method. In the intersection method, plats (maps) of intersections (Figure 10-7) are drawn on a very large scale (20 to 30 ft to an inch [6 to 9 m to a millimetre]) showing property or street lines, mains, hydrants, valves, etc. This large scale permits measurements from permanent reference points (property or street lines for example) to be entered on the intersection plat. Valves between intersections cannot be shown to scale, but they can be located by indicating measurements from intersecting streets.

Indexing intersections is accomplished by giving each intersection a number. The intersection number can then be attached to the end of the sectional plat number, which pinpoints its location. Valves and hydrants within or nearby an intersection can then be given a letter to pinpoint their location.

5th St. No. &		Taylor Avenue	433
		Wyatt Avenue	434
Ash Street	658	Miller Avenue	435
Poplar Street	715	Wood Avenue (2000 Blk)	436
Cherry Street	714	Goodnow Avenue	437
Spring Street	713	Piltz Avenue (No. of Pepper)	438
Baker Street	84	Pepper Avenue	439
Wisconsin Street	106	Clyde Avenue	561
Avon Street	29	Webb Avenue	824
Saratoga Street	126	Grove Avenue	562
		Glenwood Heights	563
5th St. So. &		Shady Lane	564
		So. of Shady Lane	825
Oak Street & E. Jackson	147		
E. Grand Avenue	151	6th St. No. &	
Peach Street	25		
Chestnut Street	824	Ash Street	659
Lee Street	203	Poplar Street	718
Dewey Street	209	Cherry Street	717
Daly Avenue	431	Spring Street	716
Strodman Avenue	432	Saratoga Street	328

Courtesy of Water Works and Lighting Comm., Wisconsin Rapids, Wis.
Figure 10-8.   Cross Index of Streets

If the size of the community is not extremely large, intersections are sometimes sequentially numbered. A cross index of streets in alphabetical order (Figure 10-8) or a print of the comprehensive map showing sectional map areas and intersection designations (Figure 10-9) can be used as a field index. If appropriate, intersection maps, like plat and list maps, can have factual information concerning the valves and hydrants. Such information can be listed in a corner or suitable open space on the intersection drawing. For more complex systems, there may be a book or a series of overlay maps and indexes to catalog this information (for example, the map book and valve book). These books may then be divided up into geographic areas for field use.

In very large systems, intersection drawings can become quite bulky. Because of this bulk, it is sometimes not feasible to equip field crews with complete records. Some systems, therefore, keep the intersection maps in the main office. However, this does defeat one of the purposes for developing the maps—field use.

Valve and hydrant maps must be updated for efficient system operation and maintenance. Working valve and hydrant map copies should be corrected daily, and original prints should be updated monthly or quarterly. After the original prints are corrected, new copies of the corrected prints should be issued to replace those copies in use.

An alternative to using a comprehensive map, a sectional map, and a valve and hydrant map is to incorporate the valve and hydrant map into the other two. Hydrant and valve locations and operating positions are generally shown on the

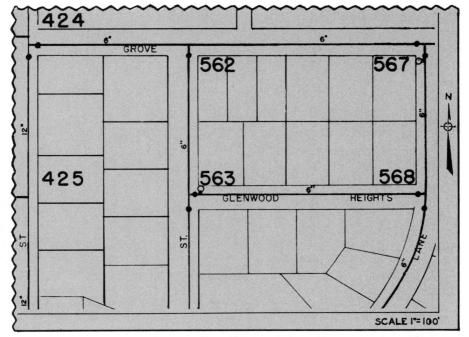

Courtesy of Water Works and Lighting Comm., Wisconsin Rapids, Wis.

Figure 10-9.   Comprehensive Map Showing Intersection Numbers

comprehensive map, with more detailed information shown on the sectional map. This may require the sectional map to be a larger scaled map than would be necessary if a hydrant and valve map were used. However, the advantage to this type of system is that one less map needs to be updated.

Plan and Profile Drawings

Occasionally water distribution system operators will deal with plan and profile drawings (Figure 10-10). These engineering drawings should be tied to section lines or property lines. They show pipe depth, pipe location (both horizontal and vertical displacements), and the correct distance from a starting reference point (stationing). Stationing is normally specified in an engineering form such as 60 + 40 meaning 6040 ft from the starting point.

Common abbreviations found on plan and profile drawings include:

- POT—Point on tangent
- POC—Point on curve
- BC or (PC)—Beginning of curve
- EC or (PT)—End of curve
- PI—Point of intersection
- EL—Elevation.

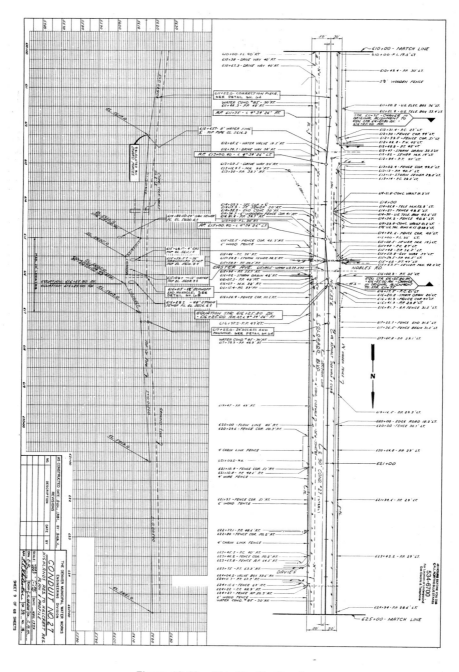

Figure 10-10.   Plan/Profile Drawings

Supplemental Mapped Records

In addition to a comprehensive map, sectional maps, and valve and hydrant maps, some systems maintain supplemental mapped records. For example, large systems may keep an arterial map and a comprehensive map at a scale of 2000 to 4000 ft to an inch (610 to 1220 m to a millimetre) showing primary distribution mains 8 in. (200 mm) or larger, for use in system analysis.

Some systems maintain pressure-zone maps (water-gradient contour maps), which record simultaneous pressure readings taken at different points throughout the system and then plot them on a print of the comprehensive map. Isopleths or contour lines can then be drawn to identify zones of equal pressure. From this map, low pressure or high pressure zones can be identified and controlled (some systems maintain pressure zones directly on the comprehensive map).

Other systems maintain LEAK FREQUENCY MAPS. On these, different colored push pins are used to identify different types of leaks on a print of the comprehensive map. Points where soil conditions or electrolysis is bad, or where faulty installation techniques or poor materials were used, show up very clearly after a few years.

Leak survey maps are used by many systems. When regular leak survey work is conducted, these maps show valves to be closed and areas to be isolated.

Card Records

A separate card record system should be maintained for pipes, valves, hydrants, and services. Key information to include on the card is the identification number, location, description, maintenance, and any noteworthy remarks. Card records should be cross referenced to maps and construction records, and they should provide supplementary descriptive, historical, and maintenance information that otherwise is not recorded on maps, drawings, and records. Typical card records include valve cards (Figure 10-11), hydrant cards (Figure 10-12), and service cards (Figure 10-13).

Information should be recorded in a legible fashion, and completed cards should be filed immediately. File systems can be arranged in a number of ways, such as by identification number of location in a cabinet or in a computer. The main criteria for a card filing system is that it be safe and efficient for future reference. When setting up a card system or filling out a new card, it is important to be thorough from the beginning. It only takes a short time to fill out a card completely, whereas years later it may take much longer to obtain the missing information.

Statistical Records

Many systems maintain statistical records (summary records) for use in preparing annual reports, audits, reports to stockholders, or taxes. Statistical records are also used in reporting construction growth or measuring the efficiency of the distribution work force. Statistical records (Figure 10-14) are commonly maintained for distribution system mains, valves, hydrants, service

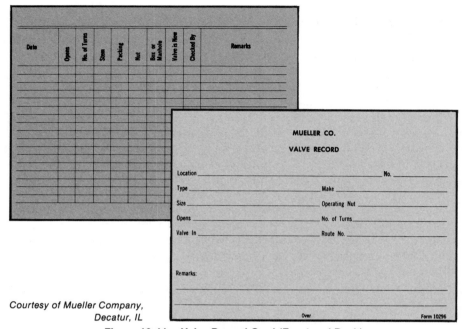

*Courtesy of Mueller Company,
Decatur, IL*

Figure 10-11.   Valve Record Card (Front and Back)

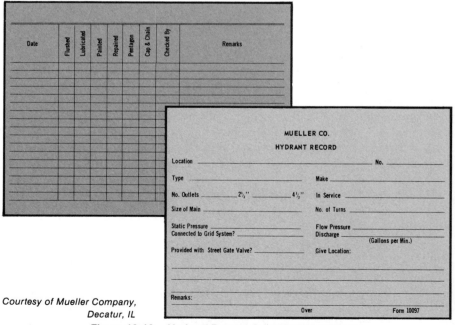

*Courtesy of Mueller Company,
Decatur, IL*

Figure 10-12.   Hydrant Record Card (Front and Back)

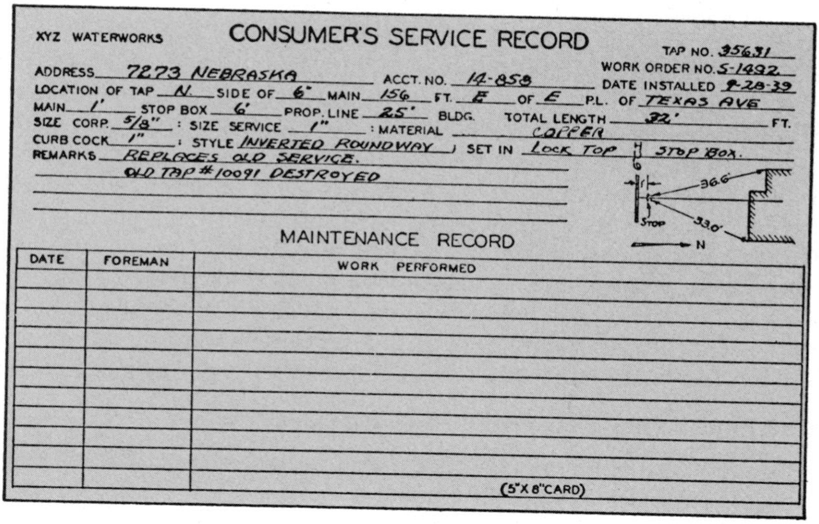
XYZ WATERWORKS
CONSUMER'S SERVICE RECORD
TAP NO. 35631
WORK ORDER NO. 5-1492
ADDRESS 7273 NEBRASKA ACCT. NO. 14-858 DATE INSTALLED 9-28-39
LOCATION OF TAP N. SIDE OF 6" MAIN 156 FT. E OF E P.L. OF TEXAS AVE
MAIN 1' STOP BOX 6' PROP. LINE 25' BLDG. TOTAL LENGTH 32' FT.
SIZE CORP. 5/8" : SIZE SERVICE 1" : MATERIAL COPPER
CURB COCK 1" : STYLE INVERTED ROUNDWAY ; SET IN LOCK TOP STOP BOX.
REMARKS REPLACES OLD SERVICE.
OLD TAP # 10091 DESTROYED
MAINTENANCE RECORD

DATE	FOREMAN	WORK PERFORMED

(5"X 8" CARD)

Figure 10-13. Service Card

Water Utility Division
Monthly Report for March 1985

Authorized Personnel 82

Now Employed 81

Item	Unit	This Month	Last Month	Same Month Last Year	This Year To Date	Last Year To Date
Water Pumped	CF	101,151,989	90,953,912	106,951,209	1,108,181,116	1,141,192,829
Water Purchased	CF	5,891,103	4,911,231	5,101,101	50,131,960	43,193,762
Total Pumped and Purchased	CF	107,043,092	95,865,143	112,052,310	1,158,313,576	1,184,386,591
Municipal Consumption	CF	1,961,204	1,013,621	2,104,136	8,261,314	7,163,412
Water Samples	EA	321	296	289	3,041	2,913
Analytical Tests	EA	1,039	987	1,011	9,536	10,431
Chemicals Used	LBS	9,291	9,931	9,604	95,461	89,362
New Services	EA	141	19	170	1,093	987
Meters Repaired	EA	27	36	12	251	232
Check Valves Tested	EA	61	32	51	279	301
Gate Valves Operated	EA	439	386	415	3,916	4,029

Figure 10-14. Statistical Record Form

pipes, and meters. Such records may be compiled monthly, semiannually, or annually, depending on reporting requirements.

Statistical records generally report quantitative information. Main statistical records report the amount of pipe installed as well as the total for the system. Valve and hydrant statistical records report the number of valves and hydrants installed, as well as the totals for each size and the system. Service statistical records report the amount (in feet) of service pipe laid for given sizes, as well as the number and size of corporation and curb stops the water system installed. Meter statistical records report the number of different size and type meters installed in the system as well as totals for each and the system.

Statistical records lend themselves well to graphing because they deal primarily with numbers. Graphing such records enables water supply personnel to compare past and present figures at a glance and identify system trends.

Record Requirements for Small and Large Systems

Distribution system records are just as important for small systems as they are for large systems. The only real difference between the two should be the number of records kept. Smaller systems, which are generally less complicated, can often consolidate certain records. Small systems should maintain the same card and statistical records as large systems, but do not need to keep as many mapped records. Generally, most small systems can consolidate their mapped records on one detailed map similar to a larger system's sectional map. Work orders should also be used by smaller systems, but the process is usually much simpler than that used in larger multidepartmental systems.

Although record keeping at smaller water distribution systems may be more consolidated than at larger ones, the problem of maintaining records is often more difficult. Small system operators frequently have additional duties beyond operating the water distribution system. Such duties compound an already busy workload and leave less time for record maintenance. In such cases, it is important that the small system operator set aside time to maintain and update distribution system records in order for these to be an accurate reference document for system operation and maintenance.

Map Symbols

Every water distribution system should adopt a set of standardized map symbols to denote different items on their mapped records. Symbols recommended by the American Water Works Association are identified in Table 10-3. These symbols were selected because of their simplicity, clarity, and acceptance in current practice.

Water distribution system employees involved in drawing or using mapped records should become familiar with their system's symbols. Familiarity with such symbols often eliminates mistakes in drawing or interpreting system maps and promotes more efficient map use. Many systems include a copy of their symbols in all their record books for easy reference.

Table 10-3.   Typical Map Symbols

Item	Job Sketches	Sectional Plats	Valve Record Intersection Sheets	Comprehensive Map & Valve Plats
3-in. & Smaller Mains				
4-in. Mains				
6-in. Mains				
8-in. Mains				
Larger Mains				
Valve	SIZE NOTED	SIZE NOTED	12" 24" 36"	12" 24" 36"
Valve, Closed				
Valve, Partly Closed				
Valve in Vault				
Tapping Valve & Sleeve				
Check Valve (Flow→)				
Regulator				
Recording Gauge				
Hydrant (2-2½-in. nozzles)	①	①	①	②
Hydrant With Steamer	①	①	①	②
Cross-Over (Two Symbols)				
Tee & Cross				
Plug, Cap, & Dead End	BSB BSBB			
	PLUG CAP			
Reducer	BS BS	12" 8"		
Bends, Horizontal	DEG. NOTED	DEG. NOTED	DEG. NOTED	
Bends, Vertical	UP DOWN	NO SYMBOL	NO SYMBOL	NO SYMBOL
Sleeve				
Joint, Bell & Spigot	BELL SPIGOT			
Joint, Dresser Type				
Joint, Flanged				
Joint, Screwed				

① OPEN CIRCLE—HYDRANT ON 4-IN. BRANCH
  CLOSED CIRCLE—HYDRANT ON 6-IN. BRANCH

② OPEN CIRCLE—4-IN. BRANCH OR NO 4½-IN. NOZZLE
  CLOSED CIRCLE—6-IN. BRANCH AND WITH
  4½-IN. NOZZLE

STEAMER NOZZLE SYMBOL IS CAPPED HORN
HOSE NOZZLE SYMBOL IS UNCAPPED HORN

① $^3/_{16}$-IN. DIAM.   ② $^3/_{32}$-IN. DIAM.

10-3.   Work Orders, Technical Information, and Recent Trends

In addition to the basic records discussed in the previous section, certain other types of records are important. Work orders are used to communicate between the office and the field. Technical information on installed or available equipment must be available for maintenance and design. This section covers those records and also discusses briefly the use of new technologies for record storage.

Work Orders

Distribution system records are only as accurate as the information recorded. Therefore, it is essential that every system have an established procedure by which field crews complete and submit accurate information. A good working relationship between field and office personnel will help ensure a reliable record system.

Many water distribution systems use a WORK ORDER to communicate field information back to the office. In addition to providing a means of communication, work orders are also used to order materials, organize field crews, and obtain easements.

A work order usually contains complete information about the location and description of the job performed; a listing of all the materials needed and used; information on the time and manpower required, equipment used, and the total cost of the work performed. A work order system should be designed to be functional rather than just additional paperwork. A sample work order is shown in Figure 10-15.

Technical Information

Water distribution systems are often swamped with technical information—information accompanying valves, hydrants, pumps, pipe, tapping machines, backhoes, and other distribution system components and equipment. Today's competitive sales market further adds to the problem by providing technical information through the mail. Technical information should not be discarded simply because of its bulk. Many technical bulletins or pamphlets are extremely important since they provide specifications for installation, operating tips, or maintenance procedures.

Technical information is generally not recopied on cards or special forms, but rather is evaluated and maintained in a file system for easy reference. Each major item within the file system should have its own section. For example, a section on hydrants could be established and subdivided by manufacturer.

A file on each manufacturer would be labeled where important information pertaining to the hydrant could be stored, along with any documents such as warranties, letters, or manufacturers' phone numbers. Also, a filing system might be established on valves, which would be subdivided by valve type and manufacturer.

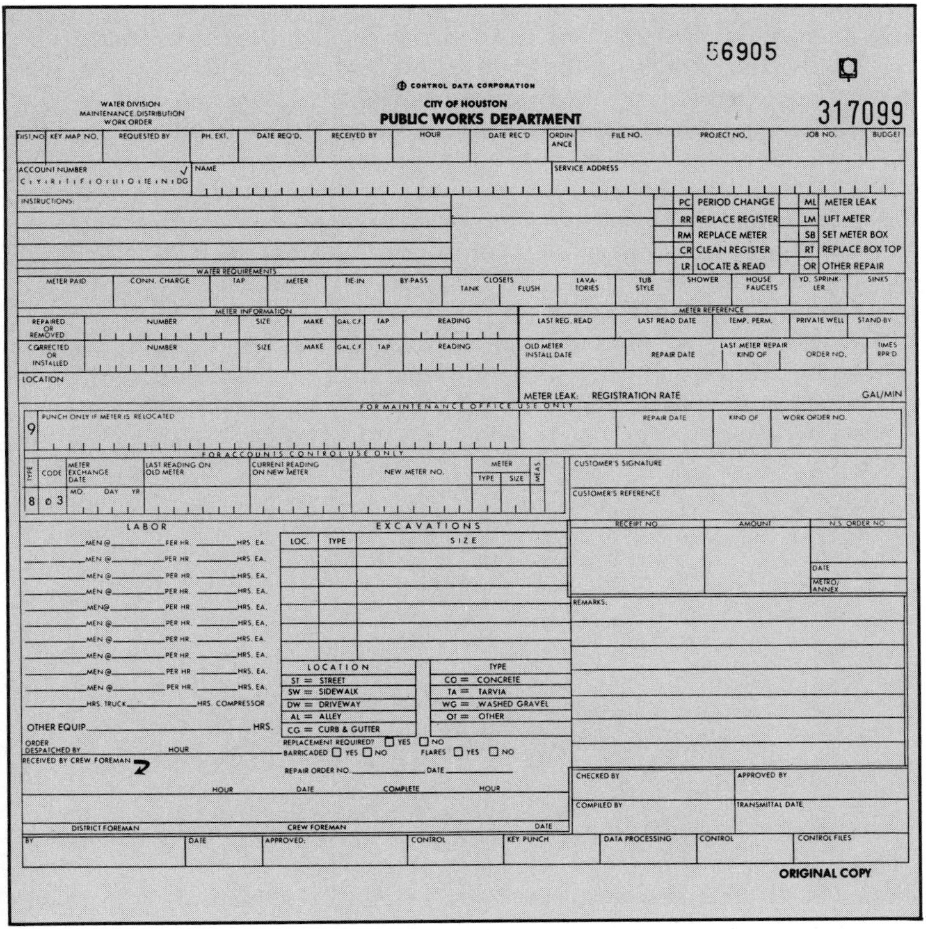

Figure 10-15. Work Order

File systems can be arranged in any number of ways. The important consideration is that every water supplier should establish an efficient means of organizing important technical information for future reference.

Recent Trends in Information Storage

Future changes in maps, drawings, and records will most likely be made in the way such information is stored, rather than in content. Variations exist from system to system in the style and format of distribution system records. However, almost all systems maintain certain basic records (a comprehensive map, sectional maps, valve and hydrant maps, card records, or statistical records) depending on their size. Changes in these basic records are not as likely to occur because of the necessity for information, but changes are taking place in how distribution records are stored. The most promising innovations seem to be microfilm and computerized information storage systems. Such systems enable information to be stored, handled, and retrieved with relative ease, at a relatively low cost, and they ensure security for permanent documents.

Microfilm systems reduce much of the bulk of large record systems. The extent to which information can be stored in computerized systems depends on the capabilities of the system, both hardware (instrumentation) and software (programming). Computerized information systems have the advantage of providing other functions in addition to record storage. As new technology evolves, computerized information storage systems will most likely become more cost effective for record storage.

The cost of microfilm and computerized information storage systems is decreasing. Smaller utilities, which were once unable to justify the expense of such systems, are now finding many microfilm and some computerized systems competitively priced compared to manual record storage systems.

Selected Supplementary Readings

Deem, R. E. Maintenance Record Systems. *Jour. AWWA*, 60:12:1345 (Dec. 1968).

Deem, R. E. The Maintenance-Record System: A Managerial Aid. *Jour. AWWA*, 66:8:447 (Aug. 1974).

Fenley, Calvin. Work Order Forms for the Field. *Jour. AWWA*, 63:12:747 (Dec. 1971).

Harris, W.S. Maps and Related Records. *Jour. AWWA*, 58:11:1453 (Nov. 1966).

Kiser, K.B. Distribution System Records. *Jour. AWWA*, 64:7:418 (July 1972).

Larkin, Donald and Grant, B. B. Planning and Using Utility Maps. *Jour. AWWA*, 63:4:213 (Apr. 1971).

Manual of Water Utility Operations. Texas Water Utilities Assoc., Austin, Texas (1979) pp. 433-434.

Operational Techniques for Distribution Systems. Proc. AWWA Distrib. System Sym. AWWA, Denver, Colo. (Feb. 1980) pp. 25-31.

Water Distribution Operator Training Handbook. AWWA, Denver, Colo. (1976) pp. XV-1 to XV-11.

Water Distribution Systems Operation and Maintenance. Water Supply and General Engrg. Sec. Minnesota Dept. Health (Sept. 1979) pp. 316-322.

Glossary Terms Introduced in Module 10

(Terms are defined in the Glossary at the back of the book.)

Arterial map
Base map
Card record
Comprehensive map
Contour lines
Coordinate system
Indexing
Intersection method
Leak frequency map
Leak survey map

List
Plan and profile drawings
Plat
Plat and list method
Pressure zone map
Sectional map
Statistical record
Valve and hydrant map
Work order

Review Questions

(Answers to Review Questions are given at the back of the book.)

1. List three reasons for maintaining maps, drawings, and records.

2. Identify the type of map that provides a clear picture of the entire distribution system.

3. What is the name of the mapped record that provides a detailed picture of a portion of the distribution system?

4. Identify two methods of showing the mapped location of valves and hydrants.

5. Identify at least two supplemental mapped records and briefly describe each.

6. Explain the purpose of maintaining a separate card record system.

7. What mapped record is usually reduced to book size for ease of handling in the field?

8. List items commonly included on a comprehensive map.

9. Identify distribution system appurtenances for which card records are typically maintained.

10. What does the list portion of a plat and list type map provide?

11. What items would be included in distribution system statistical records?

12. Explain the difference between distribution system records for small water supply systems and large water supply systems.

13. What do the following map symbols signify?

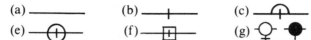

14. What function does a work order serve?

15. What information is typically included on a work order?

16. Explain the need for maintaining a file for technical information.

Study Problems and Exercises

1. You have just been appointed chief distribution operator for a water system with 4000 service connections. You learn that there are no maps or records of the system's facilities. Describe the types of maps and records you would initiate and how you would obtain the necessary information.

2. Using figures from your system, compile a set of statistical records for an upcoming month.

3. Establish an efficient file system for technical information (bulletins, pamphlets, literature, etc.) at your facility.

4. Examine your system's comprehensive map to determine where your system's largest and smallest mains are located.

5. Practice locating valves and hydrants in your system. Using your system records, have a co-worker randomly select a few valves and hydrants, then go into the field and locate them. Identify their characteristics (right-hand open, left-hand open, number of turns to open, etc.)

6. Discuss the need for updating distribution records and establish a timetable for updating your system's records.

Module 11

Public Relations

It is important to maintain the public's confidence in the quality of drinking water and the services provided by a utility. Satisfied customers will pay their bills without protest, they will provide political support for necessary rate increases or bond issues, and they will be less likely to turn to bottled water or home treatment devices of questionable quality. The PUBLIC RELATIONS activities of a utility are directed at maintaining public confidence and customer satisfaction.

Although many utilities employ public relations staff members, the main focus of this module is the role that water distribution personnel play in establishing good public relations. Frequently, the people on the "front line" can do more to ensure public confidence than all the formal media campaigns and large-scale public relations projects put together. Conversely, management's formal public relations campaigns can be undermined by an employee's poor attitude or an unwillingness to be of service beyond the immediate requirements of the job.

After completing this module you should be able to

- Explain how public relations enhances a water utility's image and builds goodwill.

- Identify specific behaviors that improve or detract from customer relations.

- Understand how informed employees are necessary to a good public relations program.

- Explain how written guidelines can assist personnel in maintaining good relations with customers.

- Identify types of formal public relations programs and explain how they benefit customer awareness and utility operations.

11-1. The Role of Public Relations

Public relations helps to create and maintain public confidence in the utility's product and organization. This requires the active cooperation of all utility personnel. Every customer contact should be viewed as an opportunity to improve communications and build GOODWILL.

Utilities depend on customer support for new budgets and for the implementation of special projects that require additional charges. It can be goodwill that tips the balance in favor of a badly needed bond issue. Customers who see their utility in a favorable light will generally vote in a like manner. They will tend to support increased service charges and new projects with long-range benefits. They will more readily tolerate problems such as temporary tastes and odors, voluntary conservation measures, or repairs that require disruption of service.

On the other hand, a utility that fosters a "them versus us" attitude can expect little customer cooperation. In fact, this approach may prompt outright antagonism from customers who already envision the utility as a faceless, heartless institution with the power to withhold a product vital to life. Everybody suffers in such cases, including water distribution personnel who may bear the brunt in terms of layoffs, lower salaries, or budget reductions affecting planned inspection or maintenance activities.

Customers must never be taken for granted. Although a utility rarely finds itself in the situation of a retail store, with competition offering lower prices or better service on the next corner, the same "best value for the dollar" approach to customer service should be applied. Higher prices must be justified by better products and services. Customer satisfaction must be a top priority.

Utility personnel in Tulsa, Okla., have maintained high community visibility. Public relations training for all personnel with an emphasis on customer communications has produced satisfied customers and a utility that is unhampered in doing its job. Figure 11-1 shows one of the devices that Tulsa uses in its public relations efforts. Tulsa officials recently reported that in the past six years water rates as a whole have climbed 150 percent, sewer rates 102 percent,

When customers call questioning the accuracy of their water meters, the Tulsa, Okla., water department tests their meter, then presents them with the card and penny shown in the photo. The penny pays for the 10 gal (40 L) of water used during the meter test, and it emphasizes what a bargain water is.

Courtesy of Tulsa, Okla., Water and Sewer Dept.

Figure 11-1. Water-Valve Awareness Card

and the average cost-per-family 69 percent. Yet during that time, additional funding in the form of a one-cent sales tax was voted twice by customers who feel that their dollars are well spent.

11-2. The Role of Water Distribution Personnel

Water distribution personnel are often in much closer contact with customers than anyone else in the utility. They may be the only contact between the utility and some customers. It may seem unfair, but a customer will remember the meter reader who took time at the doorway to wipe mud from his shoes much longer than today's favorable newspaper article about the utility. Likewise, the customer will recall the repair crew that heckled the family's cocker spaniel more vividly than the last rate hike.

Field personnel can add or detract from a utility's public image. On-the-job behavior can tell customers that the utility values their business or simply does not care. Once in the field, the utility employee is the utility. If utility employees do not respect the customer, neither, in the customer's mind, does the utility itself.

Effective public relations with customers requires three basic ingredients— communications, caring, and courtesy. Good communications means really listening to what the customer has to say, then explaining policy, answering questions, or pointing out how that customer can save water or money. Caring involves employees taking pride in themselves, their appearance, and the well-being of the customer. Finally, courtesy requires that field personnel follow the common-sense rules of polite behavior.

Meter Readers

Meter readers find themselves in the unenviable position of being naturally associated with the water bill. Whether the customer blames a high bill on a misreading of the meter or simply feels that rates are too high, the meter reader is the most available target for complaint. As if this were not enough of a burden, the meter reader must deal with unfriendly dogs, bad weather, and the general reluctance of homeowners to allow a stranger into their homes.

Irate customers are generally the exception, but an overdose of complaints can make any job difficult. Despite this, the meter reader should remain informative and polite. Meter readers are not expected to be superhuman, but a cheerful and helpful outlook goes a long way toward effective public relations, and it makes the job much easier.

Here are some basic behavioral guidelines that a meter reader can follow in performing day-to-day tasks:

- For a good first impression, maintain a neat appearance (Figure 11-2). Your uniform, if required, should be well maintained and pressed. If the utility requires no uniform, wear clean and appropriate clothing.

- Display nametags and carry credentials within easy reach. Do not make customers take your word that you are who you say you are. Many customers, especially women, are wary about admitting strangers into their homes. Anyone can claim to be a meter reader. Prove it.

- Be polite. You may not be qualified to answer all questions, but you can inform customers about inquiry or complaint procedures and provide useful information. Many customers merely want sympathy. A person who listens well and is courteous can often calm an irate customer.

- Give short, succinct answers whenever possible. A long, drawn out explanation is confusing. If the utility furnishes informative brochures, use them.

- Report any leaks found on the premises. Make sure customers know that neither you nor the utility want them to pay for unnecessary water.

- If customers read their own meters, take time to explain the procedure. This helps prevent errors or future bill adjustments.

- Show enthusiasm. Keep a smile in your voice.

- Address customers politely. Use *Miss*, *Sir*, or *Mrs. Smith*, rather than *Lady*, or *Hey You*!

- Use good judgment. For example: wipe off your muddy feet; do not kick the dog; do not smoke; do not swear; do not chat with a customer for long periods of time; walk on sidewalks, not gardens or lawns; obey all driving and parking rules.

In short, the meter reader should always try to be helpful, polite, and let customers know that he or she and the utility are on their side.

Courtesy of Colorado Springs, Colo., Water Dept.

Figure 11-2. Meter-Reader Appearance Is Important

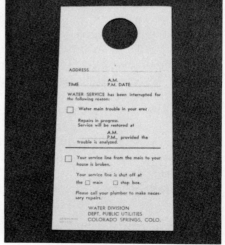

Courtesy of Colorado Springs, Colo., Water Dept.

Figure 11-3. Notification of Temporary Water Shut-off

Maintenance and Repair Crews

Like meter readers, maintenance and repair crews are highly visible to the public. It may be difficult, but field personnel who are conducting routine inspections or performing minor repairs should stay as well-groomed as possible. In most cases, maintenance or repair crews will have little, if any, face-to-face contact with customers. They may encounter a few inquiries, but usually nothing more. Personnel should answer questions politely and to the best of their ability, or refer the customer to a supervisor or the appropriate customer relations representative.

As a matter of routine, customers should be notified when service is to be temporarily discontinued (Figure 11-3), and shut-offs should be scheduled to coincide with low water-use hours. However, it may still be necessary to inform customers immediately before shut-off time and allow them a few minutes to finish a shower or collect a bottle of drinking water. The Denver, Colo., water department allows customers who are faced with water shut-offs for lack of payment a few minutes to obtain water for personal use. Customers are also allowed last-minute payment of their bill. Any such courtesy improves the utility's image and fosters good public relations.

Property should be respected and the work site kept as clean as possible. Utilities usually have policies governing lunch and breaks, but under no circumstances should field crews litter an area with wastepaper or soda cans. Smokers should not use the customer's lawn as an ashtray. Damage to lawns, sidewalks, gardens, or streets should be kept to a minimum. Customers should be forewarned if property damage is expected to occur. Repairs should be completed and property restored to its original state before leaving the site.

Vehicles should be clean and parked out of the way. They should not block driveways or alleys, nor should they hamper the flow of traffic in an area. Workers should never nap in a utility vehicle—this gives the impression that they are permitted to sleep on the job.

If streets are to be dug up, the neighborhood should be warned well in advance, with the suggestion that those affected move their vehicles. Good safety habits also promote public relations. Road barriers and warning signs communicate the utility's concern about customer and employee safety.

During the course of the job, workers should maintain a friendly attitude and a genuine desire to accomplish their work quickly and in the most inconspicuous manner possible.

Public Relations Behind the Wheel

At some time, most water distribution personnel will use a company vehicle. Public image is enhanced by good driving habits. If the driver of a utility vehicle drives 20 mph over the speed limit or tailgates the driver in front, it will not be long before a customer telephones the utility to complain. Careless driving is dangerous, makes people angry, and gives the utility a bad image.

Driving rules should always be obeyed. Vehicles should be parked so as not to impede the flow of traffic. Parked vehicles should not block driveways, intersections, or alleys, unless absolutely necessary, and then warning signals should be used on the vehicle. Where work must be performed in the street, appropriate cones, barricades, and other work-area protection methods should be employed to smoothly channel traffic around the jobsite.

The Tucson, Ariz., water utility has developed a program with its police department whereby every few years all service personnel take a four-hour safe driving course. This may not be routine for most utilities, but drivers can obtain a copy of their state's motor vehicle laws and refer to them periodically.

Public Relations and the Media

Although water distribution personnel may never be confronted with news reports, a situation could arise—a main break, for instance, or several months of water conservation restrictions—that draws the attention of local newspapers and/or television stations. The general rule for talking to reporters is, "Don't!" Large utilities maintain a public relations department whose job it is to prepare news releases and meet with the media. In other utilities, a manager or someone else is designated to take that responsibility. If approached by a reporter, distribution system operators should courteously but firmly state that they are not qualified to answer questions. Comments to the press will do little to help the situation; they may, on the contrary, do a great deal of damage.

11-3.   The Role of Informed Employees in Public Relations

When the employees do not know or understand utility procedures, they may give customers the wrong information or anger them by refusing to answer questions. Communications between the utility and its customers requires well-informed employees.

Employee orientation should be a high-priority program to familiarize new personnel with policies, regulations, and procedures. Information that relates to specific job assignments should be further explained during the initial period of employment. If an employee-orientation program is not undertaken by the utility, new personnel should acquaint themselves with the utility's procedures. Some suggestions for doing this include:

- If STANDARD OPERATING PROCEDURES (SOP) exist, read them.

- Keep a notebook and write down policies and procedures as you learn about them.

- Do not be afraid to say, "I don't know." Find out the answer, inform the customer, then keep track of the answer for future reference.

- Ask questions.

All personnel should keep abreast of new developments. After initial employment, there may be no formal effort to update personnel. Workers should read bulletin boards and memos. Attending staff meetings can provide a forum for the exchange of ideas and information. Employees should take advantage of any training that is offered.

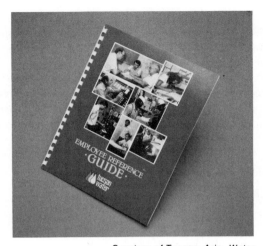

Courtesy of Tucson, Ariz., Water
Figure 11-4. Utility Procedural Manual

Procedural Manuals as a Public Relations Tool

Procedural manuals (Figure 11-4) explain the governing policies of the utility, its services, turn-on and shut-off practices, and forms. They may double as employee handbooks, providing information on sick leave, benefits, and employee procedures.

Utilities that use procedural manuals have usually devised them to assist the customer service department. Traditionally, customer service representatives spend the bulk of their time on the telephone answering questions or handling complaints. When field personnel refer customers to the utility, their calls may end up in that department. A responsive utility will be open to procedural changes and will periodically review its operating practices.

Written guidelines keep employees informed and promote a well-run organization. Procedures should be clearly stated, so employees do not need to create their own interpretations. If a clearly worded policy is quoted to a customer, no argument as to the policy's meaning is likely to occur. Customers will be far more antagonistic toward a procedure that is unclear or sounds as if it has been made up on the spur of the moment.

11-4. Formal Public Relations Programs

Although water distribution personnel play an integral part in creating a favorable image for the utility, there are many other facets of public relations that contribute to the total picture. The size of a water utility may influence the formality or scope of a public relations program, but not its importance.

Large utilities may confront political and environmental issues that affect a cross section of community interest. They may face a high volume of maintenance and repairs, billing or collection problems, rate-allocation difficulties, or growth patterns that strain current treatment and personnel levels. A larger utility, therefore, will usually maintain a full-time public relations staff whose primary concern is to promote the company's image within the community, by dispensing information to the public and working closely with the media.

Smaller utilities are not immune to political, environmental, or community issues, but by necessity they may have to operate with a one- or two-person customer relations staff to handle customer questions and complaints, plus a manager who deals with local government, community representatives, and the public in general.

Some of the functions of a larger utility's public relations efforts are discussed in the following sections.

Customer Service

CUSTOMER SERVICE representatives answer customer questions and handle complaints. They field telephone calls and complete the resulting paperwork. Some problems can be solved within seconds, while others are more complicated or involve intricate billing problems. Complex or particularly sensitive problems may be referred to a supervisor. Complaints are also dispatched to repair personnel or meter readers for prompt follow-up. In general, customer service representatives are well versed in telephone etiquette, active listening techniques, utility procedures, and persuasive speaking.

Public Information

PUBLIC INFORMATION specialists dispense information to the community. The primary goal is to project a favorable image of the organization into the public eye. Public speaking engagements, participation in civic or professional clubs,

Figure 11-5. Bill Stuffers Available from AWWA

and creating public service or special school projects consume much of the public information expert's time. Utilities that are conscious of their community image will often support local television, particularly educational or public service channels.

The public information specialist may also create and distribute literature explaining utility operations and policies, or brochures giving useful water-related information to consumers. The "bill stuffers" shown in Figure 11-5 are examples of these efforts.

Media Relations

MEDIA RELATIONS personnel specialize in communicating with newspapers, magazines, radio, and television. An ongoing program highlights utility-sponsored projects, upper management personnel changes, conservation efforts, or any newsworthy information. Bond issues, rate increases, special projects and how they affect homeowners and businesses, or emergency situations are explained to the public through the news media.

A press conference is an interview held for newsmen and conducted by a utility spokesperson, usually a public relations or media relations expert, the general manager, or the manager's designate. A press conference is usually called to explain an emergency that has some impact on the community as a whole or to make some type of announcement. Media relations experts know what to say and how to say it in order to put the utility in the best possible light.

Selected Supplementary Readings

Crawford, J.J. Effective Public Communication. *Jour. NEWWA*, 94:2:158 (June 1980).

Customer Communication Questionnaire Helps. *OpFlow*, 4:12:4 (Dec. 1978).

Dime, Rick. On-Site Customer Inspected Meter Testing. *OpFlow*, 11:8:7 (Aug. 1985).

Fenvessy, S.J. Keep Your Customers and Keep Them Happy. Dow-Jones-Irwin, Homewood, Ill. (1976).

Keeping Cool When Customers Complain. *OpFlow*, 8:11:3 (Nov. 1982).

Lakshman, B.T. Stiff Monitoring Reduces Customer Gripes in Newark. *OpFlow*, 2:6:1 (June 1976).

Maintaining Distribution-System Water Quality. AWWA, Denver, Colo. (1986).

Spangler, R.M. Public Relations Basics. AWWA, Denver, Colo. (1978).

Glossary Terms Introduced in Module 11

(Terms are defined in the Glossary at the back of the book.)

Customer service	Public information
Goodwill	Public relations
Media relations	Standard operating procedures (SOPs)

Review Questions

(Answers to Review Questions are given at the back of the book.)

1. Describe how poor customer relations can adversely affect distribution personnel.

2. List three ingredients for maintaining effective public relations with customers.

3. Identify three behaviors or practices of a meter reader that might influence a favorable public image.

4. List five behaviors that maintenance personnel should avoid displaying when on scheduled work assignments.

5. Describe how good safety practices can contribute to an overall public relations program of a water utility.

6. How can water distribution personnel familiarize themselves with safe driving regulations if the utility does not offer a formal training program?

7. What is the first and foremost rule that water distribution personnel should follow if approached by the media for an opinion or statement?

8. What are some advantages of having an operating procedures manual or other written guidelines for personnel in a water utility?

9. What is the role of a customer service representative?

10. Who is most likely to be the spokesperson for the utility at a formal press conference?

Study Problems and Exercises

1. Several public interest groups in the community and two local newspapers are opposing a proposed rate increase that your utility presented to the city council. A significant portion of the increased revenue would be used to employ additional distribution system personnel and conduct more frequent routine inspections, maintenance, and repairs. During the past three years the utility has performed only emergency or needed repairs because of budget restrictions placed on staff and replacement parts. Needless to say, repairs have not been made under the best of conditions.

   A state environmental group has attacked the utility because of a five-year plan that involves a major capital development program for a secondary water source, expanded treatment facilities, and main extension program. Local residents are sympathetic with the plan, but see it as a potential increase in the cost of water service.

   Last week, one of the distribution system operators rear-ended a $37,000 Mercedes at a traffic light, ran over Mrs. Schwartz's cat, and was seen changing a flat tire on a company pickup in front of the Flamingo Tavern.

   Your utility manager has asked for recommendations on how the organization can improve its public image and confidence. Prepare a brief report stating what you believe to be the underlying problems. Assuming the rate increase becomes effective, outline recommendations that would be appropriate for distribution personnel to follow in order to improve the utility's public relations posture.

2. Your repair crew is working in a residential neighborhood. The task requires tearing up a portion of the street, closing one lane of the two-way street, and possibly damaging several lawns. Recommend at least five precautions that will reduce customer inconvenience and foster a good public image in this neighborhood.

Appendix A

Bacteriological Sampling

Bacteriological Sampling

The first step in analyzing the bacteriological quality of water in the distribution system is the collection of samples that accurately represent the condition of the water being sampled. This appendix discusses bacteriological sample collection, preservation, and transportation. More detailed information concerning sampling and testing for bacteriological and other contaminants can be found in Volume 4, *Water Quality Analyses*.

A-1. Selecting Representative Sampling Points

One of the most common causes of error in water quality analysis is improper sampling. When a sample is tested, the test results show only what is in the sample. If those test results are to be useful, the sample must contain essentially the same constituents as the body of water from which it was taken—that is, it must be a *representative sample*.

Samples taken within the distribution system may yield water of significantly different quality than samples of finished water taken at the plant. Corrosion and cross connections are two conditions that can significantly deteriorate the quality of treated water before it reaches the customer. The selection of representative sample points within the distribution system is an important initial step in developing a sampling program that will accurately reflect water quality.

Sample Point Locations

The number and location of sample points should be established to ensure compliance with the applicable coliform (bacteria) testing requirements of the state or federal drinking water regulations. To comply with the National Interim Primary Drinking Water Regulations, every surface-water system with a population of 25 to 100 must have at least two sampling points: one for turbidity testing at the point where the water enters the distribution system, and one for coliform testing at a consumer's faucet representative of conditions within the system.

Systems drawing surface water from more than one source will require additional turbidity sampling points, and systems serving populations greater than 1000 will require more than one coliform sampling point. These sample points can also serve for other distribution system sampling, such as for other regulated contaminants or for water stability monitoring. This appendix deals primarily with the coliform sampling and testing procedures.

Two major considerations in determining the number and location of coliform sampling points are that they should be:

- Representative of each different source of water entering the system
- Representative of conditions within the system, such as dead ends, loops, storage facilities, and pressure zones.

The precise location of sampling points will depend on the configuration of the distribution system. Several detailed examples of sample-point selection are included in the Sampling module of Volume 4, *Water Quality Analyses.*

Sample Faucet Selection

Once a representative sample point has been located on the distribution system map, a specific sample faucet must be selected. These faucets can be located:

- Inside a public building, such as a fire station or school building
- Inside the home of an operator
- Inside the homes of other consumers.

Faucets selected should be on lines connected directly to the main. Only cold-water faucets should be used for sample collection. The sampling faucet must not be located too closely to a sink bottom—contaminated water or soil may be present on such faucet exteriors, and it is difficult to place a collection bottle beneath them without touching the neck interior against the outside faucet surface. Samples should not be taken from:

- Swivel faucets
- Leaking faucets, which allow water to flow out from around the stem of the valve and down the outside of the faucet
- Lines with home water-treatment units, including water softeners
- Homes with separate storage tanks
- Drinking fountains
- Faucets with aerators, strainers, or hose attachments (on nonswivel faucets, these can be removed to allow samples to be taken).

When a representative sample point has been selected, it should be marked or described so that it can be easily located for future sample collection.

A-2. Sample Collection, Preservation, and Transportation

Once sampling points have been selected, the utility can begin a regular program of sample collection and testing. Larger utilities will take and test their own samples; smaller utilities will take their own samples, but they may send them to a state or commercial lab, or to a larger utility, for testing.

Collection

Collecting the sample involves a few simple, carefully performed steps. The key to accurate sampling is to prevent contamination by allowing nothing except the sample water to enter the bottle or touch the bottle cap or neck.

1. Use only the bottles provided by the lab specifically for coliform sampling.

2. *Do not rinse the bottles.* There is a chemical in the bottles that destroys any residual chlorine in the water. The residual chlorine would otherwise kill any bacteria in the sample, yielding an incorrect test result.

3. Keep sample bottles unopened until the moment of filling. The bottles are sterile.

4. Make sure the faucet has no aerator and no swivel. Outside faucets should be prepared by squirting with concentrated chlorine solution and then flushing. Flaming the outside of the faucet is not recommended.

5. Flush the faucet for 2 to 5 min to clear any stagnant water from the service line.

6. Hold the bottle near the base; do not handle the stopper or cap and neck of the bottle.

7. When flushing is complete, without changing the flow, gently fill the bottle *without rinsing.* Leave an air space at the top.

8. Replace the cap or stopper immediately.

9. Using a separate sample, test for free chlorine residual and record the result.

10. Label the bottle, being sure to include the date and time the sample was taken, and package it for delivery to the lab.

Storage and Transportation

The number of coliform bacteria in a sample begins to change immediately after the sample is collected. These changes are slowed somewhat by the chemical supplied in the sample bottle, which neutralizes any chlorine residual; they can be further slowed by keeping the bottle cool until the time of testing. However, bacteriological testing should always be performed as soon as possible after the sample is collected—within 30 hours as a maximum.

Samples should be stored and shipped in insulated containers or insulated sample bottles. Bottle caps should be tight enough to prevent leakage, and bottles should be packed in a sturdy container with enough cushioning material to prevent breakage. If the samples are to be analyzed by a laboratory in another city, a method of shipment must be found that will ensure that the samples arrive at the lab within 30 hours (preferably less) of the time the sample was taken. Usually the US mail or a commercial package-shipping service is the best way to ship samples.

Appendix B

Construction of a Pitot Gauge

Construction of a Pitot Gauge

Accurate Pitot gauges can be purchased from several manufacturers. However, a simple gauge can also be constructed from materials commonly available in the utility shop or from local hardware stores, according to the following procedure.

B-1. Parts

The Pitot gauge (Figure B-1) consists of three main parts:

- A blade section
- An air chamber unit
- An air-pressure gauge.

An air-pressure gauge commonly has a 0–100 psi (0–690 kPa) scale. Pressure-gauge readings while testing water flow should be in the range of 10 to 30 psi (69 to 210 kPa), so it is important to select a gauge that will be easy to read in this range.

All pipe fittings should be galvanized or brass to resist corrosion. The most convenient size to use is $1/4$ in. If the gauge has a $1/8$-in. connection, it will be necessary to use a reducing tee or a $1/8$-in. reducing bushing. The wooden or plastic handle, which fits over the 3-in. nipple, is optional, but makes use of the unit more convenient.

B-2. Construction

The first step in constructing the Pitot gauge is to fabricate the blade and (if one is used) the handle. After these special parts are made, they are assembled with the other, stock, items to create the finished gauge.

Construction of the Blade

Figure B-2 shows the details of the blade section. To construct this piece, first drill a hole through the center of the top of the pipe plug to accept the $1/8$-in. OD copper tubing. Then take the 6-in. length of copper tube (make sure both openings in the ends are round and free of burrs) and sweat-solder one end into the hole in the pipe cap.

Next, take a piece of flat brass ($1^1/2$ in. × 3 in. × $1/16$ in.) and cut out a semicircle with a 3-in. diameter. With a file, taper the diameter side to a knife edge. Fit the blade to the pipe cap so that it fits tightly against the base of the copper tube, and

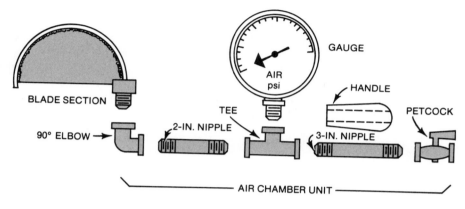

Figure B-1.   Parts of a Pitot Gauge

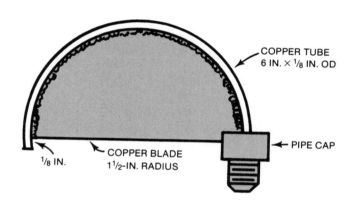

Figure B-2.   Blade Section

securely solder it in place. Next, bend the copper tube around the perimeter of the blade and solder the blade to the tubing.

Smooth the solder joints on both sides of the blade. This is important because any rough areas on the blade could cause turbulence in the water stream that will result in incorrect readings. The open end of the tube should extend out about 1/8 in. in front of the knife edge of the blade. Check to see that the openings in the copper tube are not blocked with solder or burrs, and that they are not bent out of round.

Assembly

First, thread the 90° elbow securely to the 2-in. nipple. Take this assembly and thread it securely into the tee, so the openings in the elbow and the tee point in the same direction. Now thread the 3-in. nipple securely into the tee. Install the handle, if used, over the 3-in. nipple; it should fit tightly.

Install the petcock on the open end of the 3-in. nipple so that water can be drained after each use, to prevent internal buildup of scale. It is usually best to have the petcock handle positioned so that it is on the same side as the air-pressure gauge.

Finally, install the air-pressure gauge in the top opening of the tee, and screw the blade section into the 90° elbow.

Appendix C

Flow Testing

Flow Testing

Select a hydrant at or near the location for which flow-test data is required; this will be the flowing hydrant. Select another hydrant for the residual pressure readings, or locate a hose bibb or faucet at which the readings are taken. Residual pressure should be read at a location such that the flow runs from the source to the flowing hydrant and then to the hydrant, hose bibb, or faucet at which the readings are taken. If residual pressure is read at a hose bibb or faucet, take care that there is no use of water in the building while the test is being performed.

At the start of the test, observe and record the normal system pressure. Then open the selected hydrant to allow flow. While the hydrant is flowing, determine the discharge pressure by the use of a pitot gauge, and record that pressure. Observe and record residual pressure.

Test data reflecting the flow capabilities of the system under the pressure drop induced by the test can be converted to information regarding flow capabilities under a greater pressure drop (as would be induced by fire pumping) by the use of Tables C-1, C-2, and C-3, as demonstrated in the following example.

Example:

A hydrant test gives the following results: At the location where residual pressure is to be measured, pressure before the hydrant is open is 65 psi. While the hydrant is flowing, residual pressure is 40 psi. At the flowing hydrant, a pitot reading of 27 psi is observed in the flow from a 2½-in. outlet.

Find the flow that would be produced if the residual pressure in the system were drawn down to 20 psi.

First, use Table C-1 to determine that the flow rate from a 2½-in. outlet at 27 psi is 870 gpm. (Tables relating a greater range of outlet sizes and pitot pressure readings are available in the sources listed in the supplementary readings for Module 4, Fire Hydrants.)

Next, using the formula given in Table C-3, compute what the flow Q_R would be at a residual pressure of 20 psi:

$$Q_R = \frac{Q_F \times h_R^{0.54}}{h_F^{0.54}}$$

Where: Q_R is the flow available at the desired residual pressure.

Q_F is the observed flow of 870 gpm.

h_R is the difference between the normal pressure and the desired residual pressure = 65 – 20 = 45 psi.

h_F is the actual drop in pressure observed = 65 – 40 = 25 psi.

Substituting in the formula:

$$Q_R = \frac{870 \text{ gpm} \times 45 \text{ psi}^{0.54}}{25 \text{ psi}^{0.54}}$$

From Table C-3·   45 psi $^{0.54}$ = 7.81
25 psi $^{0.54}$ = 5.69

Substituting:

$$Q_R = \frac{870 \times 7.81}{5.69}$$

$$= 1194 \text{ gpm at 20 psi}$$

If the distribution system is strong and flow from a single hydrant outlet results in only a slight change in pressure, then it may be necessary to open both outlets on a hydrant or to use several hydrants simultaneously in order to obtain a meaningful pressure drop. When several hydrants are used, the normal and residual pressures should be taken at a location between the flowing hydrants. The calculation is similar to the above example, with Q_F set equal to the sum of the hydrant flows.

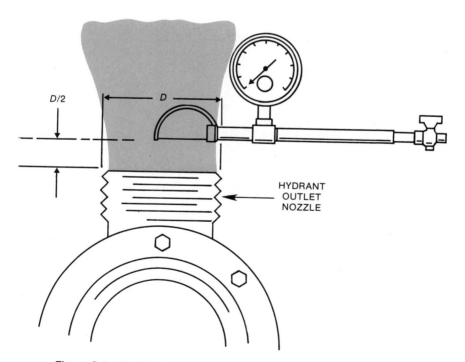

Figure C-1.   Positioning a Pitot Gauge in the Hydrant Outlet Flow

Table C-1.   Discharge Table for Circular Outlets—$2^1/_4$–$2^{11}/_{16}$ in.
(Outlet pressure measured by pitot gauge)

Outlet Pressure psi	Outlet Diameter in.							
	$2^1/_4$	$2^3/_{16}$	$2^3/_8$	$2^7/_{16}$	$2^1/_2$	$2^9/_{16}$	$2^5/_8$	$2^{11}/_{16}$
	gpm							
$^1/_4$	70	70	80	80	80	90	90	100
$^1/_2$	100	100	110	110	120	120	130	140
$^3/_4$	120	120	130	140	150	150	160	170
1	140	140	150	160	170	180	180	190
$^1/_4$	150	160	170	180	190	200	210	220
$^1/_2$	170	180	190	200	210	220	230	240
$^3/_4$	180	190	200	210	220	230	240	260
2	190	200	210	230	240	250	260	270
$^1/_4$	200	220	230	240	250	260	280	290
$^1/_2$	220	230	240	250	270	280	290.	310
$^3/_4$	230	240	250	260	280	290	310	320
3	240	250	260	280	290	310	320	340
$^1/_4$	250	260	270	290	300	320	330	350
$^1/_2$	250	270	280	300	310	330	350	360
$^3/_4$	260	280	290	310	330	340	360	380
4	270	290	300	320	340	350	370	390
$^1/_4$	280	300	310	330	350	360	380	400
$^1/_2$	290	300	320	340	360	370	390	410
$^3/_4$	300	310	330	350	370	380	400	420
5	300	320	340	360	380	390	410	430
$^1/_4$	310	330	350	370	390	400	420	440
$^1/_2$	320	340	350	370	390	410	430	450
$^3/_4$	330	340	360	380	400	420	440	460
6	330	350	370	390	410	430	450	470
$^1/_4$	340	360	380	400	420	440	460	480
$^1/_2$	350	370	390	410	430	450	470	490
$^3/_4$	350	370	390	410	440	460	480	500
7	360	380	400	420	440	470	490	510
$^1/_4$	370	390	410	430	450	480	500	520
$^1/_2$	370	390	410	440	460	480	510	530
$^3/_4$	380	400	420	440	470	490	510	540
8	380	410	430	450	480	500	520	550
$^1/_4$	390	410	440	460	480	510	530	560
$^1/_2$	400	420	440	460	490	510	540	560
$^3/_4$	400	420	450	470	500	520	550	570
9	410	430	450	480	500	530	550	580
$^1/_4$	410	440	460	480	510	540	560	590
$^1/_2$	420	440	470	490	520	540	570	600
$^3/_4$	420	450	470	500	520	550	580	600
10	430	450	480	500	530	560	580	610

Table C-1. Discharge Table for Circular Outlets—2¼–2¹¹/₁₆ in.
(Outlet pressure measured by pitot gauge) *(continued)*

Outlet Pressure psi	Outlet Diameter in.							
	2¼	2³/₁₆	2³/₈	2⁷/₁₆	2½	2⁹/₁₆	2⁵/₈	2¹¹/₁₆
	gpm							
10¼	440	460	480	510	540	570	590	620
½	440	470	490	520	540	570	600	630
¾	450	470	500	520	550	580	610	640
11	450	480	500	530	560	590	610	640
¼	460	480	510	530	560	590	620	650
½	460	490	510	540	570	600	630	660
¾	470	490	520	550	580	600	630	660
12	470	500	520	550	580	610	640	670
½	480	510	540	560	590	620	650	690
13	490	520	550	570	610	640	670	700
½	500	530	560	590	620	650	680	710
14	510	540	570	600	630	660	690	730
½	520	550	580	610	640	670	700	740
15	530	560	590	620	650	680	720	750
½	540	570	600	630	660	700	730	760
16	540	570	610	640	670	710	740	780
½	550	580	620	650	680	720	750	790
17	560	590	620	660	690	730	760	800
½	570	600	630	670	700	740	770	810
18	580	610	640	680	710	750	780	820
½	590	620	650	690	720	760	800	830
19	590	630	660	700	730	770	810	840
½	600	640	670	700	740	780	820	860
20	610	640	680	710	750	790	830	870
21	620	660	690	730	770	810	850	890
22	640	670	710	750	790	830	870	910
23	650	690	730	770	810	850	890	930
24	670	700	740	780	820	860	910	950
25	680	720	760	800	840	880	920	970
26	690	730	770	810	860	900	940	990
27	710	750	790	830	870	920	960	1010
28	720	760	800	840	890	930	980	1020
29	730	770	820	860	910	950	1000	1040
30	750	790	830	870	920	970	1010	1060
31	760	800	840	890	940	980	1030	1080
32	770	810	860	900	950	1000	1050	1100
33	780	830	870	920	970	1010	1060	1110
34	790	840	880	930	980	1030	1080	1130
35	810	850	900	940	990	1040	1090	1140
36	820	860	910	960	1010	1060	1110	1160

Source: "Fire Flow Tests"
Discharge Table for Circular Outlets
American Insurance Association

Coefficient = 0.90

Table C-2. Discharge Table for Circular Outlets—4¹/₄–4¹¹/₁₆ in.
(Outlet pressure measured by pitot gauge)

Outlet Pressure	Outlet Diameter in.							
psi	$4^{1}/_{4}$	$4^{3}/_{16}$	$4^{3}/_{8}$	$4^{7}/_{16}$	$4^{1}/_{2}$	$4^{9}/_{16}$	$4^{5}/_{8}$	$4^{11}/_{16}$
	gpm							
$^{1}/_{4}$	240	250	260	260	270	280	290	300
$^{1}/_{2}$	340	350	360	370	390	400	410	420
$^{3}/_{4}$	420	430	450	460	470	490	500	510
1	490	500	520	530	550	560	570	590
$^{1}/_{4}$	540	560	590	590	610	630	640	660
$^{1}/_{2}$	600	610	630	650	670	690	700	720
$^{3}/_{4}$	640	660	680	700	720	740	760	780
2	690	710	730	750	770	790	810	840
$^{1}/_{4}$	730	750	770	800	820	840	860	890
$^{1}/_{2}$	770	790	810	840	860	890	910	940
$^{3}/_{4}$	810	830	850	880	900	930	950	980
3	840	870	890	920	940	970	1000	1020
$^{1}/_{4}$	880	900	930	960	980	1010	1040	1060
$^{1}/_{2}$	910	940	970	990	1020	1050	1070	1100
$^{3}/_{4}$	940	970	1000	1030	1050	1080	1110	1140
4	970	1000	1030	1060	1090	1120	1150	1180
$^{1}/_{4}$	1000	1030	1060	1090	1120	1150	1180	1220
$^{1}/_{2}$	1030	1060	1090	1120	1160	1190	1220	1250
$^{3}/_{4}$	1060	1090	1120	1150	1190	1220	1250	1290
5	1090	1120	1150	1180	1220	1250	1280	1320
$^{1}/_{4}$	1110	1150	1180	1210	1250	1280	1320	1350
$^{1}/_{2}$	1140	1180	1210	1240	1280	1310	1350	1390
$^{3}/_{4}$	1170	1200	1240	1270	1310	1340	1380	1420
6	1190	1230	1260	1300	1330	1370	1410	1450
$^{1}/_{4}$	1220	1250	1290	1320	1360	1400	1440	1480
$^{1}/_{2}$	1240	1280	1310	1350	1390	1430	1470	1510
$^{3}/_{4}$	1260	1300	1340	1380	1420	1450	1490	1540
7	1290	1330	1360	1400	1440	1480	1520	1560
$^{1}/_{4}$	1310	1350	1390	1430	1470	1510	1550	1590
$^{1}/_{2}$	1330	1370	1410	1450	1490	1530	1570	1620
$^{3}/_{4}$	1350	1390	1430	1480	1520	1560	1600	1640
8	1380	1420	1460	1500	1540	1580	1620	1670
$^{1}/_{4}$	1400	1440	1480	1520	1570	1610	1650	1700
$^{1}/_{2}$	1420	1460	1500	1540	1590	1630	1680	1720
$^{3}/_{4}$	1440	1480	1520	1570	1610	1650	1700	1750
9	1460	1500	1540	1590	1630	1680	1720	1770
$^{1}/_{4}$	1480	1520	1570	1610	1660	1700	1750	1800
$^{1}/_{2}$	1500	1540	1590	1630	1680	1720	1770	1820
$^{3}/_{4}$	1520	1560	1610	1650	1700	1750	1790	1840
10	1540	1580	1630	1670	1720	1770	1820	1870

Table C-2. Discharge Table for Circular Outlets—4 ¼–4 ¹¹/₁₆ in.
(Outlet pressure measured by pitot gauge) *(continued)*

Outlet Pressure psi	Outlet Diameter in.							
	$4^{1}/_{4}$	$4^{3}/_{16}$	$4^{3}/_{8}$	$4^{7}/_{16}$	$4^{1}/_{2}$	$4^{9}/_{16}$	$4^{5}/_{8}$	$4^{11}/_{16}$
	gpm							
$10^{1}/_{4}$	1560	1600	1650	1700	1740	1790	1840	1890
$^{1}/_{2}$	1580	1620	1670	1720	1760	1810	1860	1910
$^{3}/_{4}$	1590	1640	1690	1740	1790	1830	1880	1940
11	1610	1660	1710	1760	1810	1860	1910	1960
$^{1}/_{4}$	1630	1680	1730	1780	1830	1880	1930	1980
$^{1}/_{2}$	1650	1700	1750	1800	1850	1900	1950	2000
$^{3}/_{4}$	1670	1720	1760	1820	1870	1920	1970	2020
12	1690	1730	1780	1840	1890	1940	1990	2050
$^{1}/_{2}$	1720	1770	1820	1870	1930	1980	2030	2090
13	1750	1800	1850	1910	1970	2020	2070	2130
$^{1}/_{2}$	1790	1840	1890	1950	2000	2060	2110	2170
14	1820	1870	1930	1980	2040	2090	2150	2210
$^{1}/_{2}$	1850	1910	1960	2020	2080	2130	2190	2250
15	1880	1940	1990	2050	2110	2170	2230	2290
$^{1}/_{2}$	1910	1970	2030	2090	2150	2200	2260	2330
16	1940	2000	2060	2120	2180	2240	2300	2360
$^{1}/_{2}$	1970	2030	2090	2150	2210	2270	2330	2400
17	2000	2060	2120	2180	2250	2310	2370	2440
$^{1}/_{2}$	2030	2090	2150	2220	2280	2340	2400	2470
18	2060	2120	2180	2250	2310	2370	2440	2510
$^{1}/_{2}$	2090	2150	2210	2280	2350	2410	2470	2540
19	2120	2180	2240	2310	2380	2440	2510	2580
$^{1}/_{2}$	2140	2210	2270	2340	2410	2470	2540	2610
20	2170	2240	2300	2370	2440	2500	2570	2640
21	2220	2290	2360	2430	2500	2560	2630	2710
22	2280	2350	2420	2490	2560	2620	2700	2770
23	2330	2400	2470	2540	2610	2680	2760	2830
24	2380	2450	2520	2600	2670	2740	2820	2890
25	2430	2500	2580	2650	2720	2800	2870	2950
26	2480	2550	2630	2700	2780	2850	2930	3010
27	2530	2600	2680	2750	2830	2910	2990	3070
28	2580	2650	2730	2800	2880	2960	3040	3130
29	2620	2700	2770	2850	2940	3020	3090	3180
30	2670	2740	2820	2900	2990	3070	3150	3240
31	2710	2790	2870	2950	3030	3120	3200	3290
32	2750	2830	2920	3000	3080	3170	3250	3340
33	2790	2880	2960	3040	3130	3220	3300	3390
34	2830	2920	3000	3090	3170	3260	3350	3440
35	2870	2960	3040	3140	3220	3310	3400	3490
36	2910	3000	3080	3180	3270	3360	3450	3540

Source: "Fire Flow Tests"
Discharge Table for Circular Outlets
American Insurance Association

Coefficient = 0.90

Table C-3.   Values of h to the 0.54 Power

h	$h^{0.54}$	h	$h^{0.54}$	h	$h^{0.54}$	h	$h^{0.54}$	h	$h^{0.54}$	h	$h^{0.54}$	h	$h^{0.54}$
1	1.00	26	5.81	51	8.38	76	10.37	101	12.09	126	13.62	151	15.02
2	1.45	27	5.93	52	8.44	77	10.44	102	12.15	127	13.68	152	15.07
3	1.81	28	6.05	53	8.53	78	10.51	103	12.22	128	13.74	153	15.13
4	2.11	29	6.16	54	8.62	79	10.59	104	12.28	129	13.80	154	15.18
5	2.39	30	6.28	55	8.71	80	10.66	105	12.34	130	13.85	155	15.23
6	2.63	31	6.39	56	8.79	81	10.73	106	12.41	131	13.91	156	15.29
7	2.86	32	6.50	57	8.88	82	10.80	107	12.47	132	13.97	157	15.34
8	3.07	33	6.61	58	8.96	83	10.87	108	12.53	133	14.02	158	15.39
9	3.28	34	6.71	59	9.04	84	10.94	109	12.60	134	14.08	159	15.44
10	3.47	35	6.82	60	9.12	85	11.01	110	12.66	135	14.14	160	15.50
11	3.65	36	6.93	61	9.21	86	11.08	111	12.72	136	14.19	161	15.55
12	3.83	37	7.03	62	9.29	87	11.15	112	12.78	137	14.25	162	15.60
13	4.00	38	7.13	63	9.37	88	11.22	113	12.84	138	14.31	163	15.65
14	4.16	39	7.23	64	9.45	89	11.29	114	12.90	139	14.36	164	15.70
15	4.32	40	7.33	65	9.53	90	11.36	115	12.96	140	14.42	165	15.78
16	4.47	41	7.43	68	9.61	91	11.43	116	13.03	141	14.47	166	15.81
17	4.62	42	7.53	67	9.69	92	11.49	117	13.09	142	14.53	167	15.86
18	4.78	43	7.62	68	9.76	93	11.56	118	13.15	143	14.58	168	15.91
19	4.90	44	7.72	69	9.84	94	11.63	119	13.21	144	14.64	169	15.96
20	5.04	45	7.81	70	9.92	95	11.69	120	13.27	145	14.69	170	16.01
21	5.18	46	7.91	71	9.99	98	11.76	121	13.33	146	14.75	171	16.06
22	5.31	47	8.00	72	10.07	97	11.83	122	13.39	147	14.80	172	16.11
23	5.44	48	8.09	73	10.14	98	11.89	123	13.44	148	14.86	173	16.16
24	5.56	49	8.18	74	10.22	99	11.96	124	13.50	149	14.91	174	16.21
25	5.69	50	8.27	75	10.29	100	12.02	125	13.56	150	14.97	175	16.25

Source: "Fire Flow Tests"
Discharge Table for Circular Outlets
American Insurance Association

Appendix D

Selection of Pumps and Prime Movers

Selection of Pumps and Prime Movers

The pumps and prime movers installed in a distribution system represent a major investment for a utility. If the correct equipment is selected for the application, a pump installation can provide years of efficient, reliable service. The installation of equipment that is not correct for the utility's needs, however, can result in excessive energy costs, unreliable operation, high maintenance, and rapid obsolescence of the pump station.

Because of the relatively high initial cost of a pump installation, and because of the high potential costs of improperly selected equipment, qualified engineering personnel should be consulted whenever a pump station is being designed or upgraded. The distribution system operator should also be consulted, since the operator is often the person most familiar with the demands and performance of the existing system. This appendix covers general pump and prime-mover selection criteria that an operator should be familiar with.

D-1.   Pump Selection

The first step in selecting a pump (and prime mover) is to gather data on existing system performance and needs, along with projections for the system's future needs. Information needed includes:

- Water source—surface or ground water
- Maximum capacity to be required
- Nature of the water—its temperature, pH, dissolved chemicals, gases, and suspended matter, and whether or not any suspended matter is abrasive
- Average, maximum, and minimum discharge-head conditions, including pipe sizes and length, and the shape of the system head curves
- Average pressure or suction lift, average diameter and length of suction lines
- Service—whether constant or intermittent
- Type of power available for pump drive
- Space available for pumping equipment.

Once this data is collected, various pump and prime-mover configurations can be investigated. Three general requirements for pump and motor combinations are reliability, adequacy, and economy. Reliability can be achieved by installing the best equipment available and then providing for an auxiliary power source or an emergency pump and prime mover. Adequacy can be achieved by allowing for some oversizing of pumping equipment beyond projected maximum demands. Economy can be achieved by taking into account initial cost, operating cost, life and depreciation estimates, standby charges, maintenance costs, and interest rates. With these general requirements in mind, specific attention should be given to the following areas:

- System configuration
- Pump type
- Pump capacity
- Pump speed
- Pump-station design.

System Configuration

System configuration involves consideration of the overall distribution system and of the setup of one or more pumps in an individual pump station.

Distribution system.   Figures D-1 and D-2 illustrate two typical small water system layouts. Note the considerable differences between the arrangements of supply and distribution facilities for the two systems.

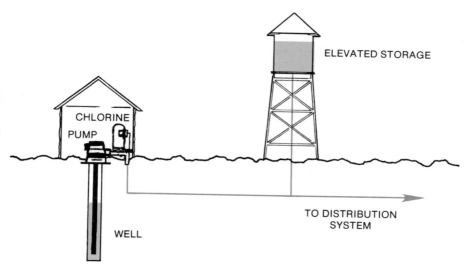

Figure D-1.   Ground-Water System

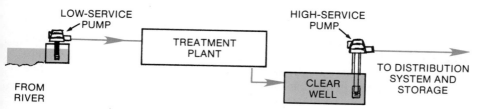

Figure D-2.   Surface-Water System

In Figure D-1, water is delivered to an elevated storage reservoir and into the distribution system with a vertical turbine or submersible pump. In many small systems, the deep-well pump would deliver water directly into the system. A standpipe or elevated tank would store excess water for peak-demand periods.

In Figure D-2, low-service pumps deliver water to the treatment plant. From the clear well, high-service pumps deliver the water to elevated storage and into the system.

A centrifugal pump should always be located as close to the source of the water supply as conditions permit, thus keeping suction piping short, and minimizing head loss.

Pump-station configuration.   To better match pumping equipment capability with demand, many stations are equipped with more than one pump. The operator can run pumps in series and/or parallel arrangements to meet varying system demand.

Series operation.   In series operation, two or more pumps are used, with each pump discharging into the suction of the next (Figure D-3A). This has the effect of increasing the discharge head. This arrangement can be used when the total head required exceeds that which is available from a single pump. Pumps in series can be used to meet heads of 250–500 ft (76–150 m).

Since pump pressure from the second pump is added to that of the first pump, care must be taken to avoid exceeding the allowable pressure in the last pump. Excessive pressures can damage flanges and shaft seals.

Parallel operation.   In parallel operation, two or more pumps discharge into a common manifold or header. This type of system is most useful when water demand fluctuates over a wide range. When demand is light, only one pump operates. When demand increases above the first pump's capacity, a second pump is turned on.

A typical arrangement of pumps and piping is shown in Figure D-3B. Both pumps take suction from a common manifold or header. Check valves prevent reverse flow through either pump when it is shut down. By having the two pumps discharging into a common system, capacity is increased for the same pressure.

Variable water demand is frequently met by installing a number of pumps with different capacities and with interconnections to allow operation in parallel. Typical pump-station configurations are shown in Figure D-4. With pumping equipment set up in this way, two or three pumps can carry the maximum load when it is necessary to remove one unit for service or repair. Generally, pumping equipment should be sized to meet maximum demand when the largest pump is out of service.

Pumps should be operated to discharge the required amount of water against the lowest possible discharge pressure. When water demand is much less than anticipated in the original design, the cost of installing an additional, smaller pump can be recovered through reduced power costs in just one to two years. The most efficient pump or pumps should be used most of the time. Automatic control systems can ensure that the most efficient pumping units come on first, go off last, and operate against the lowest aggregate discharge pressure.

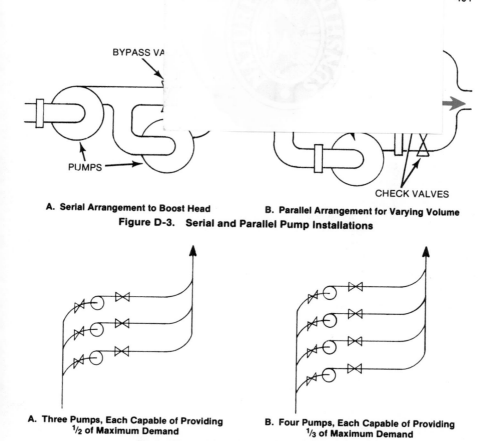

BYPASS VA

PUMPS

CHECK VALVES

A. Serial Arrangement to Boost Head   **B. Parallel Arrangement for Varying Volume**

Figure D-3.   Serial and Parallel Pump Installations

A. Three Pumps, Each Capable of Providing   **B. Four Pumps, Each Capable of Providing**
¹/₂ of Maximum Demand   **¹/₃ of Maximum Demand**

Figure D-4.   Pump-Station Configurations

Pump Type

Of the many types of pumps available, the centrifugal pump is by far the most commonly used in major water distribution systems. Centrifugal pumps are used as boosters, for raw water and high service, for backwashing filters, and sometimes as sludge pumps where sludge has a low solids content. (Advantages and disadvantages of centrifugal and other pumps are covered in detail in Module 7, Pumps and Prime Movers.)

The double-suction, horizontal split-casing centrifugal pump has generally been the pump design best suited for municipal water supply operations. A key advantage to this type of casing is that the upper half of the pump case and the entire rotating element can be serviced or changed without disconnecting any piping. At one time, end-suction pumps with single-suction impellers that had the pump and drive mounted on a common base were used. These pumps evolved into the close-coupled pump where the impeller is mounted on an extended motor shaft.

For small installations and auxiliary units, close-coupled pumps are satisfactory. In general, such pumps are easy to install, take up little space, and do not require alignment. Although less expensive in smaller sizes than double-suction pumps, caution should be used in their selection for several reasons. First, inspection of the impeller or any of the internal parts generally requires that the suction or discharge piping be disturbed. This may be complicated by the fact that piping enters and leaves the pump at right angles and in different planes, instead of in a straight line. Second, because of packing-gland location, repacking may be difficult. Finally, electric motors for these units require an extended shaft for mounting the impeller. The shaft may cause premature bearing failure.

Pump Capacity

Pump capacity for volume and total dynamic head (TDH)[1] must be selected for economic operation under present and anticipated future conditions. Pump manufacturers' representatives are generally able to give considerable guidance in this area and can supply pump characteristic curves and pump ratings for the equipment they supply. Both pump and drive performance should be known under varying conditions and guaranteed by the suppliers.

Pump characteristic curve.   Figure D-5 shows a typical head–capacity curve (also called a characteristic curve or an H-Q curve)[2], an efficiency curve (E-Q), and a power curve (P-Q) for one centrifugal pump. The horizontal scale represents the capacity in gallons per minute, and the vertical scale on the left represents the head in feet. The pump efficiency, in percent, and the brake horsepower are shown on the right vertical scale. The operation of a pump under varying conditions of head (H) and capacity (Q) can be determined from the H-Q curve. The H-Q curve in Figure D-5 shows that the pump head decreases as the flow increases. The efficiency curve indicates that the pump is most efficient at 1400 gpm.

Pump Speed

For centrifugal pumps, the speed of the edge of the impeller must be high in order to obtain high heads. Two basic methods for achieving the necessary speed are to use a larger diameter impeller and to increase impeller speed. Increasing speed usually has the lowest initial cost, but lower speeds are generally preferred on the basis of longer life for pumping-unit components. Where space is limited, it may be necessary to take advantage of higher specific speed pumps since they are generally smaller in size for a given capacity. Conversely, if high heads are necessary, high-speed and/or multistage pumps must be used.

High-speed units require more maintenance than slower-speed pumps, but because of their smaller size, replacement parts may be less costly. In most cases, a high-speed pump will prove more efficient than a low-speed pump of the same rating. Actual impeller speed can be used to determine the choice of drive unit.

---

[1] *Basic Science Concepts and Applications,* Hydraulics, Pumping Problems.

[2] *Basic Science Concepts and Applications,* Mathematics, Pump Curves.

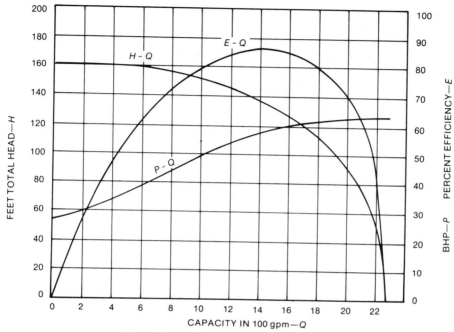

Figure D-5. Pump Performance Curves

Table D.1. Pump Comparisons Based on Speed*

High Speed (3600 rpm)	Low Speed (1800 rpm or less)
Smaller size	Larger size and/or space requirement
Lighter weight	Heavier weight to support and/or move
Less costly	Greater initial cost
Lower major repair expense	High major repair expense
Higher maintenance cost—seals, whether mechanical or packed, wear out faster	Lower maintenance cost—seals last longer
Nosier	Quieter
Vibration problems and effects more likely	Vibration effects and problems less likely due to lower frequency
Less costly foundation	More costly foundation due to larger size

*Table shows important factors to consider when weighing the use of units with high-speed drives and slower speeds for the same job. Daily inspection and maintenance costs are similar for either rpm if the pump has been properly applied.

Table D-1 summarizes several pump comparisons based on impeller speed.

Specific speed indicates the general type of pump needed. This is determined by the desired head–capacity along the approximate range of speeds of the power source. For smaller scale installations, some exceptions to these general statements may be observed, since small pumps, operating at up to 100 gpm (380 L/min) and from 50–100 ft (15–30 m) of head, will have a low specific speed with corresponding lower efficiencies.

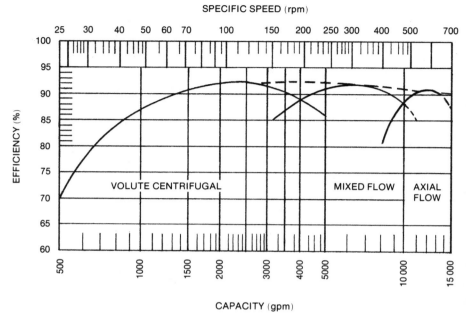

Figure D-6.   Pump Impeller Type and Specific Speed

The relationship between pump type, or impeller shape, and specific speed is shown from a different point of view in Figure D-6. Here the configuration of impeller shape and volute is shown to change from volute-centrifugal pumps on the left, to mixed-flow in the center, to axial-flow (propeller) pumps on the right.

Pump-Station Design

The pumping unit chosen for any application must fit into the available physical space. The installation must allow easy access for inspection during operation. There should also be sufficient headroom and ample space between the pumps and nearby walls for access and servicing. The pump should be located near the suction well or supply in order to minimize flow losses and in such a position that the suction and discharge piping is simplified. Where appropriate, provision should be made for an overhead crane or lifting device, with sufficient capacity to lift the heaviest part of the unit. In general, a dry, well-lit, clean room should be provided, where the temperature can be regulated.

D-2.   Prime-Mover Selection

Several types of prime movers will usually be available with the necessary power output to drive the pump selected. The type and size of prime mover selected will depend primarily on economic considerations of fuel and energy efficiency, with fuel availability and site restrictions as additional considerations.

Types of Prime Movers

The wide availability of electric power has led to the use of electrical drives for 95 percent of all water-utility pumping. In many communities, up to 7 percent of the electricity consumed is for water-supply pumping. Emergency operating concerns and increasing electrical power costs are causing many utilities to review the use of electric power. Where fuel is available, alternative gas or oil engines may be appropriate for reducing peak power demands and/or ensuring continuity of operation during power outages. An advantage of natural gas as an alternative fuel is that peak water demand occurs during the summer months, when natural gas (used for heating during the winter) is readily available. Advantages and disadvantages of several types of prime movers are discussed in detail in Module 7.

Any decision to install and operate engine-powered pumps must take into account local noise ordinances that are already in effect or that may be passed in the future. Engine units should be located where they are least likely to disturb the public, and they should be equipped with suitable mufflers or other noise-control measures to lessen their impact.

Sizing the Prime Mover

It is a good practice to have extra motor power available to meet the demands for increased volume and/or head that could become necessary in the future. Electric-motor efficiencies vary little between three-quarters and a full load. At one-half load or when overloaded, efficiency drops. Therefore, a slightly oversized motor will not use any more power than a motor operating at full load. However, it is not economical to use a motor that greatly exceeds requirements.

Another consideration in sizing the motor is speed. Pumps are usually designed to operate at speeds equivalent to full-load motor speeds. For 60-cycle operation, these are approximately 3500, 1750, and 1160 rpm. They will vary somewhat depending on the manufacturer and the size of the motor. The efficiency of a squirrel-cage type motor ranges from $1/2$- to $1 1/2$-percent higher at 1750- and 1160-rpm speeds than at higher or lower speeds. This, however, is a general rule and is not exact for all classes of motors.

Appendix E

Pump and Prime-Mover Installation

Pump and Prime-Mover Installation

This appendix gives general guidelines for pump and prime-mover installation. The equipment manufacturer should be consulted for more specific information for a given facility, and the manufacturer's representative should usually be on hand during the initial installation.

E-1. Pump Installation

Foundation and Alignment

Foundation.   The foundation of a pump-and-electric-motor unit should rest on a solid substructure. It should be sufficiently stable to absorb vibrations and to form a permanent, rigid support for the base plate, which maintains alignment. The foundation should be of sufficient area to support both the pump and driver. If the pump unit must be mounted on a structural steel frame, it should be located directly over, or as near as possible to, the main building members, beams, or walls. Base plates should be secured to the steel supports with both bolts and dowels in order to reduce distortion, prevent vibration, and retain proper alignment.

If the motor and pump are factory-mounted on a single base plate, then the unit should be placed on the foundation and carefully leveled with the aid of metal shims. Shims should also be installed to raise the base plate $1/2-3/4$ in. (13–19 mm) to allow for grouting between the base plate and foundation.

Alignment.   Proper alignment of the pump and driver is essential for trouble-free operation of a flexible coupling. Flexible couplings will compensate for slight changes in alignment caused by temperature variations that may occur during normal operation. However, couplings should not be used to correct for an improperly aligned installation. All mating surfaces on flanges and couplings should be properly cleaned before placement to avoid future operating problems.

To check the alignment of a coupling, all pins, guides, and cushions must be removed from the coupling. Figure E-1 shows two methods to use in checking coupling alignment. The first method uses a straightedge and a thickness (feeler) gauge to check the coupling for angular and parallel misalignment. These checks should be done at four positions around the coupling. It is preferable to use one half of the coupling as a base point, rotating it to the four positions 90° apart. The check is repeated using the other coupling half as the base point. The second method of testing is similar, but a dial indicator is used instead of a feeler gauge.

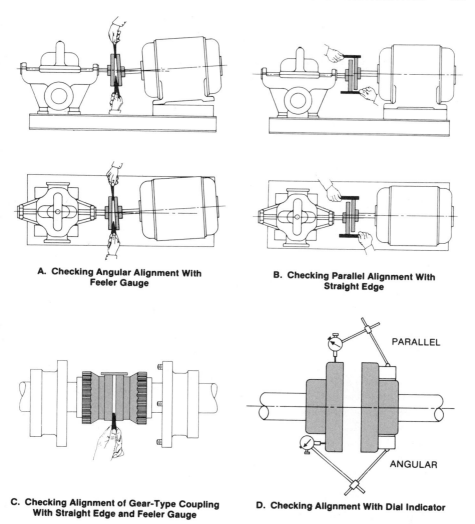

A. Checking Angular Alignment With Feeler Gauge

B. Checking Parallel Alignment With Straight Edge

C. Checking Alignment of Gear-Type Coupling With Straight Edge and Feeler Gauge

D. Checking Alignment With Dial Indicator

Courtesy of The Hydraulic Institute

Figure E-1.   Checking Coupling Alignment

On units mounted at the factory, angular and parallel misalignment can be corrected by adjusting wedges or flat shims under the base. On units not mounted by the factory, angular adjustment can be made by shimming the back motor feet (the feet opposite the coupling) in a vertical direction, or by moving the back feet in a horizontal direction, or both. Parallel adjustment can be made by moving the complete motor in a horizontal or vertical direction, or both. After each change, it is necessary to recheck alignment of the coupling halves, since adjustment in one direction may disturb adjustments already made in another direction. After the coupling is in perfect alignment, replace the pins, cushions, or guides, and tighten foundation bolts evenly, but not too firmly.

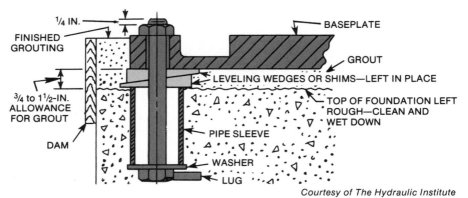

Courtesy of The Hydraulic Institute

Figure E-2.   Pump Base, Foundation, and Grout Placement

Grouting.   After the pump and driver (or base plate) are leveled, the area between the base plate and the foundations should be filled with grout, in accordance with the manufacturer's recommendations. Grouting is required to prevent lateral shifting of the base plate, not to take up irregularities in the foundation. Only nonshrink grout should be used. Cement or mortar should not be used because these materials shrink as they set.

The top of rough concrete foundations should be well saturated with water before grouting. A wooden form can be built around the outside of the base plate to contain the grout. See Figure E-2 for the placement of grout. Grout is added until the entire space under the base is filled. Grout should be poured until it reaches the required level. Grout added where the drain pocket is located should be poured to the level of the pocket, making certain that the pocket is plugged with paper or rags before pouring.

After the grout has hardened properly, and before the unit is operated, a final check should be made on alignment and direction of rotation. After running for a week or so, shaft alignment should be checked again and corrected if necessary. Both pump and motor should be doweled to the base plate at that point.

Suction and Discharge Pipes and Valves

Piping strains.   Satisfactory operation cannot be maintained if piping is not properly joined to the pump. Since pumps can be sprung and pulled out of position simply by drawing up on the bolts in the piping flanges, flanges must be brought squarely together before the bolts are tightened.

Suction and discharge pipes, along with any associated valves, should be anchored near, but independent of, the pump, so that no strain will be transmitted to the pump casing. Pipe strains can cause misalignment, hot bearings, worn couplings, and vibration. Flexible pipe connections should be used to prevent casing strain and the transmission of vibration.

Piping and valve arrangement should be properly designed and carefully installed. It is usually advisable to increase the size of both suction and discharge pipes at the pump nozzle in order to decrease the head loss due to friction. For the same reason, piping should be arranged with as few bends as possible, and those necessary should be made with a long radius whenever possible.

Suction piping.   The suction piping, in particular, should be as direct as possible. It should have a gradual rise to the pump without any high spots that could trap air, and it must be free of air leaks. Improper suction-side conditions can greatly limit pump performance. Bends in suction lines, particularly close to the pump, should be avoided. It is especially important that the joints in the suction pipe be drawn tight to prevent air leaks. Screwed or flanged pipe is recommended for the smaller sizes and flanged pipe for the larger sizes or high suction lifts. Suction lines must always pass under other interfering piping. In a new installation, dirt, pipe scale, and welding beads should be prevented from entering the pump. The suction-piping system should be thoroughly flushed before connecting it to the pump.

Valves should not be used on suction lines if they can be avoided; when used, the stem packing should be tight to prevent air leakage. If a positive head exists within the suction piping, a gate valve is required on the suction line to permit maintenance. Gate valves allow installation, removal, service, and repairs without interruption of the system. Gate valves should be locked into position during normal operation.

Discharge piping.   A check valve and a gate valve are required in the discharge line. The check valve is typically placed between the pump and gate valve, to protect the pump from any possible excessive back pressure or from reverse rotation caused by liquid running back through the casing during driver or power failure. Check valves also maintain upstream pressure when system flow ceases.

The gate valve is used in priming, starting, and shutting down the pump. In general, when an increaser is used on the discharge side to increase the size of discharge piping, the increaser should be placed between the check valve and the pump. An exception to this rule applies for some types of check valves (for example, tilting disk), which should be the same size as the pump so the valve will open fully.

The pump is equipped with a pipe tap in the bottom of the adapter for attaching a pipe to drain off the stuffing box. To give the installation a neat appearance, it is advisable to install this drain pipe. Make sure any drainage doesn't collect around the well or pump. After all the piping is completed, recheck alignment of the pump and driver.

Temperature Control

In general, the purpose of environmental control is to ensure adequate protection for equipment and operating personnel. Proper ventilation of the entire pumping station or, where indicated, of the starter and control cabinets only, will have an immediate payback in terms of reliability during periods of hot weather by reducing the likelihood of unscheduled shutdowns caused by motor and control overheating. Wherever freezing temperatures occur, provision must be made for cold-weather heating and thermostatic control. Forced-air systems usually provide for heating and cooling.

The requirements noted above for providing sufficient ventilation and heating to avoid shutdown or equipment failure are minimal. A further consideration is the need to supply clean air, relatively free of dust and corrosive fumes, to the motors to increase motor life expectancy. All of this should be considered to ensure reasonably airy and adequate working conditions for the equipment operator and maintenance or service personnel.

Pump Seals and Packing

Packing installation and adjustment.   New pumps should not need to be packed, but the packing should be checked and adjusted if necessary. Pumps that have been out of service or repaired may need repacking. New packing should be installed as shown in Figure E-3, with one or two rings of packing followed by a seal cage or lantern ring, then two or three more rings of packing. The number of rings preceding the lantern ring, if used, is determined by the location of the seal water-supply pipe or passage (Figure E-3). Rings of packing should be staggered so that the joints are 90–180° apart.

After the packing is in place, install the glands and draw them up evenly and squarely in the stuffing box. It is best to draw up the glands with a wrench to set the packing, then release the nuts about one full turn, until they can be turned with the fingers. Use only a good grade of soft packing. Hard or inferior grades will allow excessive water leakage and quickly wear grooves in the shaft sleeve. Replace all of the packing when it becomes necessary to install more than one additional ring of packing, since the seal cage should never be allowed to reach the bottom of the box.

Stuffing-box seal piping.   Generally, stuffing-box sealing liquid for potable-water pumps is piped directly from the discharge volute, with a valve for controlling the supply provided in each pipe connection. If pumped water cannot be used for sealing purposes, an independent supply of clean water should be provided at a pressure slightly higher than that of the pump suction.

Mechanical seals.   Pumps that are furnished with mechanical seals will not require any immediate attention except an inspection to make sure the hold-in bolts are tight. This will ensure that the seal is properly installed. After the pump is started and running, the seals should again be inspected to make sure that they are not leaking.

If new mechanical seals leak, turn the pump off and contact the pump manufacturer or manufacturer's representative to have the pump seal checked. Dirt under the seal face, a damaged face, or improper installation can all cause leakage, but the manufacturer's warranty should cover repairs if the pump has not been operated.

Drainage

Provision should be made to remove the water from the pump stuffing-box (seal) area. This is accomplished by connecting drain lines to the pump drainage port in the bottom of the casing. Drain piping may be directed to the nearest sloped surface of the foundation grout or directly to the drain pocket. The station or pumphouse floor drain should be adequately sized to accommodate potential momentary flows that could result from piping failures. Drainage should never be allowed to enter a well.

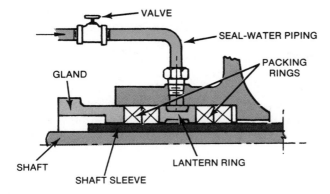

NOTE: Lantern ring must be in line with seal-water tap when packing is properly adjusted.

Courtesy of Peerless Pumps

Figure E-3.   Installation of Packing

E-2.   Prime-Mover Installation

This section covers general procedures for installation of electric motors. Similar considerations apply when installing other types of prime movers. In all cases, manufacturer's instructions should be consulted for detailed information.

Storage

When received from the supplier, a motor should be closely examined. If it is not scheduled to be installed immediately, it should be stored in a clean, dry place, preferably at a warm, constant temperature to avoid condensation of moisture.

Foundation and Alignment

It is important that the motor be installed where it will not be affected by vibration, dust, dirt (if not totally enclosed), moisture, stray oil, or grease. If the motor cannot be suitably located, provisions should be made to protect it from environmental hazards.

Motor bases or rails should be firmly bolted to a concrete platform or other suitable foundation. Standard bases usually supplied with the motor are only suitable for floor mounting; for other mountings the manufacturer should be consulted. Generally, the pump alignment, attachment, and grouting instructions (discussed in the previous section) apply to motors as well as pumps.

Correct leveling of the base is essential. Slide rails should be checked for alignment in all directions, using a straightedge and level, then firmly grouted in place for their full length. The correct alignment of the motor and driven shafts is important, since an out-of-line shaft will be subjected to deflection and eventually fail from steel fatigue. Even where flexible couplings are used, it is important that alignment be accurate, since the amount of out-of-line that can be corrected by the coupling is very limited.

Standard motor bearings will withstand only very limited end-thrust, so precautions should be taken to protect the motor bearings when end-thrust is likely to occur. When the motor is installed, the coupling should be positioned to allow for the rotor to be in its middle position when stationary.

Screen-protected or drip-proof motors require that a small space be left between the bottom of the motor and the floor or bed plate so that the ventilation of the motor is not restricted. As a general rule, the motor feet should be raised at least $1/2$ in. (13 mm) above the floor or bed plate, with either shims or slide rails.

Balance

Rotor imbalance directly contributes to vibration. Whether caused by the driven machine or transmitted from adjacent machinery, severe vibration will eventually cause motor failure. Motors are dynamically balanced during manufacture to ensure smooth running, but a badly balanced drive can cause serious vibration. It is, therefore, essential that all pulleys and couplings be balanced.

If a motor fitted with ball and roller bearings is subjected to vibration caused by adjacent machinery while the motor itself is not turning, the bearings will brinell (wear in a pattern of small dents or depressions on the bearing race) and soon fail under normal operation. The noise of a bad bearing will be especially noticeable when the motor is started up. Motor foundations should, therefore, always be arranged so that a minimum of vibration is transmitted from adjacent machinery.

Grounding

The frame of every motor must be grounded. Metallic tubing used as connecting conduit to the terminal box does not always provide a satisfactory metallic ground circuit, so a separate copper wire may also be necessary. This wire may be independently connected to the motor frame.

Insulation

Motor windings are impregnated with specially formulated oil to give maximum resistance to the effects of moisture. Nonetheless, continued exposure to moist or wet conditions will reduce the insulation resistance. In addition, windings that have become moist may tend to collect a deposit of dust from the cooling air. Windings that are heavily coated with dust will not cool properly. The way to avoid these problems is to ensure that the correct motor enclosure has been chosen and that the cooling air is cleaned where practical to reduce the presence of dust in the surrounding atmosphere. With totally enclosed motors it may be advisable to have a drain plug in the bottom of the shell to provide for the periodic removal of any moisture that has accumulated while the motor was standing in a damp atmosphere.

Before a motor is put into service, it should be examined for any foreign matter that may have become lodged in it, and its insulation resistance tested. If this is less than 1 $M\Omega$ (1 megohm), both between phases or windings and ground, then the motor should be placed in a warm, dry place until the insulation reaches this value. A motor with a low insulation resistance should not be run on full voltage since a breakdown can occur before the windings dry out.

If the motor is appreciably damp or has been subjected to water spray, then the time required for it to dry out in a warm, well-ventilated room can be excessive. In such a case, the evaporation of moisture can be speeded up by applying more heat to the windings, either by indirect heat (such as by blowing air over heaters), or by placing the motor in a steam-heated oven, or by applying a low voltage to the terminals. The temperature of the windings must not be allowed to rise higher than operating recommendations, since deterioration of the insulation occurs if a higher temperature is maintained over a prolonged period. Drying should be continued until the insulation resistance is at least 1 MΩ.

Cables

The external cables to the motor must be of sufficient cross-sectional area to carry the full-load current of the motor without excessive heating or voltage drop. Except with the smallest motors, which have straight-through connectors, cables must be soldered into the lugs. Bare cable ends should not be twisted under the lugs since they could spring loose and cause short circuits. Connections to terminals must be made with meticulous care to ensure a mechanically secure connection with low electrical resistance. Poor electrical connections will overheat.

Terminal boxes can be supplied with a variety of different glands for receiving screwed conduit or cables, depending on the motor size. In all cases, the mechanical protection for any type of cable should be connected directly to the terminal box. Where motors are mounted on slide rails to allow for drive adjustment, a short length of flexible metallic tubing can be fitted between the end of the conduit and the conduit gland on the terminal box.

Control Gear

Careful attention should be given to the location of the control gear to ensure that, although it is accessible for maintenance, it is not in the way of the machine operator or in a position where it is likely to be damaged. Emergency controls must be in a prominent position, readily accessible, and distinctively marked. The control gear should have an enclosure that will keep moisture and dirt out of switches, contactors, and relays. Total enclosure is usually recommended. Cable entries should be sealed off and covers should fit properly and be kept closed. The control-panel ventilation, if provided, should be checked periodically to ensure its adequacy and cleanliness.

Resistors must be located to allow for cooling, since considerable heat may be developed during starting and control. With equipment in which contact resistors are oil immersed, tanks should be filled to the correct level when the gear is first installed.

Fusing

In general, fuses are not a reliable means for protecting motor windings against overloads. It is difficult to grade a fuse to blow at the required overcurrent because the ambient temperature and the location of the fuse box have unpredictable effects on the operating temperature of the fuse. Also, fuses must withstand high starting currents without deterioration and must allow operation under occasional heavy loads. Fuses that will not blow under normal conditions, therefore, must be of such a high rating that they provide little protection except in the case of a short circuit on the motor wiring or starter.

For this reason it is advisable to use fuses solely for short circuit protection and install starting gear that incorporates both overload and single-phasing protection. The main cause of single-phasing is the open circuiting of one of the supply lines to the motor while it is running, as a result of the use of fuses as overload protection. A large number of burn-outs of three-phase motors are caused by single-phasing.

E-3.   Initial Start-Up Procedures

Always check the manufacturers' recommendations before initial start-up. There are several typical observations or checks that should be made before operating a pump following installation or major repair. These include:

- Turn the rotor by hand to be sure it is free.
- Check for excessive end play or lateral movement.
- Check driver rotation with the coupling unbolted.
- Make sure the coupling is properly aligned and connected.
- Open any valves in the liquid seal supply.
- Open any valves in the suction line.
- Prime the pump as required unless the water supply has sufficient head to flow into the pump and fill the casing.
- Bleed off any air from the pump casing.
- Start the driver in accordance with the manufacturer's recommendations.
- Open the discharge valve slowly after the pump is operating at full speed to avoid overloading the driver and prevent water hammer in the distribution system.

Before operating any pump, the pump casing and suction piping must be filled with water. The pump and associated piping may be filled with an ejector or vacuum pump. A wet-vacuum pump may be used in combination with a high-pressure water ejector, or exhauster equipment, to prime major units. If a foot valve is provided in the suction line, then the case and suction line can be filled from any available source.

During the first few minutes of operations, bearings and stuffing boxes should be watched closely to make sure that they do not overheat or need adjustment. It is good practice to check suction inlet and discharge pressure-gauge readings during start-up and periodically thereafter. Pump bearings should be checked for overheating at 10–15-min intervals during start-up.

Careful regulation of the valve setting is required on liquid-seal supply lines to the stuffing boxes. Stuffing box glands that have packing should be adjusted to permit a slight seepage of liquid at all times during operation to avoid excess wear on shaft sleeves. Usually 20–30 drops per minute is acceptable.

If no check valve is used, then the discharge gate valve should be closed slowly before pump shutdown to prevent water hammer in the distribution system. When a check valve is provided in the discharge line, then the pump may be shut down by stopping the drive as recommended by the manufacturer. In general, the flow control valves should be closed in the following order: discharge, suction, and sealing liquid supply, with the latter two after shutdown.

Within the first 30 days of operation, the pump and motor should be tested and complete test results recorded as a baseline for the measurement of performance in the future. Data recorded should include the following:

- Pump speed
- Fluid temperatures
- Fluid flow (gpm [L/min])
- Pump discharge head
- Pump suction (gauge at pump-shaft centerline)
- Motor data (amps, volts, power factor, kilowatts).

The standards of the Hydraulic Institute* give test procedures detailing the techniques, instruments, and calculations needed for a comprehensive evaluation. The results can be plotted and compared against the pump curve supplied by the manufacturer, to determine if the pumping unit is performing as designed.

Equipment manufacturers can supply literature to guide in the installation, operation, and maintenance of and use of the equipment, including auxiliary equipment. Installation and shop drawings should be available to maintenance personnel. Establishing a program to inspect and record operating data is the first step in planned maintenance. Every observation made on the equipment should be documented in a log book. This is not only useful from a maintenance standpoint, it is also essential for complying with equipment warranty provisions. It is important to keep a copy of any manufacturer's manuals, shop drawings, other equipment operating instructions, and related information at the point of use. The original copy of each document should be retained in a central file for record purposes.

---

* Hydraulic Institute, 712 Lakewood Center North, 14600 Detroit Avenue, Cleveland, OH 44107.

Appendix F

Pump Troubleshooting Chart

Pump Troubleshooting Chart

No Liquid Delivered

Cause	Cure
1. Lack of prime.	Fill pump and suction pipe completely with liquid.
2. Loss of prime.	Check for leaks in suction pipe joints and fittings; vent casing to remove accumulated air.
3. Suction lift too high.	If no obstruction at inlet, check for pipe friction losses. However, static lift may be too great. Measure with mercury column or vacuum gauge while pump is operating. If static lift is too high, liquid to be pumped must be raised or pump lowered.
4. Discharge head too high.	Check pipe friction losses. Large piping may correct condition. Check that valves are wide open.
5. Speed too low.	Check whether motor is directly across-the-line and receiving full voltage. Check to see if frequency is too low. Check to see if motor has an open phase.
6. Wrong direction of rotation.	Check motor rotation with directional arrow on pump casing. Check wiring for phase reversal.
7. Impeller completely plugged.	Dismantle pump and clean impeller.

Not Enough Liquid Delivered

Cause	Cure
8. Air leaks in suction piping.	Test flanges for leakage with flame or match.
9. Air leaks in stuffing box.	Increase seal lubricant pressure to above atmospheric.
10. Speed too low.	See item 5.
11. Discharge head too high.	See item 4.
12. Suction lift too high.	See item 3.
13. Impeller partially plugged.	See item 7.
14. Cavitation; insufficient net positive suction head (depending on installation).	A. Increase positive suction head on pump by lowering pump. B. Sub-cool suction piping at inlet to lower entering liquid temperature. C. Pressurize suction vessel.
15. Defective impeller.	Inspect impeller, bearings, and shaft. Replace if damaged or if vane sections are badly eroded.
16. Defective packing.	Replace packing and sleeves if badly worn.

*This chart has been excerpted from the Allis-Chalmers publication, *General Pump Instructions*.

Pump Troubleshooting Chart *(continued)*

Not Enough Liquid Delivered (cont.)

Cause	Cure
17. Foot valve too small or partially obstructed.	Area through ports of valve should be at least as large as area of suction pipe—preferably 1½ times. If strainer is used, net clear area should be 3–4 times area of suction pipe.
18. Suction inlet not immersed deep enough.	If inlet cannot be lowered, or if eddies through which air is sucked persist when it is lowered, then chain a board to suction pipe. It will be drawn into eddies, smothering the vortex.
19. Wrong direction of rotation.	Symptoms are an overloaded drive and about ⅓ rated capacity from pump. Compare rotation of motor with directional arrow on pump casing. Check wiring.
20. Too small impeller diameter. (Probable cause if none of previous causes.)	Check with manufacturer to see if a larger impeller can be used; otherwise, cut pipe losses or increase speed, or both, as needed. Be careful not to seriously overload drive.

Not Enough Pressure

Cause	Cure
21. Speed too low.	See item 5.
22. Air leaks in suction piping.	See item 8.
23. Mechanical defects.	See items 15, 16, and 17.
24. Obstruction in liquid passage.	Dismantle pump and inspect passages of impeller and casing. Remove obstruction.
25. Air or gases in liquid. (Test in laboratory, reducing pressure on liquid to pressure in suction line. Watch for bubbles.)	May be possible to overrate the pump to a point where it will provide adequate pressure despite condition. Better to provide gas separation chamber on suction line near pump and periodically exhaust accumulated gas. (See item 14.)
26. Too small impeller diameter. (Probable cause if none of previous causes.)	See item 20.

Pump Operates for Short Time, Then Stops Pumping

Cause	Cure
27. Incomplete priming.	Free pump, piping, and valves of all air. If high points in suction line prevent this, then they need to be corrected.
28. Suction lift too high.	See item 3.
29. Air leaks in suction piping.	See item 8.
30. Air leaks in stuffing box.	See item 9.
31. Air or gases in liquid.	See item 25.

(continues on next page)

Pump Troubleshooting Chart *(continued)*

Pump Requires Too Much Power

Cause	Cure
32. Head lower than rating, thereby pumping too much liquid.	Machine the impeller's OD to size advised by manufacturer.
33. Cavitation.	See item 14.
34. Mechanical defects.	See items 15, 16, and 17.
35. Suction inlet not immersed enough.	See item 18.
36. Wrong direction of rotation.	See item 6.
37. Stuffing boxes too tight.	Release gland pressure. Tighten reasonably. If sealing liquid does not flow while pump operates, replace packing. If packing is wearing too quickly, replace scored shaft sleeves and keep liquid seeping for lubrication.
38. Casing distorted by excessive strains from suction or discharge piping.	Check alignment. Examine pump for friction between impeller and casing. Replace damaged parts.
39. Shaft bent due to damage; improper shipment, operation, or overhaul.	Check deflection of rotor by turning on bearing journals. Total indicator run-out should not exceed 0.002 in. (0.05 mm) on shaft and 0.004 in. (0.10 mm) on impeller wearing surface.
40. Mechanical failure of critical pump parts.	Check bearings and impeller for damage. Any irregularity in these parts will cause a drag on the shaft.
41. Misalignment.	Realign pump and driver.
42. Speed may be too high (brake horsepower of pump varies as the cube of the speed; therefore, any increase in speed means considerable increase in power demand.)	Check voltage on motor.
43. Electrical defects.	The voltage and frequency of the electrical current may be lower than that for which the motor was built; or there may be defects in the motor. The motor may not be ventilated properly due to a poor location.
44. Mechanical defects in turbine, engine, or other type of drive exclusive of motor.	If trouble cannot be located, consult the manufacturer.

Glossary

Glossary

Words defined in the glossary are set in SMALL CAPITAL LETTERS where they are first used in the text.

Actual (cross connection), 211 Any arrangement of pipes, fittings, or devices that connects a POTABLE water supply directly to a nonpotable source at all times. Also known as a direct cross connection.

Actuator, 125 A device, usually electrically or pneumatically powered, that is used to operate valves.

Air-and-vacuum relief valve, 121 A dual-function air valve that (1) permits entrance of air into a pipe being emptied, to prevent a vacuum, and (2) allows air to escape in a pipe being filled or under pressure.

Air binding, 111 The condition in which air has collected in the high points of distribution mains, reducing the capacity of the mains.

Air gap, 222 In plumbing, the unobstructed vertical distance through the free atmosphere between the lowest opening from any pipe or outlet supplying water to a tank, plumbing fixture, or other container, and the OVERFLOW RIM of that container.

Air purging, 65 A procedure to clean mains less than 4 in. (100 mm) in diameter, in which air from a compressor is mixed with the water and flushed through the main.

Air-relief valve, 111 An air valve placed at a high point in a pipeline to release air automatically, thereby preventing air binding and pressure buildup.

Altitude-control valve, 111, 119 A valve that automatically shuts off water flow when the water level in an elevated tank reaches a preset elevation, then opens again when the pressure on the system side is less than that on the tank side.

Ambient, 273 The prevailing environmental conditions in a given area.

Ammeter, 328 An instrument for measuring AMPERES.

Ampere (Amp or A), 327 The unit of measure for electric current. One VOLT will send a current of one ampere through a resistance of one OHM.

Anaerobic, 15 The absence of air or free oxygen.

Analog, 321 Continuously variable, as applied to signals, instruments, or controls; compare DIGITAL.

Angle of repose, 81 The maximum angle or slope from the horizontal that a given loose or granular material, such as sand, can maintain without caving in or sliding; can vary considerably with changes in moisture content.

Appurtenances, 18 Auxiliary equipment, such as valves and hydrants, attached to the distribution system to enable it to function properly.

Arbor press, 261 A special tool used to force a press-fitted IMPELLER and BEARINGS off of the pump SHAFT without damaging the parts.

Armature See ROTOR.

Arterial-loop system, 3   A distribution system layout involving a complete loop of arterial mains (sometimes called trunk mains or feeders) around the area being served, with branch mains projecting inward. Such a system minimizes dead ends.

Arterial map, 348   A comprehensive map at a scale of 2000 to 3000 feet per inch, showing primary distribution mains 8 in. or larger. It is generally a supplemental mapped record that is used in system analysis.

As-built plans, 57   Plans showing how the distribution system was actually constructed, including all modifications that were made during construction.

Atmospheric vacuum breaker, 226   A mechanical device consisting of a float check valve and an air-inlet port designed to prevent BACKSIPHONAGE.

Automatic control, 336   A system in which equipment is controlled entirely by other machines or computers, without human intervention under normal conditions.

Auxiliary supply, 211   Any water source or system, other than the POTABLE water supply, that may be available in the building or premises.

Axial-flow pump, 246   A pump in which a propeller-like IMPELLER forces water out in a direction parallel to the SHAFT. Also called a propeller pump. Compare MIXED-FLOW PUMP, RADIAL-FLOW PUMP.

Backfill, 7   (1) The operation of refilling an excavation, such as a trench, after the pipeline or other structure has been placed into the excavation. (2) The material used to fill the excavation in the process of backfilling.

Backflow, 112, 210   A hydraulic condition, caused by a difference in pressures, in which nonpotable water or other fluids flow into a POTABLE water system.

Back pressure, 210, 218   A condition in which a pump, elevated tank, boiler, or other means results in a pressure greater than the supply pressure.

Backsiphonage, 210, 219   A condition in which the pressure in the distribution system is less than atmospheric pressure.

Ball valve, 119   A valve consisting of a ball resting in a cylindrical seat. A hole is bored through the ball to allow water to flow when the valve is open; when the ball is rotated 90°, the valve is closed.

Barrel, 149   The body of a fire hydrant.

Base, 153, 154   The inlet structure of a fire hydrant; it is an elbow-shaped piece that is usually constructed as a gray cast-iron casting. Also known as the shoe, inlet, elbow, or foot piece.

Base map, 350   A map used to prepare comprehensive, sectional, or other mapped records. It serves as the background for the mapped records. State agencies, such as highway departments, are a good source for base maps.

Bearing, 264   Antifriction device used to support and guide pump and motor SHAFTS.

Bedding, 38   A select type of soil used to support a pipe or other conduit in a trench.

Bellows sensor, 322   A simple, accordion-like mechanical device for sensing changes in pressure.

Bimetallic, 18   Made of two different types of metal.

Binary code, 333   A method of representing numerical values using only two signal levels—on and off (or high and low); used extensively in digital computer electronics.

Blowoff valve, 110   A valve installed in a low point or depression on a pipeline to allow drainage of the line. Also called a washout valve.

Body, 113, 118   The major part of a valve, which houses the remainder of the assembly.

Bonnet, 153   The top cover or closure on the hydrant UPPER SECTION, which is removable for the purpose of repairing or replacing the internal parts of the hydrant.

Bourdon tube, 322   A semicircular tube of eliptical cross section, used to sense pressure changes.

Braces   See TRENCH BRACES.

Breakaway hydrant, 151   A two-part, DRY-BARREL post hydrant with a coupling or other device joining the UPPER and LOWER SECTIONS. The coupling and barrel are designed to break cleanly when the hydrant is struck by a vehicle, preventing water loss and allowing easy repair.

Bronze seat ring, 153   A machined ring, mounted in the body of a hydrant or valve, against which the moving disc of the valve closes.

Brushes, 272   Graphite connectors that rub against the spinning COMMUTATOR in an electric motor or generator, connecting the ROTOR windings to the external circuit.

Bubbler tube, 324   A level-sensing device that forces a constant volume of air into the liquid whose level is being measured.

Butterfly valve, 118   A valve in which the disc rotates on a shaft as it opens or closes. In the full open position, the disc is parallel to the axis of the pipe.

Bypass, 222   (1) An arrangement of pipes, conduits, gates, or valves by which the flow may be passed around an appurtenance or treatment process. (2) In cross-connection control, any pipe arrangement that passes water around a protective device, causing it to be ineffective.

Bypass valve, 116   A small valve installed in parallel with a larger valve; it is used to equalize the pressure on both sides of the disc of the larger valve before the larger valve is opened.

Cap nut, 153   Connects a STANDARD-COMPRESSION HYDRANT valve assembly to the hydrant MAIN ROD.

Card record, 348   A card, typically 5 in. × 8 in., with a preprinted standardized format for recording descriptive, historical, and maintenance information. Card records are generally maintained for valves, hydrants, and services because it is difficult to record information of such detail on mapped records.

Carrier (of disease), 215   A human or animal who carries disease germs and can pass them to another person without getting the disease himself.

Casing, 245   The enclosure surrounding a pump IMPELLER, into which the suction and discharge ports are machined.

Cathodic protection, 16, 303   An electrical system for prevention of corrosion to metals, particularly metallic pipe.

Cavitation, 176, 257, 269   A condition that can occur when pumps are run too fast or water is forced to change direction quickly. During cavitation, a partial vacuum forms near the pipe wall or impeller blade, causing potentially rapid pitting of the metal.

Centrifugal pump, 245   A pump consisting of an IMPELLER on a rotating SHAFT enclosed by a CASING having suction and discharge connections. The spinning impeller throws water outward at high velocity, and the casing shape converts this velocity to pressure.

Channel   See TRANSMISSION CHANNEL.

Check valve, 112, 124, 212   A valve designed to open in the direction of normal flow and close with reversal of flow. An approved check valve is of substantial construction and suitable materials, is positive in closing, and permits no leakage in a direction opposite to normal flow.

Close coupled, 249, 258   A pump assembly where the IMPELLER is mounted on the SHAFT of the motor that drives the pump. Compare FRAME MOUNTED.

Closed-loop control, 336   A computerized control system that requires no operator intervention.

Combination starter, 275   A motor starter in which the safety switch (to protect the motor against short circuits) is combined with the starter.

Commutator, 272   A device that is part of the rotor of certain designs of motors and generators. The BRUSHES rub against the surface of the spinning COMMUTATOR, allowing current to be transferred between the ROTOR and the external circuits.

Compound meter, 183 A water meter consisting of two single meters of different capacities and a regulating valve that automatically diverts all or part of the flow from one meter to the other. The valve senses flow rate and shifts the flow to the meter that can most accurately measure it.

Comprehensive map, 348 A map that provides a clear picture of the entire distribution system. It usually indicates the location of water mains, fire hydrants, valves, reservoirs and tanks, pump stations, pressure zone limits, and closed valves at pressure zone limits.

Cone valve, 119 A valve in which the movable internal part is a cone-shaped rotating plug. The valve is opened by turning the plug through an angle of 90°, so fluid can pass through a port machined through it.

Continuous-feed method, 56 A method of disinfecting new or repaired mains in which chlorine is continually added to the water being used to fill the pipe, so that a constant concentration can be maintained.

Contour lines, 350 A line on a map that joins points having equal elevations.

Control equipment, 319, 335 Mechanical, electrical, and hydraulic devices used to turn on or off and to change the operating characteristics of other machines.

Control relay, 336 A device that allows low-power electrical signals to operate the on/off switch for high-power equipment.

Coordinate system, 350 Any standard system for determining the location of a particular point on the earth's surface.

Corey hydrant, 150 A type of DRY-BARREL HYDRANT in which the MAIN VALVE closes horizontally and the BARREL extends well below the connection to the pipe.

Corporation cock See CORPORATION STOP.

Corporation stop, 93, 94, 110, 170 A valve for joining a SERVICE LINE to a street water main. It cannot be operated from the surface. Also called CORPORATION COCK.

Corrosion, 12 The gradual deterioration or destruction of a substance or material by chemical action, frequently induced by electrochemical processes. The action proceeds inward from the surface.

Coupling, 265 A device that connects the pump SHAFT to the motor shaft.

Coupon, 104 In tapping, the section of the main cut out by the drilling machine.

Cross connection, 210 Any arrangement of pipes, fittings, fixtures, or devices that connects a nonpotable system to a POTABLE water system.

Crushing strength, 9 The ability of a pipe or other material to withstand external loads, such as BACKFILL and traffic.

Curb box, 110, 170 A cylinder placed around the CURB STOP and extending to the ground surface to allow access to the valve.

Curb cock See CURB STOP.

Curb stop, 110, 170 A shutoff valve attached to a water SERVICE LINE from a water main to a customer's premises, usually placed near the customer's property line. It may be operated by a VALVE KEY to start or stop flow to the water supply line. Also called CURB COCK.

Current, 327, 333 (1) The "flow rate" of electricity, measured in AMPERES. (2) In TELEMETRY, a signal whose amperage varies as the parameter being measured varies.

Current meter, 182 A device for determining flow rate by measuring the velocity of moving water; turbine meters, propeller meters, and multijet meters are common types. Compare POSITIVE-DISPLACEMENT METER.

Customer service, 376 A division of a utility that is responsible for the direct contact with customers regarding service or billing inquiries. The division is staffed with customer-service representatives.

Cut-in valve, 117 A specially designed valve used with a sleeve that allows it to be placed in an existing main.

C **value (Hazen–Williams roughness coefficient), 10**   A number used in the Hazen–Williams formula, which is used to determine flow capacities of pipelines. The *C* value depends on the condition of the inside surface of the pipe. The smoother the surface of the pipe wall, the larger the *C* value, and the greater the carrying capacity of the pipeline.

D'Arsonval meter, 327   An electrical measuring device, consisting of an indicator needle attached to a coil of wire, placed within the field of a permanent magnet. The needle moves when an electric current is passed through the coil.

Demand charge, 271   An amount charged by electrical utilities over and above the normal power rate, to take into account the extra power capacity needed to supply an electric motor's increased current demand when the motor is starting.

Detector-check meter, 186   A meter that measures daily flow but allows emergency flow to bypass the meter. Consists of a weight-loaded check valve in the main line that remains closed under normal usage and a bypass around the valve containing a POSITIVE-DISPLACEMENT METER.

Diaphragm element, 324   A mechanical SENSOR for liquid levels that uses a diaphragm and an enclosed volume of air.

Diffuser bowl, 248   The segment of a TURBINE PUMP that houses one IMPELLER stage.

Diffuser vanes, 245   Vanes installed within a pump CASING on diffuser CENTRIFUGAL PUMPS to change velocity HEAD to pressure head.

Digital, 321   Varying in precise steps, as applied to signals or instrumentation and control devices; compare ANALOG.

Direct-wire control, 337   A system for controlling equipment at a site by running wires from the equipment to the on-site control panel.

Disc, 113, 118   A circular piece of metal used in many valves as the movable element that regulates the flow of water as the valve is operated.

Dispatcher station, 337   The central location from which an operator controls equipment at one or more remote sites.

Distribution load-control center, 337   A central site with equipment for one or more operators to monitor and operate the entire distribution system.

Distribution main, 1   Any pipe in the distribution system other than a SERVICE LINE.

Distribution storage, 297   A tank or reservoir connected with the distribution system of a water supply. It is used primarily to accommodate changes in demand that occur over short periods (several hours to several days) and also to provide local storage for use during emergencies, such as a break in a main supply line or failure of a pumping plant.

Doppler effect, 186   The apparent change in frequency (pitch) of sound waves due to the relative velocity between the source of the sound waves and the observer.

Double-suction pump, 258   A CENTRIFUGAL PUMP in which the water enters from both sides of the IMPELLER. Also called a split-case pump.

Dry-barrel hydrant, 149   A hydrant with the MAIN VALVE located in the BASE. The BARREL is pressurized with water only when the main valve is opened. When the main valve is closed, the barrel drains. This type of hydrant is especially appropriate where freezing weather occurs.

Dry tap, 94, 177   A connection made to a main that is empty. Compare WET TAP.

Dry-top hydrant, 149   A DRY-BARREL HYDRANT in which the threaded end of the MAIN ROD and the revolving or OPERATING NUT is sealed from water in the BARREL when the MAIN VALVE of the hydrant is in use.

Dysentery, 214   A disease (sometimes waterborne) caused by pathogenic microorganisms and characterized by severe diarrhea with passage of mucus and blood.

Eddy hydrant   A type of DRY-BARREL HYDRANT in which the MAIN VALVE closes against pressure (downwards) and the BARREL extends slightly below the connection to the pipe. Compare STANDARD COMPRESSION HYDRANT.

Elbow, 154 See BASE (of a fire hydrant).

Electrodynamic meter, 328 A device used to measure electrical POWER (in watts or kilowatts).

Electromotive force (EMF), 327 The pressure that forces electrical current through a circuit, measured in volts.

Elevated storage, 301 In any distribution system, storage of water in a tank supported on a tower above the surface of the ground.

Elevated tank, 301 A water distribution storage tank that is raised above the ground and supported by posts or columns.

Emergency storage, 300 Storage volume reserved for catastrophic situations, such as a supply-line break or pumping-station failure.

External load, 7 Any load placed on the outside of the pipe from BACKFILL, traffic, or other sources.

Feed water, 212 Water that is added to a commercial or industrial system and subsequently used by the system, such as water that is fed to a boiler to produce steam.

Fire demand, 299 The required fire flow and the duration for which it is needed, usually expressed in gallons per minute for a certain number of hours. Also used to denote the total quantity of water needed to deliver the required fire flow for a specified number of hours.

Fire flow, 148 The rate of flow, usually measured in gallons per minute (gpm), that can be delivered from a water distribution system at a specified residual pressure for fire fighting. When delivery is to fire-department pumpers, the specified residual pressure is generally 20 psi (140 kPa).

Fire hydrant, 147 A device connected to a water main and provided with the necessary valves and OUTLET NOZZLES to which a fire hose may be attached. The primary purpose of a fire hydrant is fighting fires, but it is also used for washing down streets, filling water-tank trucks, and flushing out water mains. Sometimes called a fire plug.

Flexural strength, 9 The ability of a material to bend (flex) without breaking.

Float mechanism, 324 A simple, mechanical device to sense fluid level.

Floorstand (indicating type), 125 A device for operating a gate valve (by hand) and indicating the extent of opening.

Flow coefficient, 10 See C VALUE.

Flush hydrant, 151 A fire hydrant with the entire BARREL and head below ground elevation. The head, with OPERATING NUT and OUTLET NOZZLES, is encased in a box with a cover that is flush with the ground line. It is usually a DRY-BARREL HYDRANT.

Foot piece, 154 See BASE (of a fire hydrant).

Foot valve, 254 A check valve placed in the bottom of the suction pipe of a pump, which opens to allow water to enter the suction pipe but closes to prevent water from passing out of it at the bottom end.

Frame mounted, 258 A CENTRIFUGAL PUMP in which the pump SHAFT is connected to the motor shaft with a COUPLING. Compare CLOSE COUPLED.

Francis pump A MIXED-FLOW pump.

Full duplex, 334 Capable of sending and receiving data at the same time. Compare HALF DUPLEX, SIMPLEX.

Full voltage (motor controller), 274 An electric motor controller that uses the full line voltage to start the motor. Also called an across-the-line controller.

Galvanic cell, 18 A corrosion condition created when two different metals are connected and immersed in an electrolyte, such as water.

Galvanic corrosion, 18, 174 A form of localized corrosion caused by the connection of two different metals in an electrolyte, such as water.

Gastroenteritis, 215   An intestinal disease caused by pathogenic microorganisms, which involves inflammation of the stomach and intestines. Symptoms are diarrhea, pain, and nausea.

Gate hydrant, 150   A DRY-BARREL HYDRANT in which the MAIN VALVE is a simple GATE VALVE, similar to one side of an ordinary rubber-faced gate valve.

Gate valve, 113   A valve in which the closing element consists of a disc that slides across an opening to stop the flow of water.

Globe valve, 122   A valve having a round, ball-like shell and horizontal disc.

Goodwill, 370   An attitude of kindness, friendliness, or willingness; a good relationship, which a business enterprise should maintain with its customers.

Gooseneck, 172   A flexible coupling usually consisting of a short piece of copper pipe shaped like the letter S.

Grid system, 3   A distribution system layout in which all ends of the mains are connected to eliminate dead ends.

Ground-level tank, 301   In a distribution system, storage of water in a tank whose bottom is at or below the surface of the ground.

Half duplex, 334   Capable of sending or receiving data, but not both at the same time. Compare FULL DUPLEX, SIMPLEX.

Head (of a hydrant)   See UPPER SECTION.

Head (pressure), 152, 246   (1) A measure of the energy possessed by water at a given location in the water system, expressed in feet. Also, a measure of the pressure (force) exerted by water, expressed in feet.

Helical sensor, 322   A spiral tube used to sense pressure changes.

Hepatitis, 212   An inflammation of the liver caused by a pathogenic virus. Symptoms are jaundice (yellowing of the skin), general weakness, nausea, and presence of dark urine.

Horn, 190   See YOKE.

Hose bibb, 226   A faucet to which a hose may be attached. Also called sill cock.

Hydrant   See FIRE HYDRANT.

Hydrant map, 353   A mapped record that pinpoints the location of fire hydrants throughout the distribution system. It is generally of the PLAT-AND-LIST or intersection type.

Hydraulically applied concrete, 303   Concrete that is placed under pressure using a pneumatic gun.

Hydropneumatic system, 302   A system using an airtight tank in which air is compressed over water (separated from the air by a flexible diaphragm). The air imparts pressure to water in the tank and the attached distribution pipelines.

Hydrostatic pressure, 9   The pressure, expressed as feet of head or per unit of area (psi), exerted by water at rest (for example, in a nonflowing pipeline).

Impeller, 245   The rotating set of vanes that forces water through a pump.

Incubation period, 215   The time period that elapses between that time a person is exposed to some disease and the time that person shows the first sign or symptom of the disease.

Indexing, 352   A system in which water system maps are individually numbered for ease of reference.

Indicator, 320, 331   The part of an instrument that displays information about a system being monitored; generally either an ANALOG or DIGITAL display.

Inlet, 154   See BASE (of a fire hydrant).

In-plant piping system, 1   The network of pipes in a particular facility, such as a water treatment plant, that carry the water or wastes for that facility.

Inrush current, 271   See LOCKED-ROTOR CURRENT.

Inserting valve, 117   A shutoff valve that can be inserted by special apparatus into a pipeline while the line is in service under pressure.

Instrument, 320   Any device used to measure and display or record the conditions and changes in a system.

Instrumentation, 319   See INSTRUMENT.

Interference fit, 261   A method of joining the pump IMPELLER to the SHAFT by warming the impeller, then allowing it to cool and shrink around the shaft to provide a tight fit. Also called a shrink fit. (A slot and KEY are generally used to prevent rotational slippage of the installed impeller.)

Internal backflow, 247   In a pump, leakage around the IMPELLER from the discharge to the suction side.

Internal load, 9   The load or force exerted by the water pressure on the inside of the pipe.

Intersection method, 353   A method of preparing valve and hydrant maps. Maps of intersections are drawn on a very large scale permitting valves, hydrants, and mains to be drawn to scale. Each intersection is then given a designation for easy reference.

Iowa hydrant   See COREY HYDRANT.

Isolation valve, 110   A valve installed in a pipeline to shut off flow in a portion of the pipe, for the purpose of inspection or repair. Such valves are usually installed in the main lines.

Jet pump, 251   A device that pumps fluid by converting the energy of a high-pressure fluid into that of a high-velocity fluid.

Key, 261   A small, rectangular piece of metal used to prevent a pump IMPELLER or COUPLING from rotating relative to the SHAFT.

Kinetic pump, 245   See VELOCITY PUMP.

Lantern ring, 262   A perforated ring placed around the pump SHAFT in the STUFFING BOX. Water from the pump discharge is piped to the lantern ring. The water forms a liquid seal around the shaft and lubricates the PACKING.

Lateral, 93   Smaller-diameter pipe that conveys water from the mains to points of use.

Leak frequency map, 359   A map showing different types of leaks throughout the distribution system. Differently colored push pins are generally used to identify different leaks on a print of a comprehensive map.

Leak survey map, 348   A modified sectional or valve map showing valves to be closed and areas to be isolated during a leak survey.

Line-of-sight, 332   An open air space, unbroken by buildings or other obstructions that would prevent high-frequency radio waves (and visible light) from passing between a TRANSMITTER and a RECEIVER.

List, 355   The text portion of the PLAT AND LIST map, which provides information as to the size, make, direction to open, number of turns to open, date installed, date tested, and reference measurements for numbered valves and hydrants.

Local manual control, 335   A simple system of controlling equipment by direct human operation of switches and levers mounted on the equipment being operated.

Locked-rotor current, 271   The current drawn by the motor the instant the power is turned on while the ROTOR is still at rest. Also called the motor starting current or inrush current.

Loessial, 81   Referring to loess, which is a fine-grained fertile soil deposited mainly by the wind.

Loop system   See ARTERIAL LOOP SYSTEM.

Lower barrel, 153   The section of a hydrant that carries the water flow between the BASE and the UPPER SECTION. It is usually buried in the ground with the connection to the upper SECTION approximately 2 in. (50 mm) above ground line.

Lower main rod, 153   The lower part of a STANDARD COMPRESSION HYDRANT rod, which attaches to the MAIN VALVE ASSEMBLY and is equipped with a spring to ensure positive closure.

Lower section, 153   The part of a DRY-BARREL HYDRANT that includes the LOWER BARREL, the MAIN VALVE ASSEMBLY, and the BASE.

Lower valve plate, 153   The portion of the MAIN-VALVE ASSEMBLY in a STANDARD COMPRESSION HYDRANT that connects the valve to the LOWER MAIN ROD.

Magnetic meter, 185, 325   A flow-measuring device in which the movement of water induces an electrical current proportional to the rate of flow.

Main rod, 149   A rod, made of two sections, that connects the STANDARD COMPRESSION HYDRANT valve to the OPERATING NUT.

Main valve, 149   In a DRY-BARREL HYDRANT, the valve in the hydrant's base that is used to pressurize the hydrant BARREL, allowing water to flow from any open OUTLET NOZZLE.

Main valve assembly, 153   A STANDARD COMPRESSION HYDRANT subassembly including LOWER MAIN ROD, UPPER VALVE PLATE, RESILIENT HYDRANT VALVE, LOWER VALVE PLATE, CAP NUT, and BRONZE SEAT RING. This subassembly screws into a bronze subseat or directly into threads cut into the BASE.

Measuring chamber, 179   A chamber of known size in a POSITIVE-DISPLACEMENT METER, used to determine the amount of water flowing through the meter.

Mechanical joint, 42   A type of joint for ductile-iron pipe, using bolts, flanges, and a special gasket.

Mechanical seal, 262   A seal placed on the pump SHAFT to prevent water from leaking from the pump along the shaft; the seal also prevents air from entering the pump. Mechanical seals are an alternative to PACKING RINGS.

Media relations, 377   The methods and activities used to promote an informed public and a favorable image of an organization through good relations with print and broadcast media.

Meter box, 191   A pit-like enclosure that protects water meters installed outside of buildings and allows access for reading the meters.

Meter pit, 191   See METER BOX.

Mixed-flow pump, 246   A pump that moves water partly by centrifugal force and partly by the lift of vanes on the liquid. With this type of pump, the flow enters the IMPELLER axially and leaves axially and radially. Compare AXIAL-FLOW PUMP, MIXED-FLOW PUMP.

Mode, 336   Any of several logical schemes for determining the interaction between instruments and controls.

Motor-starting current, 271   See LOCKED-ROTOR CURRENT.

Multijet meter, 182   A type of CURRENT METER in which a vertically mounted turbine wheel is spun by jets of water from several ports around the wheel.

Multiplexing, 334   Using a single wire or channel to carry the information for several instruments or controls.

Multiplier, 194   A number noted on the meter face, such as 10^x or 100^x. The reading from the meter must be multiplied by that number to obtain the correct volume of water.

Nonrising-stem valve, 114   A GATE VALVE in which the valve stem does not move up and down as it is rotated.

Nozzle section, 152   See UPPER SECTION (of a fire hydrant).

Nutating-disc meter, 180   A type of POSITIVE-DISPLACEMENT METER that uses a hard rubber disc that wobbles (rotates) in proportion to the volume of water flowing through the meter. Also called a wobble meter.

Ohm, 327   A measure of the ability of a path to resist or impede the flow of electric current. One VOLT will send a current of one AMPERE through a resistance of one ohm.

Ohmmeter, 328   An instrument for measuring the resistance of a circuit (in ohms); usually combined with a VOLTMETER in test equipment.

On-off differential control, 336   A mode of controlling equipment in which equipment is turned fully on when a measured parameter reaches a preset value, then turned fully off when it returns to another preset value.

Open-loop control, 336   A computer-based control system that advises the operator of system status, then sequences the operation of pumps and valves in response to the operator's commands.

Operating nut, 113, 149   A nut, usually pentagonal or square, rotated with a wrench to open or close a valve or fire-hydrant valve. The nut may be a single component, or it may be combined with a weather shield.

Operating storage, 300   A tank supplying a given area and capable of storing water during hours of small demand for use when demands exceed the capacity of the pumps delivering water to the district.

Orifice meter, 185   A type of flow meter consisting of a section of pipe blocked by a disc pierced with a small hole or orifice. The entire flow passes through the orifice, creating a pressure drop proportional to the flow rate.

Outlet nozzle, 153   A threaded bronze outlet on the UPPER SECTION of a fire hydrant, providing a point of hookup for hose lines or suction hose from hydrant to pumper truck.

Outlet-nozzle cap, 153   Cast-iron covers that screw onto OUTLET NOZZLES, protecting them from damage and unauthorized use.

Outlying station, 337   A remote location of unattended distribution system equipment, such as pumps and valves, controlled by the central station.

Overflow level, 222   The maximum height that water or liquid will rise in a receptacle before it flows over the OVERFLOW RIM. Also known as flood level.

Overflow rim, 212   The top edge of an open receptacle over which water will flow. Also known as flood rim.

Packing, 113   Rings of graphite impregnated cotton, flax, or synthetic materials, used to control leakage along a valve stem or a pump SHAFT.

Packing gland, 113, 262   A follower ring that compresses the PACKING in the STUFFING BOX.

Packing ring, 262   See PACKING.

Parameter, 320   A measurable physical characteristic, such as pressure, water level, or voltage.

Peak hour demand, 299   The greatest volume of water in an hour that must be supplied by a water system during any particular time period, such as a year, to meet customer demand.

Pig, 66   Bullet-shaped polyurethane foam plug, often with a tough, abrasive external coating, used to clean pipelines. It is forced through the pipeline by water pressure.

Pilot valve, 135   The control mechanism on an automatic altitude or pressure-regulating valve.

Piston meter, 179   A water meter of the POSITIVE-DISPLACEMENT type, generally used for pipeline sizes of 3 in. or less, in which the flow is registered by the action of an oscillating piston.

Pitometer, 185   A device operating on the principle of PITOT TUBE, principally used for determining velocity of flowing fluids at various points in a water distribution system.

Pitot gauge (Pitot tube), 60, 160   A device for measuring the velocity of flowing water by using the velocity head of the stream as an index of velocity. It consists of a small tube pointed upstream, connected to a gauge on which the velocity head may be measured.

Plan and profile drawings, 348   Engineering drawings showing depth of pipe, pipe location (both horizontal and vertical displacements), and the distance from a reference point.

Plat, 351   A map showing street names, mains, main sizes, numbered valves, and numbered hydrants for the PLAT AND LIST METHOD of setting up valve and hydrant maps.

Plat and list method, 353   A method of preparing valve and hydrant maps. Plat is the map portion, showing mains, valves, and hydrants. List is the text portion, which provides appropriate information for items on the plat.

Plug valve, 119   A valve in which the movable element is a cylindrical or conical plug.

Pneumatic, 320   Operated by air pressure.

Polling, 334   A technique of monitoring several instruments over a single communications CHANNEL with a RECEIVER that periodically asks each instrument to send current status.

Positive-displacement meter, 179   A meter that measures the quantity of flow by recording the number of times a known volume is filled and emptied. The meter is primarily used for low flows. There are two common styles, the disc type and the piston type.

Positive-displacement pump, 245, 253   A pump that delivers a precise volume of liquid for each stroke of the piston or rotation of the SHAFT.

Post hydrant, 151   A fire hydrant with an UPPER SECTION that extends at least 24 in. (600 mm) above the ground.

Potable, vii   The characteristic that describes water that does not contain objectional pollution, contamination, minerals, or infective agents and is considered satisfactory for domestic consumption.

Potential (cross connection), 211   Any arrangement of pipes, fittings, or devices that indirectly connects a POTABLE water supply to a nonpotable source. This connection may not be present at all times, but it is always there potentially. Also known as an indirect cross connection.

Power, 327   A measure of the amount of work done by an electrical circuit, expressed in watts.

Power factor, 272, 279   The ratio of useful (real) POWER to the apparent power in an alternating current (AC) circuit.

Power relay, 336   See CONTROL RELAY.

Pressure differential, 325   The difference in pressure between two points in a hydraulic device or system.

Pressure-reducing valve, 111   A valve with a horizontal disc for reducing water pressures in a main automatically to a preset value.

Pressure-relief valve, 111, 121   A valve that opens automatically when the water pressure reaches a preset limit, to relieve the stress on a pipeline.

Pressure vacuum breaker, 227   A device designed to prevent BACKSIPHONAGE, consisting of one or two independently operating, spring-loaded check valves and an independently operating, spring-loaded air-inlet valve.

Pressure zone map, 348   A map showing zones of equal pressure. Pressure zone maps are sometimes called water-gradient contour maps and may be prepared on a print of a COMPREHENSIVE MAP or maintained as a supplemental mapped record.

Prestressed concrete, 303   Reinforced concrete placed in compression by a highly stressed, closely spaced, helically wound wire. The prestressing permits the concrete to withstand tension forces.

Prime mover, 243, 270   A source of power, such as an internal combustion engine or an electric motor, designed to supply force and motion to drive machinery, such as a pump.

Priming, 254   The action of starting the flow in a pump or siphon. With a CENTRIFUGAL PUMP, this involves filling the pump casing and suction pipe with water.

Propeller meter, 182, 325   A meter for measuring flow rate by measuring the speed at which a propeller spins as an indication of the velocity at which the water is moving through a conduit of known cross-sectional area.

Proportional control, 336 A mode of automatic control in which a valve or motor is activated slightly to respond to small variations in the system, but activated at a greater rate to respond to larger variations.

Proportional meter, 183 Any flow meter that diverts a small portion of the main flow and measures the flow rate of that portion as an indication of the rate of the main flow. The rate of the diverted flow is proportional to the rate of the main flow.

Public information, 376 Information that is disseminated through various communication media to attract public notice or to educate the public at large.

Public relations, 369 The methods and activities employed to promote a favorable relationship with the public.

Pulse-duration modulation (PDM), 333 An ANALOG type of TELEMETRY signaling protocol in which the time that a signal pulse remains on varies with the value of the parameter being measured.

Pumper outlet nozzle, 153 A large fire-hydrant outlet, usually 4½ in. (114 mm) in diameter, used to supply the suction hose for fire-department pumpers. Sometimes called a steamer-outlet nozzle because it was originally used to supply steam-driven fire engines.

Push-on joint, 42 The joint commonly used for ductile-iron, asbestos–cement, and PVC piping systems; one side of the joint has a bell with a specially designed recess to accept a rubber ring gasket; the other has a beveled-end spigot.

Racking, 274 A condition where a pump is subjected to frequent start–stop operations due to pressure surges affecting the pump controller. The condition can also result from a malfunctioning controller.

Radial-flow pump, 246 A pump that moves water by centrifugal force, spinning the water radially outward from the center of the impeller. Compare AXIAL-FLOW PUMP, MIXED-FLOW PUMP.

Receiver, 320, 331 (1) The part of a meter that converts the signal from the SENSOR into a form that can be read by the operator; also called RECEIVER/INDICATOR. (2) In a TELEMETRY system, the device that converts the signal from the TRANSMISSION CHANNEL into a form that the INDICATOR can respond to.

Receiver/indicator An instrument component that combines the features of a RECEIVER and an INDICATOR.

Reciprocating pump, 253 A type of POSITIVE-DISPLACEMENT PUMP consisting of a closed cylinder containing a piston or plunger to draw liquid into the cylinder through an inlet valve and force it out through an outlet valve. When the piston acts on the liquid in one end of the cylinder, the pump is termed single-action; when it acts in both ends, it is termed double-action.

Recorder, 320 Any instrument that makes a permanent record of the signal being monitored, such as a chart recorder or a magnetic tape or disc drive.

Reduced pressure backflow preventer (RPBP), 222 A mechanical device consisting of two independently operating, spring-loaded check valves with a reduced pressure zone between the check valves; it is designed to protect against both BACK PRESSURE and BACKSIPHONAGE.

Reduced-pressure-zone backflow preventer (RPZ), 222 Another term for REDUCED PRESSURE BACKFLOW PREVENTER.

Reduced voltage (motor controller), 274 An electric motor controller that uses less than the line voltage to start the motor. This type of controller is used when full line voltage may damage the electrical system.

Register, 180 That part of the meter that displays the volume of water that has flowed through a meter. Meter registers are generally either of the straight or circular type.

Regular operating delivery pressure, 9 The water pressure in the distribution system during normal operation.

Relay, 335   An electrical device in which an input signal, usually of low power, is used to operate a switch that controls another circuit, often of higher power.

Relay station, 332   A combination RECEIVER and TRANSMITTER, placed between the originator of a TELEMETRY transmission and the final receiver, that boosts the power of the signal so it can travel a greater distance.

Remote manual control, 335   A system in which personnel in a central location manually operate the switches and levers to control equipment at a distant site.

Remote-meter reading system, 194   A register that is installed and can be read at a location some distance from where the meter is located.

Reservoir, 299, 301   (1) Any tank or basin used for the storage of water. (2) A ground-level storage tank whose diameter is greater than its height.

Residual pressure, 160   The pressure remaining in the mains of a water distribution system when a specified rate of flow, such as that needed for fire-fighting purposes, is being withdrawn from the system.

Resilient   Able to return to its original shape after being compressed, a common characteristic of many synthetic rubbers and plastics.

Resilient hydrant valve, 153   A fire-hydrant valve made of RESILIENT materials to ensure effective shutoff.

Resilient-seated gate valve, 118   A gate valve with a disc that has a RESILIENT material attached to it, to allow a leak-tight shutoff at high pressures.

Resistance, 327   A characteristic of an electrical circuit that tends to restrict the flow of current, similar to friction in a pipeline; measured in ohms.

Riser, 308   The vertical supply pipe to an elevated tank.

Rotary pump, 253   A type of POSITIVE-DISPLACEMENT PUMP consisting of elements resembling gears that rotate in a close-fitting pump case. The rotation of these elements alternately draws in and discharges the water being pumped. Such pumps act with neither suction or discharge valves, operate at almost any speed, and do not depend on centrifugal forces to lift the water.

Rotor, 270   The rotating part of an electric generator or motor.

Roughness coefficient, 10   See C VALUE.

RPBP, 223   See REDUCED PRESSURE BACKFLOW PREVENTER.

RPZ, 222   See REDUCED-PRESSURE-ZONE BACKFLOW PREVENTER.

Rubber seat, 118   A valve seat made of rubber.

Saddle, 94, 177   A device attached around a main to hold the CORPORATION STOP. Used with mains with thinner walls to prevent leakage. Also called a service clamp.

Salmonellosis, 215   A disease caused by pathogenic bacteria, which affects the intestinal tract; the primary symptom is severe diarrhea.

Scanning, 334   A technique of checking the value of each of several instruments, one after another; used to monitor more than one instrument over a single CHANNEL.

Seat, 113   The portion of a valve that the disc compresses against to achieve shutoff of the water.

Sectional map, 348   A map that provides a detailed picture of a portion (section) of the distribution system. It reveals the location and valving of existing mains, the location of various fire hydrants, and the location of active service lines.

Semiautomatic control, 336   A system of controlling equipment in which many actions are taken automatically but some situations require human intervention.

Sensor, 320, 321   The part of an instrument that responds directly to changes in whatever is being measured, then sends a signal to the RECEIVER and INDICATOR or RECORDER.

Service, 1   See SERVICE LINE.

Service clamp, 94, 177   See SADDLE.

Service connection, 12   That portion of the SERVICE LINE from the utility's water main to the CURB STOP at or adjacent to the street line or the customer's property line. It includes the curb stop and any other valves, fittings, etc., that the utility may require at or between the main and the curb stop, but does not include the curb box.

Service line, 93, 169   The pipe (and all appurtenances) that runs between the utility's water main and the customer's place of use, including fire lines.

Service valve, 110   A valve, such as a CORPORATION STOP or CURB STOP, that is used to shut off water to individual customers.

Shaft, 118, 248, 261   The BEARING-supported rod in a pump, turned by the motor, on which the IMPELLER is mounted.

Shaft bearing (butterfly valve), 118   Corrosion resistant bearings that fit around the shaft on a butterfly valve to reduce friction when the shaft turns.

Shaft (butterfly valve), 118, 248, 261   The portion of a BUTTERFLY VALVE attached to the disc and a valve actuator. The shaft opens and closes the disc as the actuator is operated.

Shielding, 81   A method to protect workers against cave-ins through the use of a steel box open at the top, bottom, and ends, which allows the workers to work inside the box.

Shoe, 154   See BASE (of a fire hydrant).

Shoring, 82   A framework of wood and/or metal constructed against the walls of a trench to prevent cave-ins.

Silt stop, 308   A device placed at the outlet of water storage tanks to prevent silt or sediment from reaching the consumer.

Simplex, 334   A TELEMETRY or data transmission system that can move data through a single CHANNEL in only one direction. Compare HALF DUPLEX, FULL DUPLEX.

Single-phase (power), 271   Alternating current (AC) power in which the current flow reaches a peak in each direction only once per cycle.

Single-phasing, 283   A phase-imbalance condition that occurs in a three-phase electrical circuit when power to one leg of the circuit is interrupted.

Single-suction pump, 258   A CENTRIFUGAL PUMP in which the water enters from only one side of the impeller. Also called end-suction pump.

Siphon, 216   A bent tube or pipe that uses atmospheric pressure on the surface of a liquid to carry the liquid out over the top edge of a container. One end of the tube is placed in the liquid and the other end is placed outside the container at a point below the surface level of the liquid. The tube must be filled with the liquid (primed) before flow will start.

Slip, 246   (1) In a pump, the percentage of water taken into the suction end that is not discharged because of clearances in the moving unit. (2) In a motor, the difference between the speed of the rotating magnetic field produced by the STATOR and the speed of the ROTOR.

Sloping, 80   A method of preventing cave-ins by excavating the sides of the trench at an angle (the ANGLE OF REPOSE) so the sides will be stable.

Slug method, 57   A method of disinfecting new or repaired water mains in which a high dosage of chlorine is added to a portion of the water used to fill the pipe. This slug of water is allowed to pass through the entire length of pipe being disinfected.

Sluice gate, 124   A single, movable gate mounted in a frame, used in open channels or conduits to regulate flow.

Smoothness coefficient, 10   See C VALUE.

Solenoid, 335   An electrical device that consists of a coil of wire wrapped around a movable iron core; when a current is passed through the coil, the core moves, activating mechanical levers or switches.

Solvent, 210   A liquid that can dissolve other substances. Water is an excellent solvent.

Sonic meter, 186   A meter that sends sound pulses alternately in opposite diagonal directions across the pipe; the difference between the frequency of the sound signal traveling with the flow of water and the signal against the flow of water is an accurate indication of water's velocity.

Split tee fitting, 96   A special sleeve that is bolted around a main to allow a WET TAP to be made. Also called a TAPPING SLEEVE.

Spring line, 39   The horizontal centerline of a pipe.

Squirrel-cage induction motor, 272   The most common type of induction electric motor. The ROTOR consists of a series of aluminum or copper bars parallel to the shaft, resembling a squirrel cage. Also know as a split-phase motor.

SSU (Saybolt standard units), 267   A standard measure of the viscosity of oil and grease used for lubricating bearings.

Stability, 59   A measure of a water's tendency to corrode pipes or deposit scale in pipes.

Standard compression hydrant, 150   A type of DRY-BARREL HYDRANT in which the MAIN VALVE closes upwards with the water pressure, for a positive seal.

Standard operating procedures (SOPs), 374   Written procedures that set forth policy in the form of routine practices; commonly referred to as SOPs, meaning the accepted method for dealing with a situation involving policy.

Standpipe, 301   A ground-level water storage tank whose height is greater than its diameter.

Starter, 274   A motor-control device that uses a small push-button switch to activate a CONTROL RELAY, which sends electrical current to the motor.

Statistical record, 349   A summary record; for example, a record of the number of valves and hydrants installed, the mileage of water main laid, the amount of service pipe installed, or the number of meters installed.

Stator, 270   The stationary member of an electric generator or motor.

Steamer outlet nozzle, 153   See PUMPER-OUTLET NOZZLE (of a fire hydrant).

Steam turbine, 284   A PRIME MOVER in which the pressure or motion of steam against vanes is used for the generation of mechanical power.

Strain gauge, 322   A type of pressure sensor that is commonly used in modern instrumentation systems, consisting of a thin, flexible sheet with imbedded electrical conducting elements.

Stringer, 83   The horizontal member of a SHORING system, running parallel to the trench, to which the TRENCH BRACES are attached. The STRINGERS hold the uprights against the soil.

Stuffing box, 113, 262   A portion of the pump CASING through which the SHAFT extends and in which PACKING or a MECHANICAL SEAL is placed to prevent leakage.

Submersible pump, 249   A VERTICAL-TURBINE PUMP with the motor placed below the IMPELLERS. The motor is designed to be submersed in water.

Suction lift, 262   The condition existing when the source of water supply is below the centerline of the pump.

Superimposed load, 7   See EXTERNAL LOAD.

Supervisory control equipment, 337   Centrally located controls that transmit commands over a telemetry system to remote equipment.

Surge arrester, 276, 293   An electrical protective device installed to protect electric motors from high voltage surges in the power lines. Also called surge suppressor.

Surge pressure, 9   A momentary increase of water pressure in a pipeline due to a sudden change in water velocity or direction of flow.

Surge tank, 222   In cross-connection control, the receiving, nonpressure storage vessel immediately downstream of an air gap. Nonpotable sources can be connected to the tank without threatening the POTABLE supply.

Swab, 65   Polyurethane foam plug, similar to a PIG but more flexible and less durable.

Synchronous motor, 272   An electric motor in which the ROTOR turns at the same speed as the rotating magnetic field produced by the STATOR; this motor has no SLIP.

Synchronous speed, 270   The rotational speed, in revolutions per minute, of the magnetic field produced by an electric motor's STATOR.

Tablet method, 57   A method of disinfecting new or repaired water mains in which calcium hypochlorite tablets are placed in a section of pipe. As the water fills the pipe, the tablets dissolve, producing a chlorine concentration in the water.

Tachometer generator, 326   A sensor for measuring the rotational speed of a shaft.

Tank, 301   A structure used in a water system to contain large volumes of water or other liquids.

Tank farm, 151   A large facility for the storage of chemical or petroleum products in aboveground storage tanks.

Tapping, 93   The process of connecting LATERALS and SERVICE LINES to mains and/or other laterals.

Tapping sleeve, 96   See SPLIT TEE FITTING.

Tapping valve, 96, 117   A special shut-off valve used with a tapping sleeve.

Telemetry, 320, 331   A system of sending data over long distances, consisting of a TRANSMITTER, a TRANSMISSION CHANNEL (wire, radio, or microwave), and a RECEIVER. Used for remote instrumentation and control.

Tensile strength, 9   The ability of pipe or other material to resist breakage when it is pulled lengthwise.

Thermister, 325   A semiconductor type of SENSOR that measures temperature.

Thermocouple, 325   A SENSOR, made of two wires of dissimilar metals, that measures temperature.

Thickness class, 24   The standard wall thicknesses in which ductile-iron pipe is available.

Thin-plate orifice meter, 185   See ORIFICE METER.

Three-phase (power), 271   Alternating current (AC) power in which the current flow reaches three peaks in each direction during each cycle.

Throttling, 255   Opening or closing a valve to control the rate of flow. (Usually used to describe closing the valve).

Thrust, 45   (1) A force resulting from water under pressure and in motion. Thrust pushes against fittings, valves, and hydrants; it can cause couplings to leak or to pull apart entirely. (2) In general, any pushing force.

Thrust anchor, 47   A block of concrete, often a roughly shaped cube, cast in place below a fitting to be anchored against vertical THRUST, and tied to the fitting with anchor rods.

Thrust bearing, 118   A BEARING designed to resist THRUST in line with a turning shaft (as well as the usual force at right angles to the shaft). Often a cone-shaped roller bearing, but specially designed ball bearings are also used.

Thrust block, 47   A mass of concrete, cast in place between a fitting to be anchored against thrust and the undisturbed soil at the side or bottom of the pipe trench.

Till, 35   A type of soil consisting of a mix of clay, sand, and gravel.

Tone-frequency multiplexing, 334   A method of sending several signals simultaneously over a single CHANNEL by converting the signals into sounds (tones) and assigning a specific tone to each signal.

Totalizer, 320   A device for indicating the total quantity of flow through a flow meter. Also called integrator.

Traffic hydrant, 151   See BREAKAWAY HYDRANT.

Traffic load, 7   The load placed on a buried pipe by the traffic traveling over it.

Transducer, 320, 328   See TRANSMITTER.

Transmission channel, 320, 332   In a TELEMETRY system, the wire, radio wave, fiber-optic line, or microwave beam that carries the data from the TRANSMITTER to the RECEIVER.

Transmission line, 1   The pipeline or aqueduct used for water transmission (that is, movement of water from the source to the treatment plant and from the plant to the distribution system).

Transmitter, 320, 328   In TELEMETRY or remote instrumentation, the device that converts the signal generated by the SENSOR into a signal that can be sent to the RECEIVER/INDICATOR over the TRANSMISSION CHANNEL.

Travel-stop nut, 153   A nut used in DRY-BARREL HYDRANTS that is screwed on the threaded section of the MAIN ROD. It bottoms at the base of the packing plate, or revolving nut, and terminates downward travel (opening) of the hydrant valve.

Tree system, 3   A distribution system layout that centers around a single arterial main, which decreases in size with length. Branches are taken off at right angles with subbranches from each branch.

Trench brace, 83   The horizontal member of a SHORING system that runs across a trench, attached to the STRINGERS.

Tuberculation, 13   The growth of nodules (TUBERCULES) on the pipe interior, which reduces the inside diameter and increases the pipe roughness.

Tubercules, 11   Knobs of rust formed on the interior of cast-iron pipes by the corrosion process.

Turbine meter, 182   A meter for measuring flow rates by measuring the speed at which a turbine spins in water, indicating the velocity at which the water is moving through a conduit of known cross-sectional area.

Turbine pump, 248   (1) A centrifugal pump in which fixed guide vanes (diffusers) partially convert the velocity energy of the water into pressure head as the water leaves the impeller. (2) A regenerative turbine pump.

Unaccounted-for water, 69   Water that is pumped from the treatment plant into the distribution system but which is not delivered to the consumers or otherwise accounted for.

Underregistration, 181   A condition in which a meter records less water than is actually flowing through the meter.

Upper barrel, 152   See UPPER SECTION.

Upper section, 152   The upper part of the main hydrant assembly, including the OUTLET NOZZLES and OUTLET-NOZZLE CAPS. The upper section is usually constructed of gray-iron casting. Also known as upper barrel, nozzle section, or head.

Upper valve plate, 153   A portion of the main-valve assembly of a STANDARD COMPRESSION HYDRANT that closes against the seat.

Upright, 82   The vertical member of a SHORING system, which is placed against the trench wall.

Vacuum, 219   Any absolute pressure that is less than atmospheric (less than 14.7 psi [101 kPa] at sea level).

Vacuum breaker, 226   A mechanical device that allows air into the piping system, thereby preventing BACKFLOW that could otherwise be caused by the siphoning action created by a partial vacuum.

Valve, 109   A mechanical device installed in a pipeline to control the amount and direction of water flow.

Valve and hydrant map, 348   A mapped record that pinpoints the location of valves throughout the distribution system; it is generally of PLAT AND LIST or INTERSECTION type.

Valve box, 130   A metal or concrete box or vault set over a valve stem at ground surface to allow access to the stem so the valve can be opened and closed. A cover for the box is usually provided at the surface to keep out dirt and debris.

Valve key, 110   A metal wrench with a socket to fit a valve OPERATING NUT; the key is inserted into the valve box to operate the valve.

Valve stem, 113   The rod by means of which a valve is opened or closed.

Variable frequency, 333   A type of TELEMETRY signal in which the frequency of the signal varies as the parameter being monitored varies.

Vault, 131   An underground structure, normally made of concrete, that houses valves and other appurtenances.

Velocity meter, 182   A water meter using a spinning rotor with vanes (like a propeller) and operating on the principle that speed of the vanes accurately reflects velocity of the flowing water.

Velocity pumps, 245   The general class of pumps that use a rapidly turning IMPELLER to impart kinetic energy or velocity to fluids. The pump casing then converts this velocity head, in part, to pressure head. Also known as kinetic pumps.

Venturi, 251   An hourglass-shaped device based on the hydraulic principle that states that as the velocity of fluid flow increases, the pressure decreases. This device is used in a multitude of ways to measure flow, feed chemicals, and pump water.

Venturi meter, 184, 325   A pressure-differential meter for measuring flow of water or other fluids through closed conduits or pipes, consisting of a venturi tube and a flow-registering device. The difference in velocity head between the entrance and the contracted throat of the tube is an indication of the rate of flow.

Venturi tube   See VENTURI.

Vertical turbine pump, 248   A CENTRIFUGAL PUMP, commonly of the multistage, DIFFUSER type, in which the pump shaft is mounted vertically.

Volt, 327   The practical unit of electrical potential (electrical pressure). One volt will send a current of one AMPERE through a resistance of one OHM.

Voltage, 327, 333   (1) See VOLT. (2) In TELEMETRY, a type of signal in which the electromotive force (measured in volts) varies as the parameter being measured varies.

Voltmeter, 328   An instrument for measuring ELECTROMOTIVE FORCE (electrical pressure), measured in volts.

Volute, 245   The expanding section of pump casing (in a volute CENTRIFUGAL PUMP), which converts velocity HEAD to pressure HEAD.

Warm-climate hydrant, 151   A FIRE HYDRANT with a two-part BARREL with the MAIN VALVE located at ground level.

Water audit, 74   A procedure that combines flow measurements and listening surveys, in an attempt to give a reasonably accurate accounting of all water entering and leaving a system.

Water hammer, 9, 111, 157, 254   The potentially damaging slam, bang, or shudder that occurs in a pipe when a sudden change in water velocity (usually as a result of too-rapidly starting a pump or operating a valve) creates a great increase in water pressure.

Water meter, 169   A device installed in a pipe under pressure for measuring and registering the quantity of water passing through it.

Water purveyor, 217   Anyone who sells drinking water to the public; usually the owner of a public water system.

Watt, 327   The practical unit of electric power. In direct-current (DC) circuit, watts equals VOLTS times AMPERES. In an alternating current (AC) circuit, watts equals VOLTS times AMPERES times POWER FACTOR.

Wattmeter, 328   An instrument for measuring real power, in watts.

Wear rings, 261   Rings made of brass or bronze placed on the impeller and/or casing of a CENTRIFUGAL PUMP to control the amount of water that is allowed to leak from the discharge to the suction side of the pump.

Well point, 35   A perforated metal tube or screen, attached to a jetting or driving head. The device is jetted or driven into the earth, and ground water is withdrawn through it.

Wet-barrel hydrant, 151   A fire hydrant with no MAIN VALVE. Under normal, non-emergency conditions the BARREL is full and pressurized (as long as the lateral piping to the hydrant is under pressure and the gate valve ahead of the hydrant is open). Each outlet has an independent valve that controls discharge from that outlet. The wet-barrel hydrant is used mainly in areas where temperatures do not drop below freezing. The hydrant has no drain mechanism.

Wet tap, 94, 177   A connection made to a main that is full or under pressure.

Wet-top hydrant, 149   A DRY-BARREL HYDRANT in which the threaded end of the MAIN ROD and the revolving or OPERATING NUT are not sealed from water in the BARREL when the MAIN VALVE of the hydrant is open and the hydrant is in use.

Wobble meter, 180   See NUTATING-DISC METER.

Work order, 364   Forms used to communicate field information back to the main office. They are also used to order materials, organize field crews, and obtain easements.

Wound-rotor induction motor, 272   A type of electric motor, similar to a SQUIRREL-CAGE INDUCTION MOTOR, but easier to start and capable of variable-speed operation.

Yoke, 190   A fitting designed to assist in easy meter installation, and to maintain electrical continuity between the incoming and outgoing lines even if the meter is removed for service.

Answers
to
Review Questions

Answers to Review Questions

Module 1 Pipe Installation and Maintenance

1. The tree-type system is not recommended because only a few loops are formed; this restricts circulation, making it difficult to supply a continuous flow of good-quality water to all parts of the system.

2. The flow necessary to meet domestic, commercial, industrial, and fire-hydrant flow requirements at the necessary pressure.

3. The shock or pressure wave that travels down a pipe due to rapid opening or closing of valves or sudden starting or stopping of pumps. The increased pressure can cause extensive damage to pipes and fittings.

4. An indication of how easily water can move through a pipe. The higher the C value, the smoother the interior of the pipe is and the easier it is for water to flow.

5. (1) Internal electrochemical corrosion—aggressive water, (2) external corrosion—soil conditions, (3) galvanic corrosion—dissimilar metals, (4) stray-current corrosion—uncontrolled DC current.

6. Tuberculation is caused by the deposition of corrosion products on interior pipe walls. It causes roughness, which reduces a pipeline's carrying capacity.

7. Surface impurities, nicks in the pipe wall, and differences in the soil composition.

8. (1) Using extra thickness for pipe walls, (2) applying bitumastic or coal-tar coating, (3) wrapping in polyethylene sleeves, (4) installing cathodic protection.

9. (1) Ductile-iron pipe, (2) asbestos-cement pipe, (3) plastic pipe (primarily PVC).

10. Advantages of ductile-iron pipe are durability, strength, and ease of tapping; disadvantages include its excessive weight and susceptibility to corrosion. Advantages of asbestos–cement pipe include its light weight, low initial cost, and high C factor; disadvantages include the fact that it breaks easily and requires special care during tapping operations. Advantages of PVC pipe include its light weight, ease of installation, and high C factor; disadvantages include the fact that special care must be taken during tapping and backfilling.

11. (1) Flanged joint—aboveground installations where rigidity is required, such as in water plants and pump houses; (2) mechanical joint—above- or below-ground installation where some deflection is necessary; (3) ball-and-socket joint—above- or below-ground or below water where larger deflections are necessary; (4) push-on joint—the most common joint for normal distribution system use because it is easily installed and produces water tightness.

12. Since plastic and AC pipe are nonmetallic, they cannot be located using conventional pipe locators or thawed electrically. Metal tape or wire can be installed along the pipe to allow location and also permit nonelectrical thawing methods to be used.

13. Pipe inspection is essential to ensure that the proper type and amount of pipe and fittings have been delivered. Routine inspection procedures will most often detect any pipe damage incurred in shipping and handling.

14. Either with a derrick using a sling or rubber-covered hooks or by using skids and snubbing ropes.

15. (1) Place pipe as near to the trench as possible to avoid excessive handling; (2) string pipe alongside trench opposite the excavation material (or where excavated material will be placed); (3) secure pipe so it cannot roll; (4) safeguard pipe from traffic, blasting, and other construction hazards; (5) place bells in the direction that the work is to proceed; (6) cover the ends of the pipe to minimize contamination.

16. Climatic conditions (frost depth), soil conditions, ground-water conditions, the load that the pipe must carry, and related cost considerations.

17. They are more expensive to excavate and can considerably increase the load on the pipe.

18. The angle at which the trench wall should be excavated to prevent cave-ins. The angle varies in degree depending on the earth material being excavated.

19. (a) Water pressure in the soil; (b) external loads, such as equipment being moved close to the trench; (c) excavated soils lying too close to the trench walls; (d) trench walls that are too steep for the type of soil conditions.

20. (a) Cracks in the ground surface parallel to the trench (These are usually detected within a distance equal to $1/2$–$3/4$ of the trench depth from the edge.), (b) material crumbling off the walls, (c) settling or slumping of the ground near the trench.

21. (a) Proper slope (angle of repose) of the trench wall for type of soil, (b) brace or shore sides of trench, (c) limit the loads (equipment or excavated material) allowed near the trench wall.

22. Shielding does not prevent cave-ins. Therefore, workers are unprotected if they are outside of the shielded area.

23. The pipe must be supported uniformly along its entire length or breakage can occur when loads are imposed by backfill or traffic.

24. If any foreign material is present, the joint could leak.

25. Only the type of lubricant supplied by the manufacturer should be used. Other types, such as oil and grease, can damage the gasket and support microbiological growth.

26. Both prevent separation of the joints caused by the forces created when water abruptly changes direction or flow velocity in a pipe or fitting. Therefore, they should be used where flow will change direction at elbows, tees, and crosses, where pipelines change diameter, and wherever flow stops or is controlled, such as at valves and dead ends.

27. Yes, disturbed soils cannot be relied on to provide the proper bearing support and resistance to the thrust.

28. (1) Hand tapping, (2) mechanical tapping, (3) water settling.

29. It provides cover for the pipe and prevents settling of the surface material. Settling can allow increased surface loading and breaking of the pipe.

30. (1) Pressure and leak testing, (2) flushing, (3) disinfection, (4) bacteriological testing, (5) final inspection, (6) restoration of the excavated area.

31. (1) Continuous-feed method—adding chlorine at a constant rate while the main is filled. (2) Slug method—high rate of chlorine is applied to a slug of water moving through the main. All parts of the main should be exposed to at least 100 mg/L chlorine for 3 hours. (3) Tablet method—placing calcium hypochlorite tablets in the main as it is installed. The main is filled with water and allowed to stand for at least 24 hours.

32. This may vary depending on local and state regulations or guidelines. However, the recommendations for the three methods are as follows: (1) Continuous-feed method—25 mg/L chlorine residual during feeding operation. At least 10 mg/L residual after 24 hours. (2) Slug method—50 mg/L for 3 hours. (3) Tablet method—25 mg/L for 24 hours.

33. Illustrate how the mains, valves, hydrants, and other connections were actually installed. The records are important since final routes and installation will most often differ from the original plans.

34. They help ensure that high-quality water is being delivered to the consumer. These practices can also provide information concerning the functioning of the treatment plant, the condition of the distribution system, and whether contamination is entering the system.

35. It helps rid the pipeline of sediment, slime, and corrosion by-products, which can adversely affect water quality and the carrying capacity (C factor) of the pipe. A flushing program also allows routine maintenance to be performed on hydrants.

36. Polyurethane foam plugs used to swab or clean water mains.

37. The difference between the amount of water pumped into the distribution system and the amount billed to the consumer through a metered system. It represents lost revenue to the utility and is an indication of problems associated with leaks and underregistering of meters.

38. (1) Listening with either mechanical or electrical instruments, (2) conducting a water audit.

39. (a) Barricades, traffic cones, and flashing lights to inform the public that work is in process and to provide a safe working area; (b) flag persons may be necessary to direct traffic; (c) everyone involved with the work should wear bright reflective vests.

40. Hard hats, gloves, and steel-toed safety shoes should be worn as a minimum. Safety goggles may be necessary if jack hammers are being used.

Module 2   Tapping

1. To connect laterals or service lines to water mains.

2. A wet tap is made with the main under pressure; a dry tap is made with the main shut off.

3. (1) Water service does not need to be interrupted; (2) there is less chance of contamination.

4. A brass shut-off valve inserted into a main to connect the service line.

5. To install a corporation stop in a thin or soft pipe wall.

6. AWWA threads, which have the advantages of greater strength and less wedging pressure on the pipe.

7. (1) Service-line material, (2) type of main, (3) use of service clamp or direct insertion.

8. (1) Tee connections, (2) tapping sleeve.

9. $3/4$ in.

10. (1) Make certain the bit is clean and sharp, (2) check to determine that the cutting tool is lubricated, (3) do not force bit through pipe wall.

11. (a) Name of contractor or foreman in charge of the job, (b) date of installation, (c) exact location (address) of tap, (d) tap size, (e) main size.

12. (1) The pressure in the main can blow out a poorly installed corporation stop, (2) plastic pipe may split abruptly, (3) the trench may be suddenly flooded.

Module 3   Valves

1. Isolation, draining lines, throttling flow, regulating water-storage levels, controlling water hammer, bleeding off air and allowing air into lines, and preventing backflow.

2. Isolation: gate, butterfly, globe, plug, ball
   Drain: blow-off

Throttle: butterfly, plug, globe, ball
Regulate storage levels: altitude-control valve
Control water hammer: pressure-relief valve
Allow air in and out of lines: air-relief valve
Control backflow: check valve

3. Flanged, mechanical, push-on.

4. Provide access to the valve.

5. For large valves or valves with power actuators.

6. Prevention of cross connections.

7. Barricades or other appropriate traffic control devices should be installed; workers should wear reflective vests; confined-space entry precautions should be observed.

8. Electric motors, pneumatic, hydraulic.

9. To help equalize pressure on a large valve, making it easier to open and close.

10. Whether a valve is open or closed; position of partially closed valve.

11. Annually at least; more often for important valves.

12. Water hammer.

13. Location measurements; whether found open or closed; condition of packing, stem, operating nut, gears (if any), box or vault, box cover; number of turns to open and close.

Module 4   Fire Hydrants

1. Fire protection.

2. Flushing the water main, flushing streets and sewers, water for building contractors.

3. Increased flow in the main can stir up sediment, causing discolored or cloudy water.

4. A—wet-barrel; B—dry-barrel.
   Component parts:

a—operating nut or pentagon	h—valve
b—bonnet	i—foot piece
c—weather cap	j—valve
d—barrel	k—operating stem
e—operating stem	l—hose outlet and valve seat
f—seat ring	m—barrel
g—drain hole	n—foot piece

5. The dry-barrel hydrant has its main valve located in the base. The barrel is dry until this valve is opened. When the main valve is closed, the barrel drains. The wet-barrel hydrant has no main valve. Each outlet nozzle has an independent valve that controls its discharge. The barrel is full of water under pressure as long as the gate valve ahead of the hydrant is open and the main is under pressure.

6. The dry-barrel type must be used where freezing weather occurs.

7. (1) Be careful not to plug or block the drain holes when pouring thrust blocks, (2) make provisions for carrying off the drainage from the hydrant barrel, such as placing clean stone to a level several inches above the drain openings.

8. A color scheme is used to indicate flow capacity or main size, not pressure.

9. Toward the street.

10. (1) Set back from the road, (2) hydrant guards.

11. Hydrants should be opened and closed slowly in order to prevent pressure surges in the mains (referred to as water hammer).

12. (1) Remove outlet-nozzle cap, (2) attach auxiliary valve to outlet nozzle, (3) open main valve, (4) open auxiliary valve.

13. Remove outlet-nozzle cap from the pumper outlet nozzle and attach a fire hose or diverter.

14. (1) Check for water or ice standing in the barrel of dry-barrel hydrants, (2) check for leaks at joints, around outlet nozzle, at packing or seals, and at outlet-nozzle cap, (3) check for valve seat leakage, (4) inspect outlet-nozzle threads, (5) check ease of operation of the valve(s), (6) check outlet-nozzle cap chains, (7) check for outside obstructions that might interfere with hydrant use.

15. (1) Inspect in the fall, (2) inspect after each use in freezing weather, (3) pump out barrel if it does not drain properly.

16. Close the main valve to the position at which the drains open and allow the drains to flush for about 10 sec. Then close the valve completely.

17. (1) Valve seat leakage, (2) plugged drains, or (3) high water table. Seat leakage can be checked with a listening device. If there is no seat leakage, pump out the barrel. If water comes back into the barrel, the water table is high.

18. (1) Manufacturer, (2) type, (3) date installed, (4) location, (5) depth of bury, (6) outlet-nozzle sizes and thread types, (7) inlet size.

19. (1) Visible tag on the hydrant, (2) notification to fire department, (3) notation on hydrant record card.

20. (a) Anchor the hose end so that the hose will not whip around under pressure; and when pressure testing dry-barrel hydrants, be sure all outlet-nozzle caps are tight, (b) use a fire hose or diverter to direct the flow so that it will not flood or erode private property or cause slippery conditions on sidewalks or roads.

Module 5   Services and Meters

1. Water services provide water from the distribution main to residential, industrial, and commercial users.

2. A gooseneck is a flexible connection that provides for ease of installation and allows for any settlement of the overlying material, or expansion and contraction of the service line due to temperature variations.

3. A curb stop enables water suppliers to regulate (turn on or off) water flow in the service line before the line enters the customer's property.

4. Standardization helps simplify ordering materials and maintaining inventories, enables suppliers to take advantage of economic bulk ordering, and saves time and money in not having to train field crews on a variety of installation techniques.

5. Cost, durability, minimum AWWA standards, ease of locating, standardization.

6. Copper and plastic.

7. Lead joints are difficult to install properly and there is some question concerning the safety (in terms of water quality) of lead services. Wrought iron is rigid and requires threading, making it difficult to install. Wrought iron services may also have short lives due to corrosion.

8. Asbestos–cement, cast iron, ductile iron, and PVC plastic.

9. Use of dissimilar materials often forms a galvanic cell and causes corrosion of the pipe.

10. A wet tap is made when the water main is under pressure, while a dry tap is made when the main is empty.

11. Pressure losses and flow velocities through the pipe.

12. Frost penetration and location of other utility lines.

13. (1) Metering is a means of collecting revenue, based on service charges for the amount of water used; (2) meters encourage customers to use water wisely; and (3) meters provide an indication of the amount of water being used in the system.

14. (1) Positive displacement, (2) compound, (3) current.

15. A positive-displacement meter uses a measuring chamber of known size to determine the volume of water flowing through it. In the piston-type meter, water flows into the chamber, which houses a piston. As it flows through the chamber, the piston is displaced. The motion of the piston is then transmitted to a register, via magnets in newer models or gears in older models, that records the volume of water flowing through the meter. The nutating-disc type meter uses a measuring chamber containing a hard rubber disc instead of a piston. When water flows into and through the chamber, the disc wobbles in proportion to the volume. This motion is then transmitted to a register that records the volume of water flowing through the meter.

16. Metering residential services.

17. Where water demand varies considerably (high and low flows).

18. (1) Propeller, (2) venturi, (3) proportional, and (4) turbine.

19. (1) Expected customer demand, (2) range of flow rates, (3) pressure considerations, (4) friction loss in the service line, meter, and customer's plumbing, (5) allowable pressure losses.

20. (1) That the installation not be subject to flooding with nonpotable water; (2) that it provide an upstream and downstream shut-off valve of high quality to isolate the meter for repairs; (3) that it position the meter in a horizontal plane for optimum performance; (4) that it be reasonably accessible for service and inspection; (5) that it provide for easy reading either directly or via a remote-reading device; (6) that it be reasonably well protected against frost and mechanical damage; (7) that it not be an obstacle or hazard to customer or public safety; (8) that the meter be sealed to prevent tampering; (9) that there be proper support for large meters to avoid stress on the pipe; and (10) that there be a bypass or multiple meters on large installations.

21. A device that holds the stub ends of the pipe in proper alignment and spacing. Yokes cushion the meter against stress and strain in the pipe and provide electrical continuity if metal pipe is used.

22. Electrical continuity (bonding) around the meter reduces the chance of the worker receiving an electrical shock during meter removal due to stray current or electrical grounding to the service pipe.

23. (1) One meter can be removed for servicing by merely closing its valves, and the water supply is still left in metered operation. (2) Meters can be added and the system expanded if required in the future. (This growth can best be provided for at the time of initial installation). (3) There is no need to buy or stock parts for several different sizes of meters. All meters and valves are the same size. (4) The battery can be side-wall mounted to conserve floor space.

24. (1) Remote reading eliminates the need for entering a customer's residence, thereby avoiding inconvenience to the customer, missed readings, and call-backs. (2) Less manpower and cost is devoted to meter reading.

25. (1) Before use, (2) on removal from service, (3) after any repairs or maintenance, and (4) upon customer complaint or request.

26. (1) Running a number of different rates of flow over the operating range of the meter to determine overall meter efficiency; (2) passing known quantities of water through the meter at various test rates to provide a reasonable determination of meter registration; and (3) meeting accuracy limits on different rates for acceptable use.

27. Permanent service number, applicant's name and address, dates of application and installation, sizes of corporation and curb stop used, size and type pipe used, depth of installation, and detailed measurements of location.

28. Size, make, type, date of purchase, location, test data, and any repairs on the meter.

29. (1) Damage to the service line, plumbing, and electrical appliances, (2) stray current can cause fire or electrical shock.

Module 6   Cross-Connection Control

1. A cross connection is any connection between a potable water system and a nonpotable source through which backflow can occur. Backflow is described as a condition where a nonpotable water or other fluid flows into the potable water system through a cross connection.

2. Cross connections are found almost everywhere that potable water is used, including residential houses and apartments, office buildings, manufacturing processes, restaurants, hospitals, laboratories, etc.

3. Backsiphonage is a term used to describe backflow caused by subatmospheric or negative gauge pressure within the water distribution system. Back pressure is the condition caused by some equipment (pump, boiler, etc.) that creates pressure in the nonpotable system greater than the pressure in the potable system. In most cases, backflow occurs because of the higher pressure in the nonpotable system.

4. Waterborne diseases that have been traced to cross connections are (a) salmonellosis, (b) gastroenteritis, (c) amebic dysentery, (d) hepatitis, (e) giardiasis, (f) typhoid fever, (g) shigellosis.

5. Negative gauge pressure is a pressure that is less than atmospheric and is commonly referred to as a partial vacuum.

6. A single check valve is not positive protection against backflow since it can be easily held open by debris, corrosion products, and scale deposits.

7. The conditions that cause backsiphonage are (a) negative gauge pressure in the supply line, (b) a cross connection or link between the potable water system and a source of contamination.

8. Negative gauge pressure in a supply line can be caused by (a) fire demand, (b) a break in the distribution system at a lower level than the cross connection, (c) a booster pump, and (d) undersized piping.

9. The conditions needed for back pressure are (a) a cross connection must exist and (b) the contamination source must be at a higher pressure than the

potable supply. (In the case of back pressure, both supplies have pressure greater than atmospheric.)

10. The following installations are examples of opportunities for back pressure: (a) high pressure boilers, (b) pressurized auxiliary nonpotable water supplies such as those required for fire control or irrigation, (c) pressurized chemical feed systems.

11. An air gap.

12. Four devices for backflow prevention are (highest degree of hazard first) (a) reduced-pressure-zone backflow preventer, (b) double check-valve assembly, (c) pressure vacuum breaker, and (d) atmospheric vacuum breaker.

13. The devices that cannot be used to protect against back pressure are the atmospheric vacuum breaker and the pressure vacuum breaker.

14. A reduced-pressure-zone backflow device consists of two spring-loaded check valves with a pressure regulated relief valve between them. The check valves remain open during normal flow conditions and close under backflow pressure and backsiphonage conditions. Should either valve fail during backflow conditions, the pressure in the reduced pressure zone between the check valves will start to approach that of the inlet. When the reduced zone pressure is within 2 psi (13.8 kPa) of the inlet pressure, the relief valve will open and drain the water entering the reduced pressure zone.

15. A double check valve does not provide adequate protection against a potential health risk since the device does not drain when one or both check valves malfunction, nor does it give any indication when it is not functioning properly.

16. The atmospheric vacuum breaker must be installed downstream from the last shut-off valve because it cannot be placed under any continuing back pressure. The back pressure will force the valve to remain open even under backflow conditions.

17. When water stops flowing forward, a check valve drops, closing the water inlet and opening an atmospheric vent, letting water in the breaker body drain out and breaking the partial vacuum in that part of the system.

18. A pressure vacuum breaker has one and sometimes two spring-loaded check valves, whereas the single valve in the atmospheric breaker is not spring loaded. On a loss of internal pressure the spring action will open the

atmospheric vent in the pressure vacuum breaker, draining water from the downstream side of the check valve(s). Due to this arrangement, the pressure vacuum breaker can be installed upstream of the final shut-off valve.

19. The elements of an effective cross-connection control program are (a) an adequate cross-connection control ordinance, (b) an adequate organization with authority, (c) a systematic surveillance program, (d) follow-up procedures for compliance, (e) provision for backflow-prevention device approvals, inspection, and maintenance, (f) cross-connection control training, and (g) public awareness and information programs.

20. The agency requires reports of corrective actions taken and conducts reinspections to ensure continued compliance.

21. The reduced-pressure-zone backflow preventer should be tested at least annually.

Module 7 Pumps and Prime Movers

1. A spinning impeller accelerates water to a high velocity within a casing, which changes the high-velocity, low-pressure water to a low-velocity, high-pressure discharge.

2. Volute casing and diffuser vanes.

3. Mixed-flow pump (the design used for most vertical turbine pumps).

4. If a valve is closed in the discharge line, the pump impeller can continue to rotate for a time without pumping water or damaging the pump.

5. A multistage centrifugal pump is made up of a series of impellers and casings (housings) arranged in layers, or stages. This increases the pressure at the discharge outlet, but does not increase flow volume.

6. Shaft-type and submersible-type vertical turbines.

7. A close-coupled vertical turbine with an integral sump or pot.

8. The jet pump consists of a centrifugal pump at the ground surface and an ejector nozzle below the water level.

9. Positive-displacement pumps are generally used in water plants to feed chemicals into the water supply.

10. The foot valve prevents water from draining when the pump is stopped, so the pump will be primed when restarted.

11. The bolts holding the two halves of the casing together are removed and the top half is lifted off.

12. Wear rings prevent excessive circulation of water between the impeller discharge and suction areas. Lantern rings allow sealing water to be fed into the stuffing box.

13. (1) Packing rings are made of graphite-impregnated cotton, flax, or synthetic materials. They are inserted in the stuffing box and held snugly against the shaft by an adjustable packing gland. (2) Mechanical seals consist of two machined and polished surfaces. One is attached to the shaft, the other to the casing. Spring pressure maintains contact between the two surfaces.

14. The pump impeller is mounted directly on the shaft of the motor.

15. An experienced operator can often detect unusual vibration by simply listening or touching. Vibration, especially changes in vibration level, are viewed as symptoms or indicators of other underlying problems in foundation, alignment, and/or pump wear.

16. Racking refers to erratic operation that may result from pressure surges when the pump starts; it is often a problem when the pressure sensor for the pump control is located too close to the pump station.

17. During start-up.

18. (1) Increase system efficiency; (2) spread the pumping load more evenly throughout the day; (3) reduce power-factor charges.

19. The bearings may run hot, and excess grease or oil could run out and reach the motor windings, causing the insulation to deteriorate.

20. The abrasive material on emery cloth is electrically conductive and could contaminate electrical components.

21. Imbalance of the rotating elements, bad bearings, and misalignment.

22. A condition called single-phasing can occur, causing the motor windings to overheat and eventually fail.

23. Gasoline, propane, methane, natural gas, and diesel oil (diesel fuel).

24. Make, model, capacity, type, date and location installed, and other information for both the driver (motor) and the driven unit (pump).

25. Make sure the power to the device is disconnected. This is critical since rubber gloves, insulated tools, and other protective gear are not guarantees against electrical shock.

Module 8   Water Storage

1. (1) To equalize supply and demand, (2) to increase operating convenience, (3) to level out pumping requirements, (4) to provide water during source or pump failure, (5) to provide water to meet fire demands, (6) to provide surge relief, (7) to increase detention times, (8) to blend water sources.

2. Operating storage floats with the system; the reservoir fills when demand is low and empties when demand exceeds supply. Emergency storage is to be used only in exceptional situations.

3. Elevated tanks maintain system pressure through elevation; however, they are expensive to build. Ground-level tanks are less expensive to build, but they may require a booster pump to provide adequate pressure. The general disadvantage of standpipes is that only the upper portion of the water in the standpipe has enough pressure to be used in the distribution system.

4. Hydropneumatic systems are partially filled with pressurized air that is positioned on top of the water in the tank. This air provides the pressure needed to provide water in remote areas. A pump connected to that tank turns on only when the pressure in the tank drops below a certain point. This type of system is used in very high or remote areas with only a few customers.

5. Location is influenced by system hydraulics and water demands.

6. (1) Inlet and outlet pipes, (2) overflow pipe, (3) drain pipe.

7. Tanks should be drained, cleaned, painted (if necessary), and disinfected annually.

8. Water from the full tank must be tested for bacteriological safety before the tank is put back into service.

9. Basic information should include location, dates of inspection, conditions noted during inspection, dates maintenance was performed, and notes on the maintenance performed.

10. No, appropriate safety equipment and procedures should be used whenever work or inspections are performed in and around water storage facilities.

Module 9   Instrumentation and Control

1. To measure, display, and record conditions and changes.

2. The sensor and the indicator.

3. The transmitter, the transmission channel, and the receiver.

4. A telemetry system is required when conditions must be monitored at very distant locations, such as a remote pump station or reservoir.

5. Pressure sensors, level sensors, temperature sensors, flow sensors, speed sensors, electrical sensors.

6. The diaphragm strain gauge is the most widely used type of sensor in modern instrumentation. It consists of a section of wire fastened to a diaphragm. As the variable being measured changes, the diaphragm moves, changing the length of the wire and thus changing its resistance. The changing resistance is measured and transmitted to the receiver/indicator.

7. Electromotive force is comparable to the pressure in a water main; it is measured in volts. Current is similar to the flow rate in a water system; it is measured in amps. Resistance is similar to the friction factor of a pipe; it is measured in ohms.

8. Power, measured in watts, is also monitored. The amount of power drawn by a circuit is an indication of the amount of work performed.

9. A transmitter converts the signal from the instrument sensor into a standard signal that can be sent to a receiver/indicator in another location.

10. A receiver converts the signal sent by the transmitter to an indicator reading for the operator to monitor.

11. Analog indicators show a smooth range of values—an example is the indicator needle on a D'Arsonval type voltmeter. An analog indicator can

display any value between the minimum and the maximum that can be displayed on the instrument. Digital indicators display values as a specific number, like a digital watch. The number of values that it is possible for an instrument to display is limited by the number of digits displayed by the indicator.

12. Utility-owned cable, leased telephone lines, radio channels, or microwave systems.

13. Multiplexing, scanning, polling, duplexing.

14. Multiplexing allows a single physical channel to carry several signals simultaneously. In a tone-frequency multiplexing system, each signal is sent on a distinct frequency. Up to 21 frequencies can be sent over a single telephone line.

Scanning systems check the values of several sensors, then send those values one after another in a set order.

In a polling system, several instruments are connected to the same line, but each is given a unique number, or address. A controller is installed that sends messages over the signal line asking each instrument to send its data as needed.

Duplexing is used to allow instrument data from a remote site to pass over the same line that is used to send control signals to the remote equipment. Full duplex systems allow signals to pass in both directions simultaneously; half duplex allows signals to pass in both directions, but only in one direction at a time; simplex systems have no duplexing capability—signals can only travel in one direction.

15. In a manual control system, the operator moves the switches, levers, and other controls that change the condition of the system. In an automatic system, computers or other equipment adjust system status in response to signals received from instruments; the operator does not need to touch the controls under normal conditions. In a semiautomatic control system, both automatic and manual controls are used for different operations of a single piece of equipment.

16. Direct-wire control, used in an attended plant or pump station, has all systems wired into a control panel in the facility. Supervisory control, used for unattended facilities, requires a telemetry system to carry instrument readings and signals between the remote station and the central control point.

17. Check to be sure the instrument is not exposed to moisture, chemical gas or dust, excessive heat, vibration, or other damaging environmental factors.

18. Blow out the ports that connect to the transmitter mechanism.

19. Rubber gloves, insulated tools, rubber floor mats, and similar protective insulating devices cannot guarantee protection against electrical shock. Circuits should be de-energized (disconnected) before any maintenance is performed on them, other than tests that must be performed with the circuit energized.

20. Pneumatic failure often results in more rapid and widespread danger than hydraulic failure; however, extreme care must be exercised when dealing with either type of system when it is pressurized.

Module 10   Maps, Drawings, and Records

1. (1) Maps, drawings, and records provide a permanent record of a system's facilities. They also free water suppliers from dependence on the memory of longtime employees.
(2) Maps, drawings, and records promote efficient operation and maintenance of a distribution system and aid in developing and troubleshooting the system.
(3) Some states have statutes or regulations requiring the maintenance of maps, drawings, and records.

2. A comprehensive map.

3. A sectional map.

4. The plat and list method and the intersection method.

5. (1) Arterial map—a comprehensive map at a scale of 2000 to 3000 ft/in. showing primary distribution mains 8 in. or larger. Arterial maps are primarily used for system analysis.
(2) Leak frequency maps—different colored push pins are used to identify different types of leaks on a print of the comprehensive map.
(3) Leak survey maps—generally modified sectional or valve maps showing valves to be closed and areas to be isolated.
(4) Pressure zone maps—contour lines are drawn to identify zones of equal pressure. From pressure maps, low and high pressure zones can easily be

identified and controlled. (Some systems maintain pressure zones on their comprehensive map instead of preparing a supplemental record.)

6. Card records are usually cross-referenced to maps and construction drawings and contain detailed descriptive information not recorded on maps and other documents.

7. Valve and hydrant maps.

8. Street names   Reservoirs and tanks
   Water mains   Water source
   Sizes of mains   Scale
   Fire hydrants   Orientation arrow
   Valves   Date last corrected
   Pump stations   Pressure zone limits
   Closed valves at pressure zone limits

9. Card records are usually maintained for valves, hydrants, and services.

10. The list provides information on valves and hydrants, such as reference measurements, size and make, direction of operation and number of turns, date installed, date tested, and any remarks.

11. Water mains   Services
    Valves   Meters
    Hydrants

12. The only real difference between distribution system records for small and large water supply systems should be the number of records kept. Smaller systems can often consolidate their mapped records because of the system size and simplicity.

13. (a) 8-in. main   (e) Valve closed
    (b) Valve   (f) Valve in vault
    (c) Valve, partially closed   (g) Hydrant
    (d) Tapping valve and sleeve   (h) Check valve

14. Work orders are used to communicate field information back to the main office, where records are kept. Work orders are also used to order materials, organize work, and obtain easements.

15. A work order usually consists of a complete location and description of the job performed; a listing of all the materials needed and used; information on time, manpower, and equipment used; and the total cost.

16. A file for technical information (bulletins or pamphlets) should be established because such information often provides installation specifications, operating tips, or maintenance information.

Module 11 Public Relations

1. Poor customer relations can reduce customer support of a utility in terms of new budgets and implementation of special projects, which require additional charges. This, in turn, affects distribution personnel in terms of layoffs, lower salaries, and budget reductions affecting planned inspection or maintenance activities.

2. (1) Communications, (2) caring, and (3) courtesy.

3. (1) Maintain a neat appearance, (2) display nametags and carry credentials within easy reach, (3) be polite, (4) give short, succinct answers whenever possible.

4. (1) Tracking dirt through a customer's home, (2) abusing a customer's pet, (3) smoking on the job, (4) chatting with customers for long periods of time, (5) walking in gardens or on lawns.

5. Good safety practices, such as the use of proper road barriers and warning signs on a job, communicate the utility's concern about customer and employee safety.

6. Personnel can obtain a copy of their state's motor vehicle laws for reference.

7. Don't talk to reporters—leave that job to the utility's public relations department.

8. Procedural manuals explain the governing policies of the utility, which can be passed along to the customer when the need arises. These manuals keep employees informed and promote a well-run organization. In addition, a manual may double as an employee handbook.

9. To answer customer questions and handle complaints, including handling the resulting paperwork.

10. A public relations or media relations expert.